LASER BEAM PROPAGATION

LASER BEAM PROPAGATION

Generation and Propagation of Customized Light

Edited by
Andrew Forbes

CSIR National Laser Centre
Pretoria, South Africa

CRC Press
Taylor & Francis Group
Boca Raton London New York

CRC Press is an imprint of the
Taylor & Francis Group, an **informa** business

CRC Press
Taylor & Francis Group
6000 Broken Sound Parkway NW, Suite 300
Boca Raton, FL 33487-2742

First issued in paperback 2020

© 2014 by Taylor & Francis Group, LLC
CRC Press is an imprint of Taylor & Francis Group, an Informa business

No claim to original U.S. Government works

Version Date: 20140319

ISBN 13: 978-0-367-57619-6 (pbk)
ISBN 13: 978-1-4665-5439-9 (hbk)

Library of Congress Cataloging-in-Publication Data

Laser beam propagation : generation and propagation of customized light / edited by
   Andrew Forbes.
      pages cm
   Includes bibliographical references and index.
   ISBN 978-1-4665-5439-9 (hardcover : alk. paper)
   1. Laser beams. 2. Beam optics. I. Forbes, Andrew, editor of compilation.

QC689.5.L37L382 2014
621.36'6--dc23   2013047354

Visit the Taylor & Francis Web site at
http://www.taylorandfrancis.com

and the CRC Press Web site at
http://www.crcpress.com

Contents

SECTION I   Back to Basics

SECTION II   Generation and Characterization of Laser Beams

SECTION III Novel Laser Beams

Editor

Andrew Forbes, after earning his PhD (1998), subsequently spent several years working as an applied laser physicist, first for the South African Atomic Energy Corporation and then later in a private laser company where he was technical director. He is presently a chief researcher at the CSIR National Laser Centre in South Africa and is the research group leader for mathematical optics. Dr. Forbes sits on several international conference committees, chairs the SPIE international conference on laser beam shaping and the OSA's Diffractive Optics and Holography technical group. He holds honorary professorships at the University of Stellenbosch and the University of Kwazulu-Natal.

Preface

Laser beam propagation has been a subject of active research for over 50 years now, but as recently as 20 years ago it underwent a paradigm shift with the advent of new methods to describe and characterize general laser beams. In parallel, laser beam shaping techniques advanced due to increased computer power and the possibility of fabricating diffractive optical elements with standard lithographic techniques. In recent years, such laser beam shaping techniques have been executed with liquid crystal devices, ushering in an era of desktop creation of custom laser beams by complex light modulation. The result has been a plethora of new discoveries on laser beams and their propagation properties, including orbital angular momentum of light, the nondiffracting nature of light, light that propagates along curved paths, and reconstructs after obstacles. Such laser beams have been applied across many disciplines, from classical to quantum studies, and from physics to biology. Extending the degree of freedom from purely spatial control of the laser beam to polarization control has seen the emergence of vector beams, carrying radial and azimuthal polarization, while the advances in ultra-fast lasers have seen a concomitant interest in the coupling between spatial and temporal propagation parameters of ultra-short-pulse laser beams. This book attempts to encapsulate the field of laser beam propagation, starting with the fundamentals of how laser beams propagate, following this through to the generation of custom laser beams with modern techniques, the properties of a selection of such laser beams, and finally the application areas of these laser beams.

The book begins with introductory chapters (Section I, Back to Basics) that summarize the core concepts needed in the rest of the book, and provide the context for the chapters to follow. We begin with Optical Waves (Chapter 1), summarizing the physical optics approach to the propagation of optical waves. The chapter begins with Maxwell's equations, develops the concept of plane waves, and then leads through to the mathematical description of diffraction and Gaussian optics. Chapter 2, From Classical to Quantum Optics, considers how to adapt the concepts in Chapter 1 to the single photon level—the quantum interpretation of modes of light. Parallels between the paraxial propagation of light beams and the Schrödinger equation in quantum mechanics are explained and expanded to formulate the description of paraxial optics by means of state vectors and operators. The chapter introduces the quantization of the electromagnetic field and its application to passive linear optics, and concludes with a discussion on the concept of quantum entanglement and its counterpart in classical optics. Finally, Chapter 3, Time Domain Laser Beam Propagation, follows the analogous approach but in the time domain, where spatial diffraction is replaced with temporal dispersion and the coupling between the spatial and temporal domains is introduced.

The second part of the book (Section II, Generation and Characterization of Laser Beams) concerns methods to generate and characterize laser beams and their propagation. The content of Sections I and II would be sufficient for the lay reader to apply in the creation, propagation, and characterization of any arbitrary but coherent laser beam. This section begins with Chapter 4, Spatial Laser Beam Characterization, which outlines the application of modal decomposition to the characterization of laser beams. This modern technique is shown to completely describe any coherent field in terms of its physical parameters, such as intensity, phase, wavefront, beam quality factor, and Poynting vector. Chapter 5, Characterization of Time Domain Pulses, outlines the analogous tools for the temporal characterization of laser beams, including interference detection schemes such as FROG, STRUT, and SPIDER. Finally, the generation of arbitrary laser beams with digital holograms is considered in Chapter 6, Generation of Laser Beams by Digital Holograms. In this chapter, the use of spatial light modulators to display reconfigurable digital holograms capable of modifying and shaping laser beams is reviewed. The techniques discussed include complex amplitude modulation with spatial light modulators and include a discussion on the control of the polarization properties of light.

The heart of the book, Section III (Novel Laser Beams), covers the properties and propagation of various classes of laser beams. The selection is chosen so as to cover laser beams that are of general interest in a number of application fields (e.g., flat-top beams for materials processing), but also laser beams that have intrinsic properties of interest, for example, control of polarization (vector beams), shape invariant propagation (nondiffracting beams) and laser beams that carry orbital angular momentum (Helmholtz–Gauss beams). The section starts with the most commonly generated shaped light, flat-top beams (Chapter 7), and outlines the definitions of such beams, their propagation rules as well as means to create them in the laboratory. Chapter 8 introduces Helmholtz–Gauss beams, including various shape-invariant propagation classes of beams and outlines their general form and propagation rules through optical systems. Chapter 9, Vector Beams, is intended to offer a general overview of vector beams including the mathematical description, recently developed generation and manipulation methods, and the focusing properties of such beams. The chapter includes details of full Poincaré (FP) beams as well as generalized vector beams and their applications. Chapter 10, Low-Coherence Laser Beams, is relevant to the shaping and propagation of common classes of lasers (e.g., excimer lasers, diode lasers, and multimode fiber lasers) and is topical due to the extensive use of super-continuum and white light sources for both their spatial and spectral properties. The chapter introduces the concepts of coherence theory and then applies this to understand the propagation of low-coherence optical fields. Finally, Chapter 11, Orbital Angular Momentum of Light, outlines the recent developments in orbital angular momentum carrying fields, discussing basic properties, definitions, and applications. This chapter brings together both the classical and quantum concepts of spatial modes of light.

Contributors

Reuben Aspden
School of Physics and Astronomy
University of Glasgow
Glasgow, Scotland, United Kingdom

Jeffrey A. Davis
Department of Physics
San Diego State University
San Diego, California

Thomas Feurer
Institute of Applied Physics
University of Bern
Bern, Switzerland

Daniel Flamm
Institute of Applied Optics
Friedrich-Schiller University Jena
Jena, Germany

Andrew Forbes
CSIR National Laser Centre
Pretoria, South Africa

Sandeep Goyal
School of Physics
University of Kwazulu-Natal
Durban, South Africa

Julio C. Gutierrez-Vega
Optics Center
Tecnologico de Monterrey
Monterrey, Mexico

Thomas Konrad
School of Physics
University of Kwazulu-Natal
Durban, South Africa

Ignacio Moreno
Departamento de Ciencia
   de Materiales
Universidad Miguel
Hernandez, Spain

Miles Padgett
School of Physics and Astronomy
University of Glasgow
Glasgow, Scotland, United Kingdom

Filippus S. Roux
CSIR National Laser Centre
Pretoria, South Africa

Christian Schulze
Institute of Applied Optics
Friedrich-Schiller University Jena
Jena, Germany

Jari Turunen
Department of Physics and
  Mathematics
University of Eastern Finland
Joensuu, Finland

Qiwen Zhan
Department of Electrical and Computer
  Engineering
University of Dayton
Dayton, Ohio

I

Back to Basics

1

Optical Waves

Filippus S. Roux
CSIR National Laser Centre

Light has played a special role in many significant scientific discoveries. Right from the start, it was the light from the stars, planets, and moons reaching their eyes that gave early scientists the ability to unravel the cycles of celestial bodies. For this purpose, optical devices such as telescopes were invented. Optics is, therefore, one of the oldest fields of scientific study. Using the data from such optical observations, Kepler, Newton, and others eventually developed classical mechanics.

It was black-body radiation (the light emitted from warm objects) that eventually led to the development of quantum mechanics. When Einstein derived his special theory of relativity, the speed of light was found to play an important role.

While these examples show how light helped to increase our understanding of other fields of physics, there were also cases where we learned more about light itself. Notable among these cases was when James Clerk Maxwell unified electricity and magnetism. The resulting set of equations provides the theoretical foundation upon which our knowledge of physical optics is built and in terms of which the behavior of (classical) light can be analyzed rigorously. It also lies at the foundation of quantum electrodynamics, which described the interaction of light with matter.

It is, therefore, fitting that we start our discussion about the behavior of light with Maxwell's equations.

1.1 Maxwell's Equations

The four differential equations of Maxwell give the interrelationship among six quantities: the *electrical field strength** **E**, measured in volt per meter (V/m),[†] the *magnetic field strength* **H**, measured in ampere per meter (A/m), the *electrical flux density* **D**, measured in coulomb per square meter (C/m^2), the *magnetic flux density* **B**, measured in tesla (T),[‡] the *electrical current density* **J**, measured in ampere per square meter (A/m^2) and the *electrical charge density* ρ, measured in coulomb per cubic meter (C/m^3). All, except the last quantity, are vectorial fields. The last is a scalar field.

In SI units, Maxwell's equations are given by

$$\nabla \times \mathbf{E} = -\partial_t \mathbf{B}, \tag{1.1}$$

$$\nabla \times \mathbf{H} = \partial_t \mathbf{D} + \mathbf{J}, \tag{1.2}$$

$$\nabla \cdot \mathbf{D} = \rho, \tag{1.3}$$

$$\nabla \cdot \mathbf{B} = 0, \tag{1.4}$$

where $\partial_t = \partial/\partial t$, and the same for ∂_x, ∂_y, and ∂_z, ∇ is the nabla or del operator, defined by

$$\nabla = \partial_x \hat{x} + \partial_y \hat{y} + \partial_z \hat{z}, \tag{1.5}$$

and bold-faced symbols denote vectors

$$\mathbf{E} = E_x \hat{x} + E_y \hat{y} + E_z \hat{z}, \tag{1.6}$$

with $\hat{x}, \hat{y}, \hat{z}$ representing the orthogonal unit vectors in three-dimensional space.

---

*   The electric (magnetic) field strength is sometimes also called the electric (magnetic) field intensity.
†   The units for electric field strength can also be newton per coulomb (N/C).
‡   The units for magnetic flux density [tesla (T)] is also given as weber per square meter (W/m^2), where a weber (W) is the same as volt per second (V/s).

1.1.1 Constituent Relationships

In vacuum, the electrical flux density **D** is linearly related to the electrical field strength **E** via the constituent relation

$$\mathbf{D} = \epsilon_0\mathbf{E}, \tag{1.7}$$

where ϵ_0 is the *electrical permittivity of vacuum*. Similarly, in vacuum, the magnetic flux density **B** is linearly related to the magnetic field strength **H** via the constituent relation

$$\mathbf{B} = \mu_0\mathbf{H}, \tag{1.8}$$

where μ_0 is the *magnetic permeability of vacuum*.

Permeability is measured in henry per meter (H/m)* and the permeability of vacuum has an exact value of

$$\mu_0 = 4\pi \times 10^{-7} \text{ H/m}. \tag{1.9}$$

Permittivity is measured in farad per meter (F/m).[†] As we will see below, the *speed of light c* is related to ϵ_0 and μ_0 by

$$\mu_0\epsilon_0 = \frac{1}{c^2}. \tag{1.10}$$

The speed of light has the exact (integer) value of

$$c = 299{,}792{,}458 \text{ m/s}, \tag{1.11}$$

however, it is often approximated as 3×10^8 m/s. The value of the permittivity of vacuum ϵ_0 is then given via Equation 1.10

$$\epsilon_0 = (\mu_0 c^2)^{-1} = 8.854187817 \times 10^{-12} \text{ F/m}. \tag{1.12}$$

1.1.2 Dielectric Medium

Considering the propagation of optical beams, one often assumes that the medium through which the light propagates is a vacuum. However, the propagation of optical beams can also be considered in other media, provided that such a medium shares certain properties with vacuum. Their properties are the following:

1. The medium is *source free*. This means that $\mathbf{J} = 0$ and $\rho = 0$. In general, $\mathbf{J} = \sigma\mathbf{E}$, where σ is the *electrical conductivity*. For $\sigma = 0$, the medium does not support currents. For $\sigma \neq 0$, there is absorption in the medium. A nonzero σ is required for amplitude transmission functions.
2. It is *linear*. A process (such as propagation) is said to be linear if the output obtained when the process operates on a linear combination of inputs is equal to

---

[*] A henry (H) is the same as volt-second per ampere (Vs/A).
[†] A farad (F) is the same as coulomb per volt (C/V).

the equivalent linear combination of outputs that are obtained when the process operates on the inputs separately. In mathematical notation, one can express this requirement by

$$P\{\alpha f(x, y) + \beta g(x, y)\} = \alpha P\{f(x, y)\} + \beta P\{g(x, y)\}, \qquad (1.13)$$

where $P\{\cdot\}$ represents the linear process, $f(x, y)$ and $g(x, y)$ are input functions, and α and β are complex constants.

Nonlinear optical media will give rise to nonlinear optical effects, such as second-harmonic generation, parametric down-conversion, self-phase modulation, and four-wave mixing, to name but a few.

3. Such a medium is *isotropic*. In other words, the properties of the medium does not depend on the direction of propagation or the directions of vectors of **E** or **H**. An anisotropic medium can affect the polarization of the beam as it propagates through the medium, through phenomena that include dichroism and retardance.

4. It is also *homogeneous*. In other words, the properties of the medium are constant as functions of space:

$$\nabla \epsilon = \nabla \mu = 0. \qquad (1.14)$$

Interesting effects can appear when the medium is not homogeneous. If the refractive index varies randomly, as it does in a turbulent atmosphere, the optical field is scintillated or scattered. On the other hand, if the refractive index varies in a periodic fashion, one can find that the dispersion curves of the medium are modified, much like the dispersion curves in a crystal. Such periodic media are often called photonic crystals.

5. The medium is assumed to be *nondispersive*. This means that the flux densities, **D** and **B**, do not depend on the temporal or spatial derivatives of the field strengths, **E** or **H**. In other words, the reaction of the medium to the field strength is purely local in space and time. Dispersive media give rise to frequency-dependent refractive indices and phenomenon such as the magneto-optic effect.

6. Finally, the medium is assumed to be *nonmagnetic*. It does not have magnetic properties, which implies that $\mu = \mu_0$.

The only difference between a medium that has all the above properties and vacuum is that the permittivity of vacuum is replaced by the permittivity of the medium

$$\epsilon_0 \rightarrow \epsilon = \epsilon_r \epsilon_0 = n^2 \epsilon_0, \qquad (1.15)$$

where ϵ_r is called the *relative permittivity* or the *dielectric constant* and n is the *refractive index* of the medium. For physical media, $n \geq 1$. Apart from permittivity, everything else remains the same. We will refer to a medium that obeys these requirements as a *dielectric medium*.

1.1.3 Phasor Fields

In a dielectric medium, one can use the constituent relations for **D** and **B** in Equations 1.7 and 1.8, respectively, together with Equation 1.15, to express Maxwell's equations purely in terms of **E** and **H** (or **E** and **B**). In this case, **E** is often simply called the *electric field* and **H** (or **B**) is called the *magnetic field*.

Physical electric and magnetic fields are real-valued vector fields. It is often more convenient for the sake of calculations to work with complex-valued vector fields. Once the calculations in terms of such complex-valued vector fields are done, one can obtain the physical vector field by taking the real part of the result. The freedom to perform calculations using complex-valued vector fields is allowed when the medium is source free because then Maxwell's equations have an additional symmetry: one can transform all fields by multiplying them with the same constant phase factor, without changing the equations.

Another simplification follows due to linearity. One can express the electric and magnetic fields as linear combinations of *monochromatic* (single frequency) fields—in terms of their inverse Fourier transforms. Thanks to the linearity of the medium, the different monochromatic fields (associated with different frequencies) never couple into each other.

These simplifications now lead to a convenient way to consider electromagnetic fields, namely, as *phasor fields*. Each frequency component of, for instance, the electric field is expressed as a complex-valued vector field that only depends on the spatial coordinates. The time dependence is obtained by multiplying the phasor field with $\exp(i\omega t)$, where ω is the *angular frequency* of the monochromatic field. The physical monochromatic electric field is the real part of the product of the phasor field and the time-dependent phase factor

$$\mathbf{E}(x, y, z, t) = \mathcal{R}e\left\{\tilde{\mathbf{E}}(x, y, z)\exp(i\omega t)\right\}, \tag{1.16}$$

where $\tilde{\mathbf{E}}(x, y, z)$ represents the electric phasor field. A similar expression applies for the magnetic field

$$\mathbf{H}(x, y, z, t) = \mathcal{R}e\left\{\tilde{\mathbf{H}}(x, y, z)\exp(i\omega t)\right\}, \tag{1.17}$$

where $\tilde{\mathbf{H}}(x, y, z)$ represents the magnetic phasor field.

The Maxwell's equations for the phasor fields in a dielectric medium are given by

$$\nabla \times \tilde{\mathbf{E}} = -i\omega\mu_0\tilde{\mathbf{H}} \tag{1.18}$$

$$\nabla \times \tilde{\mathbf{H}} = i\omega\epsilon\tilde{\mathbf{E}} \tag{1.19}$$

$$\nabla \cdot \tilde{\mathbf{E}} = 0 \tag{1.20}$$

$$\nabla \cdot \tilde{\mathbf{H}} = 0. \tag{1.21}$$

From now on, we will drop the time-dependent exponential factor and only work with these phasor fields, unless stated otherwise. We will also drop the tildes in the notation for these phasor fields and we will refer to them as the electric field and the magnetic field. At the beginning of the next section, we briefly return to the full time-dependent electromagnetic fields.

1.2 Helmholtz Equation

One of the key discoveries associated with Maxwell's equations is that they lead to a wave equation in the absence of sources and currents. This wave equation, in terms of the time-dependent electric field, is obtained by taking the curl of Equation 1.1 and eliminating the time-dependent magnetic field $\mathbf{H}$, using Equation 1.2. The result is

$$\nabla \times \nabla \times \mathbf{E} + \mu_0 \epsilon \partial_t^2 \mathbf{E} = 0. \tag{1.22}$$

Then, one employs the identity

$$\nabla \times \nabla \times \mathbf{A} = \nabla(\nabla \cdot \mathbf{A}) - \nabla^2 \mathbf{A}, \tag{1.23}$$

together with Equation 1.3 to obtain

$$\nabla^2 \mathbf{E} - \mu_0 \epsilon \partial_t^2 \mathbf{E} = 0. \tag{1.24}$$

This differential equation has the form of a wave equation with a wave velocity given by

$$v = \frac{1}{\sqrt{\mu_0 \epsilon}}. \tag{1.25}$$

In vacuum, $\epsilon = \epsilon_0$ and the resulting wave velocity turns out to be the *speed of light in vacuum*

$$c = \frac{1}{\sqrt{\mu_0 \epsilon_0}} \tag{1.26}$$

given in Equation 1.11. In a dielectric medium, the speed of light is reduced by the refractive index n

$$\frac{1}{\sqrt{\mu_0 \epsilon}} = \frac{1}{\sqrt{\mu_0 n^2 \epsilon_0}} = \frac{c}{n}. \tag{1.27}$$

So, one can write the wave equation of Equation 1.24 in a dielectric medium as

$$\nabla^2 \mathbf{E} - \frac{n^2}{c^2} \partial_t^2 \mathbf{E} = 0. \tag{1.28}$$

Following a similar procedure, one can obtain an identical expression for the magnetic field $\mathbf{H}$ (or $\mathbf{B}$).

Starting from Equations 1.18 through 1.21, one can derive the equivalent of the wave equation for the phasor fields. The result has a similar form as Equation 1.28, but instead of a time-derivative, one finds a term linear in the electric field

$$\nabla^2 \mathbf{E} + \omega^2 \mu_0 \epsilon \mathbf{E} = 0. \tag{1.29}$$

The angular frequency ω is related to the speed of light via

$$c = \frac{\omega}{k}, \tag{1.30}$$

where

$$k = \frac{2\pi}{\lambda} \tag{1.31}$$

is the *wave number* in vacuum. Using Equations 1.30 and 1.27, one can express Equation 1.29 as

$$\nabla^2 \mathbf{E} + n^2 k^2 \mathbf{E} = 0. \tag{1.32}$$

The differential equation in Equation 1.32 is known as the *Helmholtz equation*. The equivalent equation for the magnetic field is obtained by replacing $\mathbf{E}$ in Equation 1.32 with $\mathbf{H}$ (or $\mathbf{B}$). Solutions of Equation 1.32 represent the three-dimensional vector fields for the electric (phasor) field of the electromagnetic waves in a dielectric medium.

Note that the Helmholtz equation, Equation 1.32 (as well as the wave equation in Equation 1.28), operates independently on each component of the vector field $\mathbf{E}$. As a result, Equation 1.32 represents three different equations, one for each vector component. Under some conditions, the overall vector field may have (approximately) the same direction at all points in space. For such a case, the three components are proportional to each other. Therefore, one can remove the direction of the vector and treat the electric field as a scalar field. This leads to the scalar form of the Helmholtz equation:

$$\nabla^2 U + n^2 k^2 U = 0, \tag{1.33}$$

where $\mathbf{E} = U\hat{n}$, with U being the complex-valued scalar electric (phasor) field and $\hat{n}$ being a constant unit vector.

1.2.1 Phase Functions

Any complex-valued function $g(x, y, z)$ can be written as

$$g(x, y, z) = A(x, y, z) \exp[i\theta(x, y, z)], \tag{1.34}$$

where $A(x, y, z)$ is the *amplitude function* given by the magnitude or absolute value of the complex function and $\theta(x, y, z)$ is the *phase function* or argument of the complex function.

Phase is a relative concept and is only defined in terms of a reference value, which may be a predefined phase at some point in space and time. Although the phase can take on any real value, phase values that differ by an integer multiple of 2π radian represent the same phase. As a result the function value of a phase function at any point in space and time is uniquely defined by the modulo 2π value of the phase function at that point. The set of all distinct phase values can be represented by the points on the circumference of the unit circle.

A phase function is a mapping from some domain, which in this case is the three-dimensional space, to the unit circle. The range of a phase function is the unit circle, which differs from the range of the more common conventional functions—the real line. As a result, phase functions have interesting properties that are not shared by conventional functions. These properties are related to the nontrivial topology of its compact range. In the language of topology, it is said that the second homotopy group [1] is nontrivial. What it means is that if one considers the mapping of a closed curve in space onto the unit circle, one will find situations where this mapping wraps itself around the unit circle an integer number of times. When the closed curve is contracted to a single point in space, the mapping of the contracting curves cannot continuously contract to a point on the unit circle. At some stage, the mapped points will have to jump from one side of the circle to the opposite side. Physically, this means that the contracting curve in space crossed a phase singularity. A singularity is a point where the quantity is undefined. Hence, *phase singularities* are special points in a phase function where the phase is undefined.

In optical fields, one can find such points of singular phase where the optical field becomes zero. In fact, all first-order zeros in a complex function represent phase singularities. In optical fields, such singularities are called *optical vortices* due to the helical structure of the phase function in the region surrounding the singularity. Optical vortices play an important role in optical beams and are intrinsic in the expressions for some of the sets of laser modes: for example, in Helmholtz–Gauss beams (Chapter 8), vector beams (Chapter 9), and light fields carrying orbital angular momentum (Chapter 11).

1.2.2 Solutions of the Helmholtz Equation

One often refers to the solutions of the Helmholtz equation as *modes*. When the Helmholtz equation is solved for a given system, with specific boundary conditions, one generally obtains multiple possible solutions. Each solution that can exist independently in that system (i.e., satisfies all boundary conditions of the system without having to excite additional fields) is a mode of that system. Different modes are normally associated with different values for certain parameters, which can be used to distinguish the modes. These parameters may be discrete or continuous variables. In the former case, one often refers to the parameter as the *mode index*. Different modes (associated with different parameters) are orthogonal to each other, with respect to some orthogonality condition that is defined for the specific system. As a result, all the modes together form a complete orthogonal basis in terms of which all possible electric fields in the system can be expressed.

One can expedite the calculation of the solutions of the Helmholtz equation by a convenient choice of coordinate system. The boundary conditions of the structure under consideration should preferably coincide with surfaces where one or more of the coordinates have a constant value. We will now consider some important special cases, where the coordinate systems are, respectively, the Cartesian coordinates and the cylindrical coordinates. Here, we will only obtain the general solutions, that is without imposing any boundary conditions. In all these cases, the differential equations are solved using the *separation-of-variables technique*.

1.2.2.1 Cartesian Coordinates: Plane Waves

The simplest case is obtained when considering the *Cartesian coordinate system*. The Laplacian operator for the Cartesian coordinate system is given by

$$\nabla^2 U = \partial_x^2 U + \partial_y^2 U + \partial_z^2 U. \tag{1.35}$$

Following the separation-of-variables technique, one separates the three spatial coordinates as a product of functions, by assuming that the solution has the form $U(x, y, z) = X(x)Y(y)Z(z)$. Substituting this anzats into the scalar Helmholtz equation Equation 1.33 and dividing it by $U(x, y, z)$, one obtains

$$\frac{\partial_x^2 X(x)}{X(x)} + \frac{\partial_y^2 Y(y)}{Y(y)} + \frac{\partial_z^2 Z(z)}{Z(z)} + k^2 = 0. \tag{1.36}$$

Here, we only consider the solutions in vacuum, for which $n = 1$. Since the first three terms can only depend on x, y, and z, respectively, the only way that they can satisfy the equation is when each of them is equal to a constant. Furthermore, when these constants are added to k^2, the result must be zero. So, we will assume that the terms in Equation 1.36 can be replaced by

$$\frac{\partial_x^2 X(x)}{X(x)} = -k_x^2, \tag{1.37}$$

$$\frac{\partial_y^2 Y(y)}{Y(y)} = -k_y^2, \tag{1.38}$$

$$\frac{\partial_z^2 Z(z)}{Z(z)} = -k_z^2, \tag{1.39}$$

respectively, which then leads to the requirement that

$$k_x^2 + k_y^2 + k_z^2 = k^2. \tag{1.40}$$

Obviously, not all three of these constants need to be negative. Some may be positive. In general, one can say that either the three values k_x, k_y, and k_z are all positive real-valued quantities, or one or two of them are positive real-valued quantities while the rest are imaginary values. The latter case represents an *evanescent wave*, which will be discussed more thoroughly later.

The three decoupled differential equations in Equations 1.37 through 1.39, associated with the three coordinates in the Cartesian coordinate system, can be solved separately to give the solutions

$$X(x) = \exp(-ik_x x),\tag{1.41}$$

$$Y(y) = \exp(-ik_y y),\tag{1.42}$$

$$Z(z) = \exp(-ik_z z).\tag{1.43}$$

Note that these solutions could also have had a positive sign in their exponents. We deliberately picked the negative exponent. The reason for this will become clear later.

The general solution can now be constructed out of the three solutions for the three spatial coordinates. It is given by

$$U(x, y, z) = U(\mathbf{x}) = \exp\left(-i\mathbf{k} \cdot \mathbf{x}\right).\tag{1.44}$$

where we expressed the argument as the dot-product between the *position vector*,

$$\mathbf{x} = x\hat{x} + y\hat{y} + z\hat{z},\tag{1.45}$$

which is a vector pointing from the origin of the coordinate system to the point denoted by (x, y, z), and the *propagation vector* (or just k-vector),

$$\mathbf{k} = k_x\,\hat{x} + k_y\,\hat{y} + k_z\,\hat{z},\tag{1.46}$$

which denotes the direction of propagation as we will show below.

The magnitude of a vector is given by the square root of the dot-product of the vector with itself. It thus follows from Equation 1.40 that the magnitude of the propagation vector is the wave number

$$\sqrt{\mathbf{k} \cdot \mathbf{k}} = \sqrt{k_x^2 + k_y^2 + k_z^2} = k.\tag{1.47}$$

For this reason, one can express the k-vector as $\mathbf{k} = k\hat{k}$, where $\hat{k}$ is a unit vector that points in the direction of propagation.

The general solution of the Helmholtz equation given in Equation 1.44 is a solution for the scalar phasor field. To form the time-dependent field, one needs to multiply it with the time-dependent frequency factor

$$U(\mathbf{x}, t) = \exp[i(\omega t - \mathbf{k} \cdot \mathbf{x})].\tag{1.48}$$

This expression represents a plane wave that propagates in the (positive) direction of the k-vector. Different k-vectors represent different plane waves.

1.2.2.2 Cylindrical Coordinates: Bessel Modes

The Laplacian in cylindrical coordinates (ρ, ϕ, z) is given by

$$\nabla^2 U = \partial_\rho^2 U + \frac{1}{\rho}\,\partial_\rho U + \frac{1}{\rho^2}\,\partial_\phi^2 U + \partial_z^2 U. \tag{1.49}$$

For the separation of variables, we use the anzats

$$U(\rho, \phi, z) = R(\rho)P(\phi)Z(z). \tag{1.50}$$

The decoupled differential equations are given by

$$0 = \partial_\rho^2 R(\rho) + \frac{1}{\rho}\,\partial_\rho R(\rho) + \left(k^2 - \beta^2 - \frac{m^2}{\rho^2}\right)R(\rho), \tag{1.51}$$

$$0 = \partial_\phi^2 P(\phi) + m^2 P(\phi), \tag{1.52}$$

$$0 = \partial_z^2 Z(z) + \beta^2 Z(z), \tag{1.53}$$

where we introduced the constants β^2 and m^2 for the z- and ϕ-dependent terms, respectively. Continuity requires that m is an integer. The solution of these three differential equations are

$$R(\rho) = \begin{cases} \left.\begin{matrix} J_m\left(\rho\sqrt{k^2 - \beta^2}\right) \\[2mm] Y_m\left(\rho\sqrt{k^2 - \beta^2}\right) \end{matrix}\right\} & \text{for } \beta^2 \le k^2 \\[6mm] \left.\begin{matrix} K_m\left(\rho\sqrt{\beta^2 - k^2}\right) \\[2mm] I_m\left(\rho\sqrt{\beta^2 - k^2}\right) \end{matrix}\right\} & \text{for } \beta^2 > k^2 \end{cases} \tag{1.54}$$

$$P(\phi) = \exp(im\phi), \tag{1.55}$$

$$Z(z) = \exp(-i\beta z), \tag{1.56}$$

where $J_m(\cdot)$ and $Y_m(\cdot)$ are the mth-order Bessel functions [2] of the first and second kinds, respectively, and $I_m(\cdot)$ and $K_m(\cdot)$ are the mth-order modified Bessel functions of the first and second kinds, respectively.

A physical field is continuous at the origin and has a finite energy, which implies that the field goes to zero at infinity. Therefore, the general solution for a physical electric field in cylindrical coordinates is given by

$$U(\rho, \phi, z) = J_m(\rho k_T)\exp(im\phi)\exp(-i\beta z), \tag{1.57}$$

where $k_T = \sqrt{k^2 - \beta^2}$.

1.3 Plane Waves

Since plane waves form an important building block of light propagating through free space we will discuss them in depth. For the purpose of the discussion on plane waves, we will briefly return to time-dependent electric fields. As we saw above, the electric field of a plane wave can be expressed as a complex function of Cartesian coordinates and time

$$\mathbf{E}(\mathbf{x}, t) = \mathbf{E}_0 \exp[i(\omega t - \mathbf{k} \cdot \mathbf{x})], \tag{1.58}$$

where $\mathbf{E}_0$ is a constant vector that is orthogonal to the direction of propagation.

1.3.1 Sign Convention

Note that we could also have expressed the plane waves as

$$\mathbf{E}'(\mathbf{x}, t) = \mathbf{E}_0 \exp[-i(\omega t - \mathbf{k} \cdot \mathbf{x})]. \tag{1.59}$$

The two expressions, Equations 1.58 and 1.59, both have the same real part and are, therefore, equivalent. We shall, as far as possible, use the **sign convention that phase increases with time**. This means that we will use Equation 1.58.

1.3.2 Phase Function of a Plane Wave

The instantaneous local phase of a plane wave is given by the argument of the expression for the plane wave in Equation 1.58:

$$\Theta(\mathbf{x}, t) = \omega t - \mathbf{k} \cdot \mathbf{x}. \tag{1.60}$$

It is a function that gives the phase of the plane wave at every point in space and time. The unit (or dimensions) of this argument is the unit of phase—radians. So the angular frequency ω is measured in radian per second (rad/s) and the propagation vector $\mathbf{k}$ carries units of radian per meter (rad/m).

If we fix the time of the phase function of the plane wave Equation 1.60 at a specific instant in time, for instance, at $t = 0$, we have

$$\Theta(\mathbf{x}, 0) = -\mathbf{k} \cdot \mathbf{x} = -k(\hat{k} \cdot \mathbf{x}) = -2\pi \frac{\hat{k} \cdot \mathbf{x}}{\lambda}, \tag{1.61}$$

where k is the wave number Equation 1.31, representing the length of the propagation vector, and $\hat{k}(= \mathbf{k}/k)$ is a unit vector in the direction of propagation.

The dot-product between $\hat{k}$ and the position vector $\mathbf{x}$ computes the projection of the position vector onto $\hat{k}$, or the component of the position vector along the propagation direction. Every set of position vectors for which these projections are equal describe

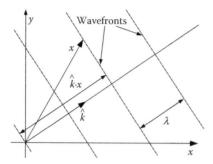

FIGURE 1.1 Wavefront of a plane wave.

a plane that is perpendicular to $\hat{k}$, as shown in Figure 1.1. Since the projections for all the points on such a plane are equal, all these points have the same phase value. Such equiphase surfaces in space are called *wavefronts*. Hence, we see that the wavefronts of a plane wave are planar surfaces perpendicular to the propagation vector. Due to the negative sign in Equation 1.61, the phase decreases linearly in the direction of the propagation vector. This is a consequence of the sign convention.

Two points, x_1 and x_2, for which $\hat{k} \cdot x$ differ by an integer value times λ represent the same phase because

$$\mathbf{k} \cdot \mathbf{x}_1 - \mathbf{k} \cdot \mathbf{x}_2 = \frac{2\pi}{\lambda}(\hat{k} \cdot \mathbf{x}_1 - \hat{k} \cdot \mathbf{x}_2) = \frac{2\pi}{\lambda}(m\lambda) = m2\pi, \tag{1.62}$$

where m is an integer. Different planes (wavefronts) represent the same phase if their phase values, modulo 2π, are equal. Consecutive planes that represent the same phase are always separated by a distance equal to an integer multiple of the wavelength λ.

1.3.3 Direction of Propagation

When the time increases (becomes larger than zero, $t > 0$), we add a positive value of ωt to the phase value of Equation 1.61. As a result, the wavefront associated with a specific phase value moves forward, in the direction of the propagation vector. In fact, the entire phase function that we had for $t = 0$ is now shifted in the direction of the propagation vector by an amount equal to $(\omega/k)t = ct$. So we see that the wave *propagates* in the direction of the propagation vector at the speed of light. In Figure 1.2, it is shown how a one-dimensional (real-valued) wave shifts in the positive x-direction with time.

Note that if the sign of the temporal and the spatial parts in the phase function were the same, it would represent a plane wave propagating in a direction opposite to that of the propagation vector.

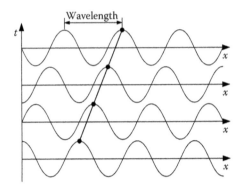

FIGURE 1.2   One-dimensional real-valued wave shifts in the positive x-direction with increasing time.

1.3.4 Direction Cosines

Each component of the propagation vector is obtained by projecting the vector onto the associated coordinate axis. For example

$$k_x = \mathbf{k} \cdot \hat{x} = k\cos(\alpha), \tag{1.63}$$

where α is the angle between the propagation vector and the x-axis. Similar angles can be defined for the other axes as well. In Figure 1.3, all such angles are shown for a specific vector. Using these angles, one can write the propagation vector as

$$\mathbf{k} = k\left[\cos(\alpha)\,\hat{x} + \cos(\beta)\,\hat{y} + \cos(\gamma)\,\hat{z}\right]. \tag{1.64}$$

These cosines are referred to as the *direction cosines* of the plane wave. Since $\mathbf{k} \cdot \mathbf{k} = k^2$, the direction cosines obey the property

$$\cos^2(\alpha) + \cos^2(\beta) + \cos^2(\gamma) = 1. \tag{1.65}$$

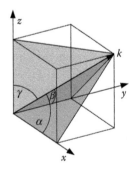

FIGURE 1.3   Direction cosine angles.

1.3.5 Spatial Frequency

In analogy to the temporal frequency ν, which is equal to the inverse of the period, the spatial frequency of a plane wave is equal to $1/\lambda$. Hence, the wave number k represents the "angular spatial frequency." There is also directional information associated with spatial frequencies. For this reason, it is convenient to express the propagation vector in terms of the spatial frequencies associated with the respective coordinate axes. This follows by substituting $k = 2\pi/\lambda$ into Equation 1.64 and pulling out a factor of 2π

$$\mathbf{k} = 2\pi \left[\frac{\cos(\alpha)}{\lambda} \hat{x} + \frac{\cos(\beta)}{\lambda} \hat{y} + \frac{\cos(\gamma)}{\lambda} \hat{z} \right] = 2\pi (a\,\hat{x} + b\,\hat{y} + c\,\hat{z}). \tag{1.66}$$

Here, a, b, and c represent the respective spatial frequency components associated with the x-, y-, and z-axes. These spatial frequency components can be combined into a *spatial frequency vector*

$$\mathbf{a} = a\,\hat{x} + b\,\hat{y} + c\,\hat{z}. \tag{1.67}$$

The dot-product in the expression for the plane wave can now be written as

$$\mathbf{k} \cdot \mathbf{x} = 2\pi (ax + by + cz) = 2\pi (\mathbf{a} \cdot \mathbf{x}), \tag{1.68}$$

and one can write the plane waves as

$$\exp[i2\pi (\nu t - \mathbf{a} \cdot \mathbf{x})] = \exp[-i2\pi (ax + by + cz - \nu t)]. \tag{1.69}$$

Dropping the explicit time dependence and returning to phasor fields, we write the plane waves as

$$\Phi(\mathbf{x}; \mathbf{a}) = \exp(-i2\pi\, \mathbf{a} \cdot \mathbf{x}) = \exp[-i2\pi (ax + by + cz)]. \tag{1.70}$$

To visualize the concept of a spatial frequency, one can consider a number of equispaced parallel lines on a piece of paper, as shown in Figure 1.4. The magnitude of the spatial frequency is given by the inverse of the distance between adjacent lines—that

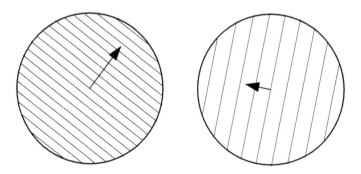

FIGURE 1.4   Spatial frequencies.

is the number of lines per distance. The closer the lines are to each other, the higher the spatial frequency. The orientation of the lines gives the direction of the spatial frequency vector. The direction of the spatial frequency vector, indicated by the arrow in Figure 1.4, is always perpendicular to the lines and its length (magnitude) is inversely proportional to the distance between the lines.

1.3.6 Propagating Plane Waves

Due to the relationship among the direction cosines, given in Equation 1.65, the three spatial frequencies are not independent, but are related by

$$a^2 + b^2 + c^2 = \frac{1}{\lambda^2}. \tag{1.71}$$

This is the expression for a sphere with radius $1/\lambda$. Hence, for a given wavelength, the set of all the different plane waves is represented by the surface of a sphere. Each point on the sphere represents a specific plane wave and the coordinates of the point are the spatial frequencies a, b, and c that define the plane wave.

In practice, only the plane waves associated with one half of this sphere (hemisphere) are considered. As we will see below, one can use this set of plane waves to construct an arbitrary beam of light propagating toward this hemisphere. The plane waves of the other hemisphere propagate in the general opposite direction. The surface of the upper hemisphere (where a and b are measured along the two horizontal axes and c is given on the positive vertical axis) can be expressed as a function of c in terms of a and b:

$$c(a, b) = \sqrt{\frac{1}{\lambda^2} - a^2 - b^2} \ \text{ for } \ a^2 + b^2 \leq \frac{1}{\lambda^2}. \tag{1.72}$$

(For the lower hemisphere, c is given by the negative square root.) Thus, we see that one can specify each propagating plane wave uniquely by specifying the values of a and b.

1.3.7 Evanescent Waves

Above, we have only considered propagating plane waves for which $a^2 + b^2 \leq 1/\lambda^2$. When $a^2 + b^2 > 1/\lambda^2$, the value of c is imaginary

$$c(a, b) = \pm i\sqrt{a^2 + b^2 - \frac{1}{\lambda^2}} = \pm i\frac{\alpha}{2\pi}, \tag{1.73}$$

where α is a decay constant (not to be confused with one of the direction cosine angles). Such cases also give solutions of the scalar Helmholtz equation in Cartesian coordinates. For such a case, the expression of the plane wave becomes

$$\Phi(\mathbf{x}; \mathbf{a}) = \exp(-\alpha z) \exp[-i2\pi(ax + by)]. \tag{1.74}$$

This describes a wave that decays* along the z-direction with a spatial frequency in the xy-plane that is larger than $1/\lambda$. These waves are called *evanescent waves*. They form part of the reactive field in the vicinity of a physical boundary lying in the xy-plane.

Evanescent waves are only found in the vicinity of a physical structure (or scatterer). The latter sets up a boundary condition that cannot be satisfied by a superposition of propagating waves alone. Therefore, evanescent waves must also be excited. In general, both evanescent waves and propagating waves are required to satisfy the boundary conditions, especially when there are features in the structure that are smaller than the wavelength. Only the propagating waves can carry away power. The evanescent waves store energy in the vicinity of the boundary.

1.3.8 Spatial Frequency Plane

Together, the propagating and evanescent waves are represented by all the points on an infinite two-dimensional plane with coordinates a and b. This is the complete spatial frequency plane, as shown in Figure 1.5. Points inside the circle with radius $1/\lambda$ denote propagating waves and points outside of it denote evanescent waves.

1.3.9 Plane Wave Slices

On occasion, it will be necessary to restrict the plane waves to the xy-plane, where $z = 0$:

$$\Phi(x, y; a, b) = \exp[-i2\pi(ax + by)]. \tag{1.75}$$

We will refer to these two-dimensional functions as *plane wave slices*. Note the difference in notation: $\Phi(\mathbf{x}; \mathbf{a})$, as defined in Equation 1.70, denotes a three-dimensional plane wave, while $\Phi(x, y; a, b)$, as defined in Equation 1.75, denotes a two-dimensional plane wave slice. The latter is obtain from the former by setting $z = 0$.

A number of comments about plane wave slices are in order.

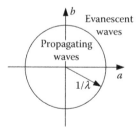

FIGURE 1.5 The *ab*-plane.

---

* One needs to select the correct sign for the decaying function. The other option gives a growing function, which is nonphysical.

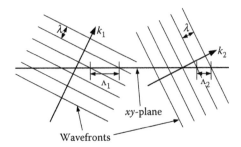

FIGURE 1.6   In-plane wavelengths of plane wave slices.

1. The spatial frequency components a and b can take on any real value.
2. The *in-plane wavelength* $\Lambda = (a^2 + b^2)^{-1/2}$ can vary from zero to infinity, even though the three-dimensional field is monochromatic (λ is fixed). (See Figure 1.6.)

    If the three-dimensional plane wave propagates in the z-direction (perpendicular to the xy-plane) the in-plane wavelength is infinite. If the propagation direction of the 3D plane wave is perpendicular to the z-direction (parallel to the xy-plane), the in-plane wavelength is equal to the wavelength λ. In-plane wavelengths smaller than λ are associated with evanescent waves.

3. Without specifying λ, it is not possible to say whether a particular plane wave slice is part of a propagating plane wave or of an evanescent wave. It is only once λ is specified that one can determine whether $a^2 + b^2$ is larger or smaller than $1/\lambda^2$. For propagating waves: $a^2 + b^2 \leq 1/\lambda^2$ ($\Rightarrow \Lambda \geq \lambda$).
For evanescent waves: $a^2 + b^2 > 1/\lambda^2$ ($\Rightarrow \Lambda < \lambda$).

4. One can reproduce the full three-dimensional plane wave by multiplying the plane wave slice by $\exp[-i2\pi zc(a, b)]$, where $c(a, b)$ is given by Equation 1.72 for propagating waves, or by Equation 1.73 for evanescent waves. For a specific λ, each three-dimensional plane wave is uniquely determined by a particular plane wave slice.

5. The plane wave slices are not to be confused with plane waves in two dimensions. The latter have a fixed wavelength, whereas the in-plane wavelength of plane wave slices can vary.

1.3.10 Orthonormal Basis

One of the most important properties of plane waves and, in particular, plane wave slices is that they form a complete orthogonal basis. Their orthogonality follows from the fact that plane waves are the solutions of the Helmholtz equation—a property that was briefly mentioned in Section 1.2.2. First, we need to explain what is meant by saying that functions are *orthogonal*.

To determine whether mathematical objects are orthogonal, one first needs to define an inner product between these objects. An *inner product* between two objects A and B is represented by $\langle A, B \rangle$. When $\langle A, B \rangle = 0$, the two objects A and B are said to be orthogonal to each other. An inner product is a bilinear operation, operating on A and B with the following properties:

- $\langle A, A \rangle = |A|^2 \geq 0$
- $\langle A, A \rangle = 0 \quad \Leftrightarrow \quad A = 0$
- $\langle aA, B \rangle = a\langle A, B \rangle$, where a is any complex-value constant
- $\langle A, B \rangle = \langle B, A \rangle^*$, where $*$ denotes the complex conjugate
- $\langle A + B, C \rangle = \langle A, C \rangle + \langle B, C \rangle$

The dot-product for vectors is an example of an inner product. For functions, one can define the inner product as

$$\langle f, g \rangle = \int fg^* \, d\mathcal{D}, \tag{1.76}$$

where $\mathcal{D}$ represents the domain on which the functions are defined.

Functions have the property that one can multiply them with arbitrary complex constants and again end up with functions. Another property that they have is that one can add them together to end up with functions yet again. Sets of mathematical objects that have these properties are generally called *vector spaces*. These properties imply that one can represent all the functions as a linear combination of a special set of functions. This special set of functions is called a *basis*. To have the smallest number of elements in the basis, the basis elements must be *linearly independent*, which means that none of the basis elements can be formed in terms of a linear combination of the other basis elements.

If one can define an inner product for the elements in a vector space, then it is called an *inner product space*. With the aid of an inner product one can find an orthogonal basis for the space, such that the inner product between any two different basis elements gives zero. In other words

$$\langle \Phi_n, \Phi_m \rangle = \lambda_m \delta_{n,m}. \tag{1.77}$$

where Φ_n represents the basis elements, λ_m is an orthogonality constant, and $\delta_{n,m}$ is the Kronecker delta, which is equal to 1 if $n = m$ and equal to zero otherwise. If all the orthogonality constants are equal to 1, the basis elements are normalized and the basis is called *orthonormal*. The expression in Equation 1.77 is called an *orthogonality condition*.

A basis is called a *complete* orthogonal basis if, in addition to the orthogonality condition, it also satisfies a completeness condition. The latter ensures that the expansion of an object in terms of the basis is exact. In other words, the expansion reproduces the object exactly.

The benefit of an orthogonal basis lies in the ease with which one can determine the coefficients in the expansion of a particular object as a linear combination of basis elements. Assume that the expansion for the object is given by

$$g = \sum_n \alpha_n \Phi_n, \tag{1.78}$$

where α_n denotes the coefficients and Φ_n are the orthonormal basis elements. Then

$$\langle g, \Phi_m \rangle = \sum_n \alpha_n \langle \Phi_n, \Phi_m \rangle = \sum_n \alpha_n \delta_{n,m} = \alpha_m. \tag{1.79}$$

The coefficient associated with a particular basis element is directly obtained by taking the inner product between the object and that basis element.

1.3.11 Plane Wave Slices as a Complete Orthonormal Basis

The plane wave slices form a complete orthonormal basis for all finite-energy complex-valued functions on the xy-plane. In other words, plane wave slices satisfy both an orthogonality condition and a completeness condition. The inner product for the plane wave slices is defined as an integral on the two-dimensional xy-plane

$$\langle f(x,y), g(x,y) \rangle = \iint_{-\infty}^{\infty} f(x,y) g^*(x,y) \, dx \, dy. \tag{1.80}$$

Using this inner product and the general expression for plane wave slices, given in Equation 1.75, one can show that plane wave slices are orthonormal

$$\langle \Phi(x,y;a,b), \Phi(x,y;a',b') \rangle = \iint_{-\infty}^{\infty} \exp\{-i2\pi[(a-a')x + (b-b')y]\} \, dx \, dy$$

$$= \delta(a-a')\delta(b-b'), \tag{1.81}$$

where the last expression follows from the inverse Fourier transform of a Dirac delta function. The fact that the Dirac delta functions are not multiplied by some constant means that the plane wave slices are normalized.*

The completeness condition is similar to the orthogonality condition in Equation 1.81. The only difference is that the integration is done over the spatial frequency components instead of the spatial coordinates

$$\iint_{-\infty}^{\infty} \Phi(x,y;a,b)\Phi^*(x',y';a,b) \, da \, db$$

$$= \iint_{-\infty}^{\infty} \exp\{-i2\pi[(x-x')a + (y-y')b]\} \, da \, db$$

$$= \delta(x-x')\delta(y-y'). \tag{1.82}$$

The fact that the result is a Dirac delta function for each coordinate implies that the basis can reproduce the two-dimensional functions exactly and is, therefore, a complete basis.

---

* This is actually a consequence of expressing them in terms of the spatial frequency components a and b, instead of the components of the propagation vectors k_x and k_y. In the latter case, an orthogonality constant of $(2\pi)^2$ would appear.

When one makes an expansion of an arbitrary function in terms of a complete basis then the expansion reproduces the function exactly.

1.4 Propagation and Diffraction

1.4.1 Defining Diffraction

One can say that any *deflection* experienced by a beam of light that is not due to *reflection* or *refraction* is the result of *diffraction*. According to this definition, deflection is the collective noun for reflection, refraction, and diffraction. However, diffraction differs from the other two in an essential way. Reflection and refraction are not wave phenomena—they are geometrical optics effects. Diffraction, on the other hand, is a result of the wave properties of light.

1.4.2 Diffraction and Scattering

Often, one may think of diffraction purely as the result of sharp boundaries of an obstacle in the path of a light beam. However, although such boundaries have an effect on the way a beam diffracts, the process of diffraction is what happens to the beam of light as it propagates. At one point, the beam may have a specific intensity distribution, while some distance further the intensity distribution has changed. This change is the result of the diffraction process. The effect of the obstacle is to set up boundary conditions, which in turn affects the electrical field in the immediate vicinity of the obstacle. Diffraction is what happens when this electrical field in the immediate vicinity of the obstacle propagates further. To distinguish between these two processes, one can refer to the effect of the obstacle on the electrical field around it as *scattering*, while *diffraction*, as stated before, is what happens to the electrical field as it propagates away from the obstacle.

Here, we are not discussing scattering as such. We will only consider the propagation of beams of light away from any obstacle or scatterer. The terms propagation and diffraction are used interchangeably.

1.4.3 Two-Dimensional Complex-Valued Functions

The functions that we are interested in are all those that can represent physical beams of light—electrical fields in a dielectric medium. The candidate functions are scalar three-dimensional complex-valued functions $g(\mathbf{x})$. To represent a physical beam of light, such a function must obey the scalar Helmholtz equation Equation 1.33—the equations of motion for this particular situation.

Due to the equations of motion, one only needs to specify the function on a two-dimensional plane, for instance, at $z = 0$: $g(x, y) = g(x, y, z = 0)$. The rest of the function in the three-dimensional space is then uniquely determined via the equations of motion. In other words, the function on the two-dimensional plane serves as a boundary condition that uniquely specifies the rest of the function. (This is why a hologram can reproduce a three-dimensional image.) In practice, the rest of the function is found by propagating it away from the two-dimensional plane.

Another requirement is that the beam of light has a finite amount of power (or energy). This means that the modulus square of the two-dimensional function on the cross section of the beam integrates to a finite amount

$$\int |g(x,y)|^2 \, dx \, dy < \infty, \tag{1.83}$$

which also implies that $g(x,y) \to 0$ for $x, y \to \pm\infty$.

It thus follows that the set of all scalar two-dimensional, finite energy, complex-valued functions $g(x,y)$ completely specify all the possible (scalar) beams that can exist in free space.

1.4.4 Propagation as a Process

The basic free-space optical system associated with beam propagation is shown in Figure 1.7. We will assume that we have a beam propagating in the general direction of the z-axis. Define a plane at $z = z_1$ and call it the *input plane*. (For convenience, we usually set $z_1 = 0$.) The transverse coordinates in the input plane are x and y. Consider another plane at $z = z_2$, where $z_2 > z_1$, and call it the *output plane.** The transverse coordinates in the output plane are u and v. The electric field of a beam of light in the input plane is given by a two-dimensional complex-valued function $g(x,y)$, which we call the *input function*. For a given input function we compute the complex-valued function $h(u,v)$ for the electric field in the output plane, and call it the

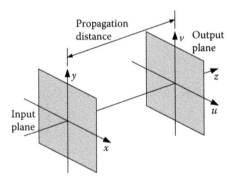

FIGURE 1.7   Free-space optical system associated with beam propagation.

---

*   Sometimes one may also want to consider *backward propagation*, for which $z_2 < z_1$.

output function. This process defines the propagation of a beam of light from $z = z_1$ to $z = z_2$.

Note that the input and output functions are both parts of a continuous complex-valued three-dimensional function $f(\mathbf{x})$ for the electric field of the light. In the input plane, $g(x, y) = f(x, y, z = z_1)$, and in the output plane, $h(u, v) = f(x = u, y = v, z = z_2)$. Hence, one can consider the propagation process as the process through which the rest of the function $f(\mathbf{x})$ is uncovered.

It is important to note that the free-space optical system, as shown in Figure 1.7, does not include any optical components, such as lenses. The free-space optical system of Figure 1.7 is assumed for the entire discussion on diffraction in Section 1.4.

1.4.5 Approach

Traditionally, the diffraction of light is discussed within the framework of the work done by Kirchhoff and Sommerfeld [3]. As such it usually, more or less, follows the historical development of the topic. This is not the approach that will be followed here. Instead, we will use an approach based on the orthogonality of plane waves, which is just as rigorous as the Kirchhoff–Sommerfeld approach, easier to implement and conceptually much clearer.

In essence, the plane wave approach follows these three basic steps:

1. Use the orthogonality of the plane wave slices to expand the input function in terms of plane wave slices in the input plane.
2. Extend the plane wave slices to full three-dimensional plane waves and determine what they look like in the output plane.
3. Reconstruct the output function in the output plane by using the same expansion that was obtained in the input plane but replacing the input plane wave slices, by their equivalents in the output plane.

Although we only consider scalar diffraction here, the approach can be extended to *vector diffraction*—the diffraction theory for the propagation of vector fields [4].

1.4.6 Plane Wave Expansion

Any beam of light can be expressed in terms of a linear combination of plane waves. Such a *plane wave expansion* is expressed by

$$
g(\mathbf{x}) = \iint_{-\infty}^{\infty} G(a, b) \Phi(\mathbf{x}; \mathbf{a}) \, da \, db
$$

$$
= \iint_{-\infty}^{\infty} G(a, b) \exp(-i2\pi [ax + by + c(a, b)z]) \, da \, db, \tag{1.84}
$$

where $g(\mathbf{x})$ is the three-dimensional complex amplitude of the beam and the function $G(a, b)$ represents the "coefficients" in the expansion. Although the spatial frequency components a and b are continuous variables, they act like the "indices" for the plane waves.

In the two-dimensional xy-plane at $z = 0$, the plane wave expansion becomes an expansion in terms of plane wave slices

$$g(x, y) = \iint_{-\infty}^{\infty} G(a, b)\Phi(x, y; a, b) \, da \, db = \iint_{-\infty}^{\infty} G(a, b) \exp(-i2\pi[ax + by]) \, da \, db.$$

(1.85)

To compute $G(a, b)$ from the input function $g(x, y)$, one uses the orthogonality of the plane wave slices. Substituting the plane wave expansion into the inner product and using Equation 1.81, we obtain

$$\langle g(x, y), \Phi(x, y; a, b)\rangle = \iint_{-\infty}^{\infty} G(a', b')\langle \Phi(x, y; a', b'), \Phi(x, y; a, b)\rangle \, da' \, db' = G(a, b).$$

(1.86)

Hence, $G(a, b)$ is given by

$$G(a, b) = \langle g(x, y), \Phi(x, y; a, b)\rangle$$

$$= \iint_{-\infty}^{\infty} g(x, y)\Phi^*(x, y; a, b) \, dx \, dy$$

$$= \iint_{-\infty}^{\infty} g(x, y) \exp(i2\pi[ax + by]) \, dx \, dy.$$

(1.87)

The final integral expression in Equation 1.87 is a *two-dimensional Fourier transform*

$$G(a, b) = \mathcal{F}\{g(x, y)\}.$$

(1.88)

For this reason, $G(a, b)$ is called the *angular spectrum*. The spatial frequency components a and b determine the direction (or "angle") of propagation of each plane wave. The plane wave expansion in Equation 1.85 is the *two-dimensional inverse Fourier transform* of the angular spectrum

$$g(x, y) = \mathcal{F}^{-1}\{G(a, b)\}.$$

(1.89)

The Fourier transform always exists for physical beams because they are given by continuously differentiable functions and they have finite energy. We often use unphysical functions such as Dirac delta functions, trigonometric functions, exponential functions, and so on to do calculations. However, these functions can never represent physical beams and are usually only valid in the sense of *functionals* or *distributions* under integrals.

It is important to note that if a beam of light has a specific angular spectrum in a plane at $z = z_1$, then this beam will have the same angular spectrum in any other plane, provided that the reference frame (the coordinate system) remains the same. Therefore, for an arbitrary value of z, we have

$$G(a, b) = \langle g(\mathbf{x}), \Phi(\mathbf{x}; \mathbf{a}) \rangle$$

$$= \iint_{-\infty}^{\infty} g(\mathbf{x}) \exp(i2\pi [ax + by + c(a, b)z]) \, dx \, dy, \tag{1.90}$$

where $c(a, b)$ is given in Equation 1.72. Furthermore, one can expand a beam of light at any value of z using this angular spectrum, as is implied in the plane wave expansion in Equation 1.84.

1.4.7 Propagation Phase Factor

If the same angular spectrum is used to reconstruct the beam at different values of z, why would the beam ever look different at different values of z? The difference comes from the z-dependent part of the three-dimensional plane waves

$$\Phi(\mathbf{x}; \mathbf{a})|_{x=y=0} = \exp[-i2\pi c(a, b)z] = \psi(a, b, z). \tag{1.91}$$

This is the phase factor that a plane wave slice must be multiplied with to extend it to the three-dimensional plane wave

$$\Phi(\mathbf{x}; \mathbf{a}) = \exp(-i2\pi [ax + by + c(a, b)z]) = \exp(-i2\pi [ax + by])\psi(a, b, z). \tag{1.92}$$

It relates the two-dimensional Fourier transforms at different values of z to each other. It also indicates how the relative phase among the different plane waves change as they propagate along the z-direction. The change in the phase is a result of the fact that, although the plane waves all have the same wavelength (in three dimensions), they do not propagate in the same direction. We will call $\psi(a, b, z)$ the *propagation phase factor*.

For propagation from $z = z_1$ to $z = z_2$, the propagation phase factor is $\psi(a, b, \Delta z)$, where $\Delta z = z_2 - z_1$. The difference between the plane wave expansions at $z = z_1$ and $z = z_2$ is given by a relative factor of $\psi(a, b, \Delta z)$. For propagation over a zero distance ($\Delta z = 0$), the propagation phase factor equals unity: $\psi(a, b, 0) = 1$. Therefore, it has no effect when evaluated at the reference plane (input plane).

1.4.8 Steps of the Propagation Process

The plane wave approach to propagation, as set out in Section 1.4.5, can now be expressed in terms of mathematical expressions. Thus, the computational propagation process from

$z = 0$ to an arbitrary value of z for a complex-valued input function $g(x, y)$ at $z = 0$ is performed by the following three steps:

1. Compute the angular spectrum by computing the Fourier transform of the input function:

$$G(a, b) = \mathcal{F}\{g(x, y)\} = \iint_{-\infty}^{\infty} g(x, y) \exp[i2\pi(ax + by)] \, dx \, dy. \qquad (1.93)$$

2. Multiply the angular spectrum with the appropriate propagation phase factor:

$$G(a, b)\psi(a, b, z) = G(a, b) \exp\left[-i2\pi zc(a, b)\right]. \qquad (1.94)$$

3. Reconstruct the output function at z by computing the inverse Fourier transform of the previous result:

$$g(\mathbf{x}) = \mathcal{F}^{-1}\{G(a, b)\psi(a, b, z)\}$$

$$= \iint_{-\infty}^{\infty} G(a, b)\psi(a, b, z) \exp[-i2\pi(ax + by)] \, da \, db. \qquad (1.95)$$

1.4.9 Propagation Kernel

The expressions for the three steps in Equations 1.93 through 1.95 can be combined into one expression for the entire propagation process:

$$g(\mathbf{x}) = \mathcal{F}^{-1}\{\mathcal{F}\{g(x, y)\}\,\psi(a, b, z)\} = g(x, y) \otimes \mathcal{F}^{-1}\{\psi(a, b, z)\}, \qquad (1.96)$$

where the last expression follows from the Fourier theorem that states that the (inverse) Fourier transform of the product of two functions is equal to the convolution of the (inverse) Fourier transforms of the respective functions. Hence, the propagation process can be expressed as a convolution integral

$$g(\mathbf{x}) = \iint_{-\infty}^{\infty} g(u, v)\Psi(x - u, y - v, z) \, du \, dv \qquad (1.97)$$

with

$$\Psi(\mathbf{x}) = \mathcal{F}^{-1}\{\psi(a, b, z)\} = \iint_{-\infty}^{\infty} \exp(-i2\pi[ax + by + c(a, b)z]) \, da \, db. \qquad (1.98)$$

The function $\Psi(\mathbf{x})$ represents the propagation process in free space and is called the *propagation kernel*.

According to linear optical systems theory, the propagation kernel $\Psi(\mathbf{x})$ is the *impulse response* of the propagation process in free space. The fact that this process can be expressed as a convolution, means that free-space propagation is *shift invariant*, as expected. In the Fourier domain, the *transfer function* (i.e., the Fourier transform of the impulse response) for the free-space propagation process is given by the propagation phase factor Equation 1.92.

Propagation over zero distance must give back the original input function. This follows immediately from the fact that

$$\Psi(x, y, 0) = \iint_{-\infty}^{\infty} \exp(-i2\pi[ax + by]) \, da \, db = \delta(x)\delta(y), \tag{1.99}$$

where $\delta(\cdot)$ is the Dirac delta function.

The propagation kernel is expressed in terms of an integral—one that is difficult to solve analytically. By solving it numerically, one can see that it looks like a spherical wave radiating outward from the origin. This is reminiscent of the *Huygens principle*.

1.5 Fresnel and Fraunhofer

1.5.1 Paraxial Approximation

Often, a beam of light propagates predominantly in a specific direction. The angular spectrum of such a beam is narrow compared to the circular region of propagating plane waves. The dominant part of the angular spectrum is located in a region of the ab-plane for which

$$\frac{1}{\lambda} \gg \sqrt{a^2 + b^2}. \tag{1.100}$$

This is called the *paraxial (or Fresnel) approximation*, which means that all the plane waves in the beam propagate in a direction close to the axis of the beam. As a result of this approximation, the expression for $c(a, b)$ can be simplified by

$$c(a, b) = \sqrt{\frac{1}{\lambda^2} - a^2 - b^2} \approx \frac{1}{\lambda} - \frac{\lambda}{2}(a^2 + b^2). \tag{1.101}$$

The propagation phase factor then becomes

$$\psi(a, b, z) = \exp[-i2\pi c(a, b)z] \approx \exp(-ikz) \exp\left[i\pi \lambda z(a^2 + b^2)\right]. \tag{1.102}$$

The first exponential function in the final expression represents the phase retardation due to propagation over a distance z. (Note that the sign of the exponent is negative because of our sign convention, which implies that phase decreases with distance.) The second exponential contains a quadratic phase function. It is reminiscent of a Gaussian function, but with an imaginary argument.

Upon substituting the paraxial propagation phase factor of Equation 1.102 into the propagation kernel Equation 1.98, one finds

$$\Psi(\mathbf{x}) = \mathcal{F}^{-1}\left\{\psi(a, b, z)\right\}$$

$$\approx \exp\left(-ikz\right) \iint_{-\infty}^{\infty} \exp(-i2\pi[ax + by]) \exp\left[i\pi\lambda z(a^2 + b^2)\right] \, da \, db. \quad (1.103)$$

This is the inverse Fourier transform of a quadratic phase factor, which is again a quadratic phase factor, as one can see from the Fourier transform of a Gaussian function

$$\int_{-\infty}^{\infty} \exp\left(-\alpha t^2\right) \exp\left(-i2\pi ut\right) \, dt = \sqrt{\frac{\pi}{\alpha}} \exp\left(-\frac{\pi^2 u^2}{\alpha}\right). \quad (1.104)$$

Therefore

$$\Psi(\mathbf{x}) \approx \exp\left(-ikz\right) \frac{i}{\lambda z} \exp\left[-i\frac{\pi}{\lambda z}\left(x^2 + y^2\right)\right]. \quad (1.105)$$

1.5.2 Fresnel Diffraction Integral

Now we are ready to substitute the paraxial approximation of the propagation kernel back into the propagation integral. After substituting Equation 1.105 into Equation 1.97, we have

$$g(\mathbf{x}) = \frac{i}{\lambda z} \exp(-ikz) \iint_{-\infty}^{\infty} g(u, v) \exp\left\{-i\frac{\pi}{\lambda z}\left[(x - u)^2 + (y - u)^2\right]\right\} \, du \, dv.$$

$$(1.106)$$

This is the *Fresnel diffraction integral*. The paraxial approximation of the propagation kernel in Equation 1.105 is called the *Fresnel kernel*.

One can find a slightly different expression for the Fresnel diffraction integral by expanding the squares in the exponent of the Fresnel kernel. This gives

$$g(\mathbf{x}) = \frac{i}{\lambda z} \exp(-ikz) \exp\left[-i\frac{\pi}{\lambda z}\left(x^2 + y^2\right)\right] \quad (1.107)$$

$$\times \iint_{-\infty}^{\infty} g(u, v) \exp\left[-i\frac{\pi}{\lambda z}\left(u^2 + v^2\right)\right] \exp\left[i\frac{2\pi}{\lambda z}\left(xu + yv\right)\right] du \, dv.$$

From the above expression, we can see that one can express the Fresnel diffraction process in terms of a Fourier transform. One multiplies the input function with a quadratic phase function before and after performing the Fourier transform. Hence

$$g(\mathbf{x}) = \frac{i \exp(-ikz)}{\lambda z} Q(x, y, z)\mathcal{F}\left\{g(u, v)Q(u, v, z)\right\}, \quad (1.108)$$

where

$$Q(x, y, z) = \exp\left[-i\frac{\pi}{\lambda z}\left(x^2 + y^2\right)\right]. \tag{1.109}$$

Due to the paraxial approximation, Fresnel diffraction is not as accurate as the rigorous computations. However, in practice, rigorous computation is only used where Fresnel diffraction breaks down, such as over short distances. For longer distances, Fresnel diffraction is accurate enough. The rigorous procedure requires two Fourier transforms (actually a Fourier transform and an inverse Fourier transform), whereas the Fresnel procedure only requires one Fourier transform.

1.5.3 Fraunhofer Diffraction

When the distance z increases relative to the size of the input plane (input aperture), then at some point the quadratic phase factor under the integral can be neglected, because for $z \gg \pi(x^2 + y^2)/\lambda$, we have

$$\exp\left[-i\frac{\pi}{\lambda z}\left(x^2 + y^2\right)\right] \approx 1. \tag{1.110}$$

This is the *Fraunhofer approximation*, which implies that Fresnel diffraction integral reduces to

$$g(\mathbf{x}) = \frac{i\exp(-ikz)}{\lambda z} \exp\left[-i\frac{\pi}{\lambda z}\left(x^2 + y^2\right)\right] \tag{1.111}$$

$$\times \iint_{-\infty}^{\infty} g(u, v) \exp\left[i\frac{2\pi}{\lambda z}\left(xu + yv\right)\right] du\, dv.$$

Apart from an overall complex constant, the output function is the Fourier transform of the input function times a quadratic phase factor. If one is only interested in the intensity of the output function, one only needs to take the Fourier transform of the input and compute the modulus square of it.

The Fraunhofer approximation is only valid at large distances. Typically, for $z \gg D^2/\lambda$, where D is the size of the input function.

1.6 Thin Lens

Perhaps the most important optical element in most optical systems is the thin lens. With modern technology, lenses are constructed in so many different ways that it almost defies any attempt to provide a general definition of a lens. However, one can regard any optical devise with the ability to change the radius of curvature of the wavefront as a lens. Conventionally, a (positive)* lens is a piece of glass (or some other dielectric material) with (spherically) curved surfaces, giving it the property to focus light. If a plane wave is

---

* A negative lens increases divergence.

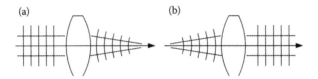

FIGURE 1.8   Positive lens (a) focusing a plane wave and (b) collimating a spherical wave.

incident on a positive lens along its symmetry axis, which is also called the *optical axis*, it will focus the light (cause it to converge) to a *focal point* located at a distance of one *focal length* behind the lens, as shown in Figure 1.8a. Likewise, if a point source is located at one focal length in front of a positive lens, the light, diverging from the point source, will be *collimated* by the lens, creating a plane wave behind the lens, as shown in Figure 1.8b. One can thus define a *front focal plane* and a *back focal plane*, respectively, located at distances of one focal length in front of and behind a lens.

The transmission function of a thin lens (see Ref. [3]) can be expressed as

$$t_{lens}(x, y) = A(x, y) \exp\left(\frac{i\pi r^2}{\lambda f}\right), \tag{1.112}$$

where $A(x, y)$ is an aperture function, which is usually circular, $r^2 = x^2 + y^2$, and f is the focal length.

1.6.1 Fourier Transforming Properties of the Thin Lens

One of the most powerful properties of a thin lens is the fact that one can use it to perform a two-dimensional Fourier transform instantaneously. To understand how this is possible, consider an arbitrary beam passing though a lens. One can express the beam in terms of its plane wave expansion, as in Equation 1.84. Each plane wave passing through the lens forms a focal point in the back focal plane. If the plane wave passes obliquely through the lens, the focal point is shifted to the side by a distance proportional to the angle of incidence. Since the amplitude of each plane wave in the beam is given by the angular spectrum of the beam, the distribution in the back focal plane will correspond to the angular spectrum. Hence, the lens breaks up the beam into its Fourier components—it performs a two-dimensional Fourier transform on the beam.

Now, we will give a quantitative analysis to show that the lens performs a Fourier transform on the input distribution $g(x, y)$, located in the front focal plane. To compute the distribution in the back focal plane, we perform a three-step process: The first step is just the Fresnel transform of the complex distribution, as given in Equation 1.108, with $z = f$. The next step is to multiply the result after the first Fresnel transform $g_1(x, y)$ with the transmission function of the lens, given in Equation 1.112. Note that the expression for the transmission function of the lens is equal to the complex conjugate of the phase factor in Equation 1.109 for $z = f$. The result is that the lens removes the $Q(x, y, f)$-factor

$$g_2(x, y) = g_1(x, y)Q^*(x, y, f) = \frac{i\exp(-ikf)}{\lambda f} \mathcal{F}\{g(x', y')Q(x', y', f)\}. \tag{1.113}$$

The third and final step is another Fresnel transform with $z = f$

$$g_3(u, v) = \left(\frac{i\exp(-ikf)}{\lambda f}\right)^2 Q(u, v, f)\mathcal{F}\{\mathcal{F}\{g(x', y')Q(x', y', f)\}Q(x, y, f)\}, \tag{1.114}$$

which becomes a convolution

$$g_3(u, v) = \left(\frac{i\exp(-ikf)}{\lambda f}\right)^2 Q(u, v, f)\left([g(x', y')Q(x', y', f)] \otimes \mathcal{F}\{Q(x, y, f)\}\right). \tag{1.115}$$

The Fourier transform of the quadratic phase function is proportional to the complex conjugate of a quadratic phase function

$$\mathcal{F}\{Q(x, y, f)\} = -i\lambda f\, Q^*(u, v, f), \tag{1.116}$$

which follows from Equation 1.104. The result then becomes

$$g_3(u, v) = \frac{i\exp(-i2kf)}{\lambda f} Q(u, v, f) \iint_{-\infty}^{\infty} g(x', y')Q(x', y', f)$$

$$\times \exp\left(\frac{i\pi}{\lambda f}[(u + x')^2 + (v + y')^2]\right) dx'\,dy'. \tag{1.117}$$

One can now expand the argument of the resulting exponential function to find that some of the factors cancel the remaining quadratic phase functions. The result is

$$g_3(u, v) = \frac{i\exp(-i2kf)}{\lambda f} \iint_{-\infty}^{\infty} g(x, y) \exp\left[\frac{ik}{f}(ux + vy)\right] dx\,dy. \tag{1.118}$$

Hence, the distribution in the back focal plane $g_3(u, v)$ is the Fourier transform of the distribution in the front focal plane $g(x, y)$, up to an overall complex constant.

1.7 Gaussian Beam Optics

1.7.1 Paraxial Wave Equation

The output distribution that one obtains from the Fresnel diffraction integral, given by Equation 1.106, always contains a factor of $\exp(-ikz)$. It suggests that the scalar field of an optical beam under the paraxial approximation generally has the form

$$U(\mathbf{x}) = A(\mathbf{x}) \exp(-ikz). \tag{1.119}$$

One can substitute this form into the Helmholtz equation to see what it produces. For this purpose, the Laplacian operator is separated into the z-dependent part and the transverse part:

$$\nabla^2 = \nabla_T^2 + \partial_z^2. \tag{1.120}$$

The resulting Helmholtz equation, divided by $\exp(-ikz)$, becomes

$$\nabla_T^2 A(\mathbf{x}) + \partial_z^2 A(\mathbf{x}) - i2k\partial_z A(\mathbf{x}) = 0. \tag{1.121}$$

Under the paraxial approximation, the variation of $U(\mathbf{x})$ in the z-direction is very close to $\exp(-ikz)$. Once this z-dependent phase factor is removed, the remaining function $A(\mathbf{x})$ has a very slow variation in the z-direction. As a result, one can assume that $\partial_z^2 A(\mathbf{x}) \approx 0$. Hence, under the paraxial approximation the equation becomes

$$\nabla_T^2 A(\mathbf{x}) - i2k\partial_z A(\mathbf{x}) = 0, \tag{1.122}$$

and is called the *paraxial wave equation*. (In analogy with quantum mechanics, it is also sometimes referred to as a (2+1 dimensional) *linear Schrödinger equation*. In a (nonlinear) Kerr medium, the equation becomes $\nabla_T^2 A - i2k\partial_z A + \xi|A|^2 A = 0$, where ξ represents the coupling strength of the nonlinear term. This equation is called the *nonlinear Schrödinger equation*.)

1.7.2 Gaussian Beam

The solutions of the paraxial wave equation form an orthogonal basis, just like those of the Helmholtz equation, discussed in Section 1.2.2. Before we discuss the general solution of the paraxial wave equation, we first consider a simple but important solution. It is given by

$$A(\mathbf{x}) = \frac{1}{z - C} \exp\left[\frac{-ikr^2}{2(z - C)}\right], \tag{1.123}$$

where C is an unspecified complex-valued constant and $r^2 = x^2 + y^2$. The real part of C simply represents a shift in the function along the z-axis. We can remove it by redefining the z-axis. That still leaves an imaginary part, which we define as $-iz_R$. The resulting expression becomes

$$A(\mathbf{x}) = \frac{1}{z + iz_R} \exp\left[\frac{-ikr^2}{2(z + iz_R)}\right]. \tag{1.124}$$

To satisfy the requirement that $A \to 0$ for $x, y \to \pm\infty$, we find that z_R must be positive. The parameter z_R is called the *Rayleigh range*.

When we set $z = 0$, we find

$$A(x, y, 0) = \frac{1}{iz_R} \exp\left[-\frac{k}{2z_R}r^2\right] = \frac{-i}{z_R} \exp\left[-\frac{\pi}{\lambda z_R}r^2\right]. \tag{1.125}$$

This is a (rotationally symmetric) Gaussian function in terms of the radial coordinate r. Apart from the $-i$ factor, the function at $z = 0$ is real-valued. The phase is, therefore, constant over the xy-plane at $z = 0$.

It is customary to define the width of this Gaussian function with a parameter ω_0, such that the power density at $r = \omega_0$ is $\exp(-2)$ times the peak value. As a result, the Gaussian function at $z = 0$ is given by

$$A(x, y, 0) = \frac{-i}{z_R} \exp\left[-\frac{r^2}{\omega_0^2}\right]. \tag{1.126}$$

The shape of this function in shown in Figure 1.9. Comparing the two different expressions for the Gaussian function in Equations 1.125 and 1.126, one finds a relationship between the Rayleigh range z_R and the beam radius ω_0, given by

$$z_R = \frac{\pi \omega_0^2}{\lambda}. \tag{1.127}$$

For $z \neq 0$, the function becomes complex-valued and the phase is not constant anymore. To separate the amplitude and phase of the function in Equation 1.124, one must separate the real and imaginary parts of its exponent. Thus, one finds

$$\frac{-ikr^2}{2(z + iz_R)} = \frac{-kr^2 z_R}{2(z^2 + z_R^2)} + \frac{-ikr^2 z}{2(z^2 + z_R^2)}. \tag{1.128}$$

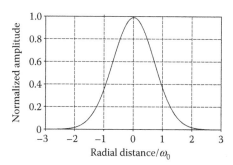

FIGURE 1.9 Gaussian function.

Hence, the solution of the paraxial wave equation can be written as that of a Gaussian function times a quadratic phase factor

$$A(\mathbf{x}) = \frac{1}{z + iz_R} \exp\left[\frac{-r^2}{\omega^2(z)}\right] \exp\left[\frac{-i\pi r^2}{\lambda R(z)}\right], \tag{1.129}$$

where

$$\omega(z) = \omega_0 \sqrt{1 + \frac{z^2}{z_R^2}} \tag{1.130}$$

is the radius of the beam as a function of z and

$$R(z) = z\left(1 + \frac{z_R^2}{z^2}\right) \tag{1.131}$$

is the radius of curvature of the parabolic wavefront as a function of z. This radius of curvature follows from a comparison with the paraxial approximation of a spherical wave,

$$\exp\left[-ik\left(R - \sqrt{R^2 - r^2}\right)\right] \approx \exp\left(\frac{-i\pi r^2}{\lambda R}\right). \tag{1.132}$$

The beam width and the radius of curvature are shown as functions of the propagation distance in Figures 1.10 and 1.11, respectively.

We can now interpret these expressions in terms of the following observations:

- The smallest beam radius exists at $z = 0$, where it is $\omega(z = 0) = \omega_0$. This is called the *waist* of the beam. At $z = z_R$, the beam radius becomes $\omega(z = z_R) = \sqrt{2}\omega_0$.
- Apart from the $1/(z + iz_R)$-factor, the function for the Gaussian beam in Equation 1.129 is symmetric upon inverting the z-axis (i.e., for $z \to -z$). Therefore, the width of the beam at z is equal to the width of the beam at $-z$ and the same applies for the magnitude of the radius of curvature of the wavefront.
- At $z = 0$, the radius of curvature is infinite (the wavefront is flat). It decreases to its minimum value $R_{min} = 2z_R$ at $z = \pm z_R$ and then increases again gradually.

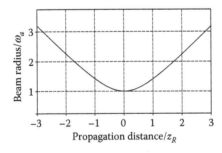

FIGURE 1.10   Gaussian beam radius as a function of propagation distance.

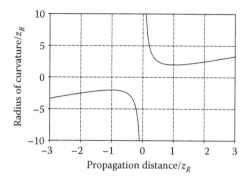

FIGURE 1.11   Gaussian beam wavefront radius of curvature as a function of propagation distance. (A negative radius of curvature implies the wavefront curve in the opposite direction.)

- The Rayleigh range is a scale parameter that determines the form of the Gaussian beam as a function of z. For $z \ll z_R$, the beam is effectively collimated—the wavefront is approximately flat and the beam radius remains approximately constant. For $z \gg z_R$, the beam is diverging—it is close to a spherical wave with a radius of curvature $R(z) \approx z$.
- The *numerical aperture* (NA) of the beam is determine by the beam divergence angle, θ_B for $z \gg z_R$. The numerical aperture is defined by $\mathrm{NA} = \sin\theta_B$. Under the paraxial approximation, which implies $\sin\theta_B \approx \theta_B \approx \tan\theta_B$, one can define the NA of a Gaussian beam as

$$\sin\theta_B = \mathrm{NA} \approx \frac{\omega(z)}{z} = \frac{\omega_0}{z}\sqrt{1 + \frac{z^2}{z_R^2}} \approx \frac{\omega_0}{z_R} = \frac{\lambda}{\pi\omega_0}. \qquad (1.133)$$

1.7.3 Gouy-Phase

The complex-valued factor $(z + iz_R)^{-1}$ can be written as

$$\frac{1}{z + iz_R} = \frac{\exp[-i\arctan(z/z_R)]}{\sqrt{z^2 + z_R^2}}. \qquad (1.134)$$

As a result, there is an additional z-dependent phase shift coming from this factor. This z-dependent phase

$$\gamma(z) = \arctan(z/z_R) \qquad (1.135)$$

is called the *Gouy-phase*. The Gouy-phase is shown as a function of propagation distance in Figure 1.12.

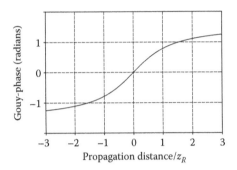

FIGURE 1.12   Gouy-phase as a function of propagation distance.

1.7.4 ABCD Matrix Method

The action of a thin lens on a Gaussian beam can be calculated with the aid of the Fresnel diffraction integral, after multiplying the Gaussian beam profile in the plane of the lens with the transmission function of the lens. One finds that the beam thus produced is again a Gaussian beam, but with a different Rayleigh range, different beam radius, and a different location of the beam waist. A simple method to determine the parameters of the Gaussian beam behind the lens is based on the ABCD matrices that are used to calculate the propagation of rays in geometrical optics. The rays are denoted by two-dimensional vectors. There is an ABCD matrix for every basic transformation that a ray can undergo as it propagates through an optical system. Such transformations include those that are produced by free-space propagation, passing though a lens or being refracted by a dielectric interface, to name but a few [5].

For Gaussian beams, one still uses the same ABCD matrices, but instead of performing vector-matrix multiplications, one specifies the Gaussian beam in terms of a q-parameter, given by

$$q = z + iz_R \tag{1.136}$$

and then calculates the transformation of the q-parameters with the formula

$$q' = \frac{Aq + B}{Cq + D}, \tag{1.137}$$

where A, B, C, and D are the elements from the ABCD matrix. At $z = 0$ (i.e., at the waist of the Gaussian beam), the q-parameter is just iz_R.

The ABCD matrix for free-space propagation is given by

$$\begin{bmatrix} A & B \\ C & D \end{bmatrix} = \begin{bmatrix} 1 & z \\ 0 & 1 \end{bmatrix} \tag{1.138}$$

After propagating a distance z (from $z = 0$), the q-parameter is given by the expression in Equation 1.136.

A thin lens with a focal length of f is represent by the ABCD matrix

$$\begin{bmatrix} A & B \\ C & D \end{bmatrix} = \begin{bmatrix} 1 & 0 \\ -1/f & 1 \end{bmatrix}. \tag{1.139}$$

This transforms the q-parameter in Equation 1.136 into

$$q' = \frac{(z + iz_R) + 0}{-1/f(z + iz_R) + 1} = \frac{fz(f - z) - fz_R^2 + if^2 z_R}{(f - z)^2 + z_R^2}. \tag{1.140}$$

For $z = f$ (i.e., when the waist is located one focal length in front of the lens), q' becomes

$$q' = -f + i\frac{f^2}{z_R}, \tag{1.141}$$

which means that the new Gaussian beam's waist is located one focal length behind the lens and the new Rayleigh range is given by

$$z_R' = \frac{f^2}{z_R} = \frac{f^2 \lambda}{\pi \omega_0}. \tag{1.142}$$

The radius of the new Gaussian beam's waist is

$$\omega_0' = \frac{f \lambda}{\pi \omega_0}. \tag{1.143}$$

1.7.5 Hermite–Gaussian Modes and Laguerre–Gaussian Modes

The Gaussian function given above is not the only solution of the paraxial wave equations. One can find a set of solutions that represents a complete orthogonal basis for arbitrary paraxial beams.

The set of orthogonal modes that one obtains depends on the coordinate system that is used. For Cartesian coordinates, one obtains the Hermite–Gaussian modes, given by

$$U_{mn}^{HG} = \mathcal{N} \frac{(1 + it)^{N/2}}{(1 - it)^{N/2+1}} H_m \left(\frac{\sqrt{2}u}{\sqrt{t^2 + 1}} \right) H_n \left(\frac{\sqrt{2}v}{\sqrt{t^2 + 1}} \right) \exp \left(\frac{-(u^2 + v^2)}{1 - it} \right), \tag{1.144}$$

where we use normalized coordinates $u = x/\omega_0$, $v = y/\omega_0$, and $t = z/z_R$, $\mathcal{N}$ is a normalization constant, H_x represents the Hermite polynomials [2], and m and n are integers such that $m + n = N$, with N being the total order of the polynomial.

In (normalized) polar coordinates, one obtains the Laguerre–Gaussian modes

$$U_{p\ell}^{LG}(\rho, \phi, t) = \mathcal{N} \frac{\rho^{|\ell|} \exp(i\ell\phi)(1 + it)^p}{(1 - it)^{p+|\ell|+1}} L_p^{|\ell|} \left(\frac{2\rho^2}{1 + t^2} \right) \exp \left(\frac{-\rho^2}{1 - it} \right), \tag{1.145}$$

where $L_p^{|\ell|}$ represents the associate Laguerre polynomials, p is the radial index (a nonnegative integer), ℓ is the azimuthal index (a signed integer), and $\mathcal{N}$ is a normalization constant given by

$$\mathcal{N} = \left[\frac{2^{|\ell|+1} p!}{\pi (p + |\ell|)!} \right]^{1/2}. \tag{1.146}$$

The polar coordinates (ρ, ϕ) are related to the normalized Cartesian coordinates by $u = \rho \cos \phi$ and $v = \rho \sin \phi$. The topological charge associated with each Laguerre–Gaussian mode is given by the integer ℓ. The order of the polynomial prefactor of these modes is given by $N = 2p + |\ell|$. The propagation distance t is defined relative to the waist where the beam has its smallest width. Note that these modes are nonrotating apart from the phase factor, and their intensity distributions are rotationally symmetric. The Gouy-phase is contained in the t-dependence in Equation 1.145. Further classes of laser beams will be covered in the later chapters.

References

1. Hocking, J. G. and Young, G. S. 1961. *Topology*. Toronto: Dover.
2. Abramowitz, M. and Stegun, I. A. 1972. *Handbook of Mathematical Functions*. Toronto: Dover.
3. Goodman, J. W. 1996. *Introduction to Fourier Optics*. 2nd Ed. New York: McGraw-Hill.
4. Clarke, R. H. and Brown, J. 1980. *Diffraction Theory and Antennas*. Chichester: Ellis Horwood.
5. Saleh, B. E. A. and Teich, M. C. 1991. *Fundamentals of Photonics*. New York: John Wiley & Sons.

2

From Classical to Quantum Optics

Thomas Konrad
University of Kwazulu-Natal

Sandeep Goyal
University of Kwazulu-Natal

2.1 Introduction

The aim of this chapter is to give an introduction to methods of quantum mechanics that can be applied to optics as well as an introduction to quantum optics (QO). The purpose of the latter is to grant a fast access for the reader who is familiar with optics and wishes to extend her/his studies to the realm where quantum effects determine the behavior of light.

It is well known that physical optics and quantum mechanics partly describe similar physics, most prominently, interference phenomena. Because of the close relationship, methods from quantum mechanics can be used to solve problems in optics in an elegant and efficient way. Here, we focus on the transfer from quantum mechanics to optics. Of course, the converse direction is not less fruitful and has already left a deep imprint in quantum mechanics, which is the younger of the two disciplines. For example, the transfer of Huygens' principle can be seen to give rise to the formulation of quantum mechanics in terms of Feynman's path integrals [15].

In the first part, we review the application of the Dirac formalism and operator algebra to the optics of paraxial light beams that are governed by an evolution equation that resembles the Schrödinger equation in quantum mechanics. In particular, we formulate the description of paraxial optics by means of state vectors and operators in terms of postulates to compare them to those of quantum mechanics and find—remarkably—great similarity except for measurements.

The second part describes the quantization of the electromagnetic field and its application to passive linear optics. We conclude this chapter by discussing the difference between quantum entanglement and its counterpart in classical optics.

2.2 Paraxial Optics in the Dirac Formalism

2.2.1 Paraxial Wave Equation

Monochromatic light of frequency ω with linear polarization is characterized by the Helmholtz equation (cf. Chapter 1, Equation 1.33).

$$\nabla^2 g + k^2 g = 0, \tag{2.1}$$

where $k = \omega/c$ is the magnitude of the wave vector and g is the scalar part of the electric field $\vec{E} = g\vec{n}$ (here $\vec{n}$ is a unit-length polarization vector). Light beams that propagate with small divergence along the z-direction satisfy the paraxial approximation (cf. Chapter 1, Equation 1.100)

$$k^2 \gg k_x^2 + k_y^2, \tag{2.2}$$

where x and y denote directions orthogonal to the z-axis. By writing $g(x, y, z) = u(x, y, z) \exp(-ikz)$ and neglecting the second derivative of the slowly varying envelop function u with respect to z, we obtain the paraxial wave equation,

$$-2ik\frac{\partial}{\partial z}u(x, y, z) = \left(\frac{\partial^2}{\partial x^2} + \frac{\partial^2}{\partial y^2}\right)u(x, y, z), \tag{2.3}$$

which determines in good approximation the electric field of the beam along the z-direction. In other words, given the scalar part $u(x, y, z_0)$ of the electric field in a transversal plane at $z = z_0$, the solution of the paraxial wave equation yields its full z-dependence $u(x, y, z)$. Therefore, the state of a monochromatic light beam can be represented by the amplitudes and phases that the corresponding electric field, assumes of a single transversal plane. All complex-valued functions $u(x, y, z_0)$ on such a plane form a vector space since the linear combination of two such functions yields another function. Moreover, this vector space is endowed with an inner product that measures the "overlap" between any two functions u and v on the plane

$$\langle u(z_0), v(z_0) \rangle = \int_{-\infty}^{\infty} \int_{-\infty}^{\infty} dx \, dy \, u^*(x, y, z_0) v(x, y, z_0) \tag{2.4}$$

and induces a norm, the square of which is proportional to the total intensity of the electromagnetic wave integrated over the plane

$$||u(z_0)||^2 \equiv \langle u(z_0), u(z_0) \rangle = \int_{-\infty}^{\infty} \int_{-\infty}^{\infty} dx \, dy \, |u(x, y, z_0)|^2 = E_0^2. \tag{2.5}$$

Please note that we followed the physicists' convention to define the inner product linear in its second argument and that we only considered functions $u(z_0)$ with finite norm since the integrated intensity of the light beam is always finite. Hence, these functions form a Hilbert space of square-integrable functions on a real plane $\mathcal{L}^2(\mathbb{R}^2)$.

2.2.2 States of Paraxial Beams

Instead of using the scalar part of the electric field $u(x, y, z)$ as a function of the points on the transversal plane, we rather consider it as a vector denoted in Dirac's notation $|u(z)\rangle$ in an abstract infinite-dimensional Hilbert space $\mathcal{H}$. Thus, a coordinate-free representation is obtained. Here, z is viewed as a parameter rather than a coordinate. The concrete functional dependence $u(x, y, z)$ can at any point of the calculation be recovered from the abstract vector $|u(z)\rangle$ as its coordinate representation with respect to the so-called position basis.

Not only is the electric field determined by its values on a single transversal plane represented by a vector $|u(z_0)\rangle$, this vector is also related to the vector $|u(z)\rangle$ representing the field values on any other transversal plane by a linear transformation because of the linearity of the differential equation (2.3) as demonstrated below. This guarantees the consistency of the following postulate that is similar to the postulate on states of quantum systems in quantum mechanics.

Postulate 1

The state of a paraxial light beam is represented by a vector $|u\rangle$ in an infinite-dimensional Hilbert space isomorphic to $\mathcal{L}^2(\mathbb{R}^2)$. The square of the norm of $|u\rangle$ is proportional to the total intensity of the light beam on any transversal plane $||u||^2 \equiv \langle u \,|\, u \rangle = E_0^2$.

2.2.3 Dirac Formalism

In the first postulate, we used the Dirac notation of an inner product $\langle u \,|\, u \rangle$ of a vector with itself. This can also be seen as a linear mapping $\langle u |$, corresponding to the vector $|u\rangle$ acting on $|u\rangle$. In general, $\langle u |$ can transform any other vector $|v\rangle$ in the Hilbert space $\mathcal{H}$, into a complex number by means of the inner product $\langle u |(|v\rangle) \equiv \langle u \,|\, v \rangle \in \mathbb{C}$. The linearity of this mapping is inherited from the linearity of the inner product:

$$\langle u | (\alpha | v_1 \rangle + \beta | v_2 \rangle) = \alpha \langle u | v_1 \rangle + \beta \langle u | v_2 \rangle, \tag{2.6}$$

where the coefficients α, β are complex numbers. With this notation the orthogonal projection of a vector $| v \rangle$ onto a state $| w \rangle$ can be written as,

$$| v \rangle \rightarrow | w \rangle \langle w | (| v \rangle) \equiv | w \rangle \langle w | v \rangle = \alpha | w \rangle \quad \text{with } \alpha = \langle w | v \rangle. \tag{2.7}$$

This leads to an elegant way to find representations with respect to orthonormal bases. Let us assume that the sequence of vectors $| 1 \rangle, | 2 \rangle, \ldots$ forms a complete orthonormal system (orthonormal basis), that is, $\langle i | j \rangle = \delta_{ij}$ and if any vector $| u \rangle \in \mathcal{H}$ can be written as a linear combination

$$| u \rangle = \sum_{i=0}^{\infty} \alpha_i | i \rangle \tag{2.8}$$

with unique coefficients $\alpha_i = \langle i | u \rangle$, then

$$| u \rangle = \sum_{i=0}^{\infty} \alpha_i | i \rangle = \sum_{i=1}^{\infty} | i \rangle (\langle i | u \rangle) = \left(\sum_{i=0}^{\infty} | i \rangle \langle i | \right) | u \rangle \quad \text{for all } | u \rangle \in \mathcal{H} \tag{2.9}$$

$$\Rightarrow \sum_{i=0}^{\infty} | i \rangle \langle i | = \mathbb{I} \tag{2.10}$$

which means that the sum of the projectors $| i \rangle \langle i |$ onto orthonormal basis vectors is equal to the identity operator $\mathbb{I}$. Equation 2.10 is the condition to check whether a set of orthonormal vectors is complete, that is, it spans the whole Hilbert space and thus forms a basis. Using the completeness relation (2.10), one also finds the basis representation of linear operators A mapping vectors $| u \rangle \in \mathcal{H}$ onto vectors $| w \rangle = A | u \rangle \in \mathcal{H}$:

$$A = \mathbb{I} A \mathbb{I} = \left(\sum_{i=1}^{\infty} | i \rangle \langle i | \right) A \left(\sum_{j=1}^{\infty} | j \rangle \langle j | \right) = \sum a_{ij} | i \rangle \langle j |, \tag{2.11}$$

where $a_{ij} \equiv \langle i | A | j \rangle$ are the matrix elements of A with respect to the basis $| 1 \rangle, | 2 \rangle, \ldots$. If there exists an orthonormal basis of eigenvectors $| a \rangle$ of A, that is, vectors with the property $A | a \rangle = a | a \rangle$, the basis representation of A simplifies:

$$A = \sum_a a | a \rangle \langle a |, \tag{2.12}$$

where the numbers $a \in \mathbb{C}$ are known as eigenvalues or spectral values of A. The matrix of A is then diagonal ($\langle b | A | a \rangle = a \, \delta_{ab}$). A spectral decomposition (2.12) of A exists if and only if A is a "normal" operator, that is, the commutator $[A, A^\dagger] \equiv AA^\dagger - A^\dagger A$ equals zero [12]. Here, $A^\dagger \equiv \sum_i a_{ji}^* |i\rangle\langle j|$ is the adjoint of A with respect to any orthonormal basis ($|i\rangle$). For example, unitary operators ($U^\dagger = U^{-1}$) and self-adjoint operators ($O^\dagger = O$) possess spectral decompositions with complex phases $\exp(i\varphi)$ and real numbers, respectively, as spectral values. An important feature of operators on Hilbert spaces is the definition of a function of an operator. For any function $f : \mathbb{C} \rightarrow \mathbb{C}$, the corresponding function of a normal operator A is defined via its spectral decomposition as

$$f(A) = \sum_a f(a) |a\rangle\langle a|. \tag{2.13}$$

2.2.4 Observables of Paraxial Light

After these mathematical preliminaries, we can represent the state $|u(z)\rangle$ of an electric field with respect to the position states $|x, y\rangle$. These states represent an electric field that has only a nonvanishing intensity in a single point (x, y) on a particular transversal plane and can, therefore, only be prepared approximately. They can be treated as an orthonormal basis satisfying a continuous version of the completeness relation (2.10), that is, $\int dx\, dy\, |x, y\rangle\langle x, y| = \mathbb{I}$. Hence

$$|u(z)\rangle = \mathbb{I}\, |u(z)\rangle = \int dx\, dy\, |x, y\rangle\langle x, y| u(z)\rangle = \int dx\, dy\, u(x, y, z) |x, y\rangle. \tag{2.14}$$

This means that we recover the scalar function $u(x, y, z)$ in the position representation as the coordinate function of our abstract state vector $|u(z)\rangle$

$$u(x, y, z) = \langle x, y| u(z)\rangle. \tag{2.15}$$

Physically, expansion (2.14) represents the electric field in the plane as generated by the superposition of light from point sources. For example, a plane wave $|k_x, k_y\rangle$ with a transversal wave vector (k_x, k_y) has the position representation $\langle x, y| k_x, k_y\rangle = \exp\{-i(k_x x + k_y y)\}/2\pi$. Since these functions also form an orthonormal basis with $\int dk_x dk_y\, |k_x, k_y\rangle\langle k_x, k_y| = \mathbb{I}$, we can write any state $|u(z)\rangle$ as

$$|u(z)\rangle = \int dk_x dk_y\, |k_x, k_y\rangle\langle k_x, k_y| u(z)\rangle \tag{2.16}$$

with the coefficient function

$$\langle k_x, k_y| u(z)\rangle = \langle k_x, k_y| \mathbb{I}| u(z)\rangle = \int dx\, dy\, \langle k_x, k_y| x, y\rangle\langle x, y| u(z)\rangle$$

$$= \frac{1}{2\pi} \int dx\, dy\, e^{i(k_x x + k_y y)} u(x, y, z) \equiv \tilde{u}(k_x, k_y, z), \tag{2.17}$$

where we have used the antisymmetry of the inner product $\langle k_x, k_y \,|\, x, y \rangle = (\langle x, y \,|\, k_x, k_y \rangle)^*$. The function $\tilde{u}(k_x, k_y, z)$ is the Fourier transform of the scalar function $u(x, y, z)$ with respect to the coordinates x, y. Note that the Dirac formalism comes with built-in Fourier transform and is, therefore, *predestined* to represent Fourier optics as becomes obvious below.

To take full advantage of the situation, the Lie algebra of operators corresponding to the position and momentum representation of the electromagnetic field is essential. Moreover, from an operational point of view, it is necessary to represent measurements that can be carried out on light. Light is detected by means of photodetectors often after some manipulation using filters that project onto certain components of the electromagnetic field. The simplest measurement is the direct detection of the intensity profile of a beam by means of an array of photodetectors or photography. Lenses are employed to measure the transversal spatial frequencies in the focal plane behind the lens. Spatial light modulators serve to select certain light modes by transforming them into a Gaussian mode, which can then be coupled into an optical fiber and detected. For example, in this way, Laguerre–Gaussian (LG) modes are detected, which carry orbital angular momentum (for more details, see Chapters 4 and 11) and the angular momentum spectrum of a beam is measured. Polarization can be detected by analyzers that only transmit or reflect electromagnetic fields that oscillate in a given direction. The common denominator of many spectroscopic methods is to map different light modes into spatially separated detectors.

The outcome of the measurement of an intensity profile in a (transverse) plane can be modeled by an observable, which detects signals at points (x, y) (or small areas, depending on the spatial resolution of the detector used). The intensity of the signal at each point is proportional to

$$I(x, y) = |u(x, y, z)|^2 = \langle (\,|\, x, y \rangle \langle x, y \,|\,) \rangle_{u(z)} = \langle u(z) \,|\, x, y \rangle \langle x, y \,|\, u(z) \rangle.$$

In practice, only a time average of $I(x, y)$ can be measured with the current detector technology. This observable is represented by the operator

$$\vec{X} = \int d^2\vec{x}\, \vec{x} \,|\, \vec{x} \rangle \langle \vec{x} \,|, \tag{2.18}$$

where we introduced the notation $\vec{x} \equiv (x, y)$. A measurement of this observable will lead in general to a detection of signals from many points simultaneously. The measurement of the transversal wave vector $\vec{k} \equiv (k_x, k_y)$ is represented by the "momentum" operator

$$\vec{P} = \int d^2\vec{k}\, \vec{k} \,|\, \vec{k} \rangle \langle \vec{k} \,|. \tag{2.19}$$

The expectation value of this observable $\langle \vec{P} \rangle = \langle u(z) \,|\, \vec{P} \,|\, u(z) \rangle$ is proportional to the momentum of the electromagnetic wave in the transversal plane at z. Both the position operator and the momentum operator act on an abstract state vector $|u(z)\rangle$ by first transforming it into the appropriate spectral representation. This can be seen at the example of the action of one component P_x of the momentum operator:

$$|u'(z)\rangle \equiv \vec{P}|u(z)\rangle = \int d^2\vec{k}\,\vec{k}\,|\vec{k}\rangle\langle\vec{k}|u(z)\rangle. \qquad (2.20)$$

Note that $\langle\vec{k}|u(z)\rangle = \tilde{u}(k_x, k_y, z)$ is the Fourier transform of the scalar field $u(x, y, z)$ (cf. Equation 2.17). What is the position representation of the resulting vector $|u'(z)\rangle$ and how is it obtained from $u(x, y, z)$?

$$\langle\vec{x}|\vec{P}|u(z)\rangle = \int d^2\vec{k}\,\vec{k}\langle\vec{x}|\vec{k}\rangle\langle\vec{k}|u(z)\rangle = \frac{1}{2\pi}\int d^2\vec{k}\,\vec{k}\,e^{i\vec{k}\cdot\vec{x}}\langle\vec{k}|u(z)\rangle$$

$$= \frac{1}{2\pi}\int d^2\vec{k}\,(-i\nabla)e^{i\vec{k}\cdot\vec{x}}\langle\vec{k}|u(z)\rangle = -i\nabla\int d^2\vec{k}\,\langle\vec{x}|\vec{k}\rangle\langle\vec{k}|u(z)\rangle$$

$$= -i\nabla\langle\vec{x}|u(z)\rangle, \qquad (2.21)$$

where we used the completeness of the momentum eigenstates to arrive at the expression in the last line. Thus the action of the momentum operator on the scalar field is that of a differential operator, for example, P_x acts as $u(x, y, z) \rightarrow -i\frac{d}{dx}u(x, y, z)$. Using the action of X and P in the position representation, it is now easy to check that not all components of these observables do commute:

$$[x_l, -i\frac{d}{dx_k}]\,u(x, y, z) = i\,\delta_{l,k}u(x, y, z) \;\Leftrightarrow\; [X_l, P_k] = i\,\delta_{lk} \quad \text{for } l, k = x, y. \qquad (2.22)$$

Instead, they reproduce the Heisenberg algebra known from the position and momentum observables of quantum mechanical systems.

For any spectral analysis of the light, that is, decomposition into a set of orthonormal modes and measurement of their intensities, a corresponding observable is naturally defined in the way done above. Two such sets are given by the Hermite–Gaussian and the LG modes produced by lasers (see Chapter 1 for the definition and Chapter 8 for more examples of modes in other geometries), which are solutions of the paraxial wave equation. For example, consider the observables

$$\vec{R} = \sum_{m,n}\binom{m}{n}|m, n\rangle\langle m, n|, \quad L_z = \sum_{l,p}l\,|l, p\rangle\langle l, p|. \qquad (2.23)$$

While $\vec{R}$ represents a detection of the intensities of a decomposition into Hermite–Gaussian modes $\langle x, y|m, n\rangle \equiv u_{mn}^{HG}$ (cf. Chapter 1, Equation 1.47), L_z is an example of a degenerate observable that distinguishes only the l- but not the p-index of LG modes $\langle\rho, \phi|l, p\rangle \equiv u_{lp}^{LG}$ (cf. Chapter 1, Equation 1.48), where ρ and ϕ are polar coordinates. Since $u_{lp}^{LG} = u_{lp}(\rho)\exp(il\phi)$, the same method as for momentum leads to the position representation of L_z as the differential operator $-id/d\phi$. $L_z = XP_y - YP_x$ represents the z-component of the orbital angular momentum of a paraxial light beam [1]. The considerations in this section can be summarized as follows:

Postulate 2

Each observable of a monochromatic paraxial light beam can be represented by a self-adjoint operator acting on the Hilbert space of the system. A measurement of the observable $O = \sum_o o \, |o\rangle \langle o|$ yields the spectral values o of the operator together with the intensities given by $I = |\langle o \, | u(z)\rangle|^2$, with $|u(z)\rangle$, the state of the electric field in the transversal plane where the intensity is measured.

Although not discussed here, based on the electric field in only a single transversal plane, it is possible to define three–dimensional position and momentum operators $\vec{X} = \int dx \, dy \vec{x} \, | x, y\rangle \langle x, y|$ and $\vec{P} = \int dk_x dk_y \vec{k} \, | k_x, k_y\rangle \langle k_x, k_y|$ with $\vec{x} = (x, y, z)$, where z is defined by the plane in which the light is detected and $\vec{k} = (k_x, k_y, (k^2 - k_x^2 - k_y^2)^{1/2})$.

For any two different sets of orthonormal modes, one obtains noncommuting observables. Usually, it is assumed that all observables in classical physics commute. For example, all observables of a particle in classical mechanics can be written as functions of its position and momentum values at a given time and thus commute. The difference in optics is that optics is a (classical) field theory with states that are functions on a transversal plane and not just points in phase space as in classical mechanics.

Postulate 2 in the context above is only dealing with observables of the spatial part of the electromagnetic field. It can be generalized to incorporate polarization by considering $|u(z)\rangle$ as a vector in the tensor product space that represents the set of pure states of the spatial and polarization part of light; see Section 2.6.

Unlike quantum mechanics, there is no mention of a collapse of the wave function due to a measurement for optics. In optics, the measurement rather yields different spectral values of the observable with nonvanishing intensities simultaneously. However, the integrated signal from a measurement of classical light in a certain mode per time unit is the same as would be obtained from repeated measurements on single photons, that is, elementary excitations of the same mode.

2.3 Propagation of Monochromatic Light

The propagation of a paraxial beam of monochromatic light is represented by the paraxial wave equation. In the Dirac formalism, it reads

$$i\frac{\partial}{\partial z}\langle x, y \, | u(z)\rangle = -\frac{1}{2k}\nabla^2 \langle x, y \, | u(z)\rangle = -\frac{1}{2k}\langle x, y \, |\vec{P}^2 \, | u(z)\rangle \qquad (2.24)$$

$$\Leftrightarrow \quad i\frac{\partial}{\partial z} \, | u(z)\rangle = -\frac{1}{2k}\vec{P}^2 \, | u(z)\rangle, \qquad (2.25)$$

where we used Equation 2.21 in the first line and the completeness of the basis ($|x, y\rangle$) to infer from the paraxial equation in position representation (first line) the abstract paraxial equation. The solution of the paraxial Equation 2.25 is given in terms of an operator that propagates the state of the electric field on an initial plane at z_0 to the plane at z:

$$|u(z)\rangle = U(z - z_0)\,|u(z_0)\rangle \quad \text{with } U(z - z_0) = \exp\left(-i\frac{1}{2k}\vec{P}^2(z - z_0)\right). \tag{2.26}$$

Note that U is unitary since $(U(z - z_0))^\dagger = (U(z - z_0))^{-1} = U(z_0 - z)$ and can be written in its spectral decomposition:

$$U(z - z_0) = \int d^2\vec{k}\,\exp\left(\frac{-i}{2k}(k_x^2 + k_y^2)(z - z_0)\right)\,|\vec{k}\rangle\,\langle\vec{k}|. \tag{2.27}$$

The paraxial equation resembles the Schrödinger equation of a free massive particle moving in a plane where time replaces the propagation coordinate z and Planck's constant assumes the value $\hbar = 1$. Indeed, a Gaussian electric field with width Δr on the transversal plane z_0 disperses just like the wave function of a free particle to a Gaussian with width $\Delta r' = |z - z_0|/(2k\Delta r)$ at z, which can be determined from $u(x, y, z) = \langle x, y|U(z - z_0)|u(z)\rangle$.

Moreover, it should be kept in mind that the paraxial wave equation is just an approximation of the Helmholtz equation (2.1) under assumption (2.2). The Helmholtz equation is a linear differential equation too and allows to use the same formalism introduced for the paraxial wave equation. Also, in this case, we can represent the scalar part of the electric field on a plane as a vector $|g(z)\rangle$ with its propagation along the orthogonal z-axis governed by the differential equation

$$\frac{\partial^2}{\partial z^2}|g(z)\rangle = -(k^2 + P_x^2 + P_y^2)|g(z)\rangle \quad \text{and thus} \tag{2.28}$$

$$|g(z)\rangle = U_H\,|g(z_0)\rangle \tag{2.29}$$

$$\text{with } U_H = \exp\left(-i\sqrt{k^2 + P_x^2 + P_y^2}\,(z - z_0)\right) \tag{2.30}$$

$$= \int d^2\vec{k}\,\exp\left(-i\sqrt{k^2 + k_x^2 + k_y^2}\,(z - z_0)\right)\,|\vec{k}\rangle\langle\vec{k}|. \tag{2.31}$$

That $|g(z)\rangle$ in Equation 2.29 is the solution of the Helmholtz equation (2.28) that can be checked by direct insertion. Therefore, in the general case, the transversal electric field at

z is obtained from the transversal field at z_0 from (cf. Chapter 1, Equation 1.96).

$$g(x, y, z) = \langle x, y | U_H | g(z_0) \rangle$$
$$= \int dx' dy' \langle x, y | U_H | x', y' \rangle \langle x', y' | g(z_0) \rangle$$
$$= \int dx' dy' \langle x, y | U_H | x', y' \rangle g(x', y', z_0). \tag{2.32}$$

Note that under the paraxial approximation $k^2 \gg k_x^2 + k_y^2$, the square root in the spectral values of U_H in Equation (2.31) can be approximated by

$$\sqrt{k^2 + k_x^2 + k_y^2} \approx k \left(1 + \frac{1}{2k} \left(k_x^2 + k_y^2 \right) \right) \tag{2.33}$$

and the paraxial propagator U given in Equation 2.27 is recovered:

$$U_H \approx e^{ikz} U. \tag{2.34}$$

Postulate 3

The electric field of monochromatic light in a transverse plane can be obtained from the electric field on another transverse plane by a unitary operator depending on the positions (z, z_0) of the planes, that is, $|u(z)\rangle = U(z, z_0)|u(z_0)\rangle$.

This postulate resembles the quantum mechanical postulate for the time evolution of closed quantum systems, which are also given in terms of unitary operators.

2.4 Imaging Properties of Thin Lenses Using Operator Algebra

The power of the Dirac formalism in classical optics can be seen at the example of thin lenses as shown by Stoler [18]. For the sake of simplicity, we do not take into account finite apertures. The action of a thin lens on the transverse plane of the light beam is represented by the transmission function (cf. Chapter 1, Equation 1.114):

$$t(x, y) = \exp \left(-i \frac{k}{2f} (x^2 + y^2) \right), \tag{2.35}$$

where k is the wave number and f is the focal length of the lens. A beam with the field amplitude $u(x, y, z_0)$ passing through the lens placed in the transversal plane at z_0 transforms as follows:

$$\tilde{u}(x, y, z_0) = t(x, y) u(x, y, z_0). \tag{2.36}$$

In Dirac notation, the action of the lens can be expressed by means of the unitary operator T:

$$T = \exp\left(-i\frac{k}{2f}(X^2 + Y^2)\right), \qquad (2.37)$$

with X and Y being the components of the position operator defined in Equation 2.18. Equation 2.36 can then be expressed as

$$|\tilde{u}(z_0)\rangle = T|u(z_0)\rangle, \qquad (2.38)$$

which can be seen by multiplying $\langle x, y|$ on both sides of the equation from the left. The mapping properties of the lens can now be explored by considering an object that generates an electric field $|u_1\rangle \equiv |u(0)\rangle$ in the transversal plane at $z = 0$. A thin lens with focal length f is positioned in a transversal plane at a distance s_1 (object distance, cf. Figure 2.1). Then the electric field $|u_2\rangle \equiv |u(s_2 + s_1)\rangle$ at a distance s_2 behind the lens is obtained by the consecutive action of the free propagator U given by Equation 2.27, the action of the lens operator T, and another free propagator U again.

$$|u_2\rangle = G|u_1\rangle \quad \text{with } G \equiv U(s_2)\,T\,U(s_1). \qquad (2.39)$$

For the sake of simplicity, we suppress the coordinate y in the following. The problem is symmetric with respect to rotations about the propagation axis z and the same results are obtained for the x and y coordinates. Thus, the operators U and T read

$$U(s) = \exp\left(-i\frac{s}{2k}P^2\right) \quad \text{and} \quad T = \exp\left(-i\frac{k}{2f}X^2\right), \qquad (2.40)$$

where P is the x-component of the momentum operator $\vec{P}$ (2.19).

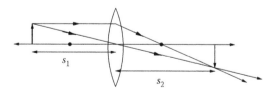

FIGURE 2.1 Principal ray diagram for a thin converging lens. In this diagram, the object is placed at a distance s_1 in front of a lens with the focal length $f > 0$ that generates a real image at distance s_2 behind the lens.

The task at hand is to find the kernel $g(x, x') = \langle x|G|x'\rangle$ since

$$|u_2\rangle = G|u_1\rangle,$$

$$\Leftrightarrow \quad \langle x|u_2\rangle = \int_{x'} \langle x|G|x'\rangle\langle x'|u_1\rangle dx',$$

$$\Leftrightarrow \quad u_2(x) = \int_{x'} g(x, x')u_1(x'). \tag{2.41}$$

To determine $g(x, x')$, we need to calculate $G^\dagger X G$. Using the Baker–Campbell–Hausdorff formula [16]

$$e^A Be^{-A} = B + [A, B] + \frac{1}{2!}[A, [A, B]] + \frac{1}{3!}[A, [A, [A, B]]] + \cdots \tag{2.42}$$

One finds

$$GXG^\dagger = \left(1 - \frac{s_1}{f}\right)X - \frac{s_1 s_2}{k}\left(\frac{1}{s_1} + \frac{1}{s_2} - \frac{1}{f}\right)P. \tag{2.43}$$

Let us choose the object distance s_1 and the distance s_2 of the second plane from the lens, such that

$$\frac{1}{s_1} + \frac{1}{s_2} = \frac{1}{f}. \tag{2.44}$$

As shown below, we obtain subject to this condition an image of the electromagnetic field from the object plane in the second plane. This gives rise to the famous lens Equation 2.44, where s_2 is the distance of the image from the lens and f is the focal length of the lens. Note that if s_1 and s_2 do not satisfy this equation, the image will not be positioned at s_2 and hence, the light distribution at s_2 generates a blurred image of the original distribution.

Once s_1 and s_2 are chosen to satisfy the lens equation (2.44), the term corresponding to momentum in Equation 2.43 will vanish and we are left with

$$GXG^\dagger = \left(1 - \frac{s_1}{f}\right)X,$$

$$= \frac{1}{m}X, \tag{2.45}$$

with $m \equiv -s_2/s_1$. Interestingly, Equation 2.45 indicates that the position operator X is mapped by G to a multiple of itself. This corresponds to the so-called Heisenberg picture, where observables change but states are considered constant (cf. Ref. [16]). It implies that

$G|x'\rangle$, the image of a point source at x', is an eigenvector of X with eigenvalue mx', as can be seen by the following calculation:

$$GXG^\dagger = \frac{1}{m}X,$$

$$\Leftrightarrow \quad GX = \frac{1}{m}XG, \tag{2.46}$$

$$\Rightarrow GX|x'\rangle = \frac{1}{m}XG|x'\rangle,$$

$$\Leftrightarrow \quad X\left(G|x'\rangle\right) = mx'\left(G|x'\rangle\right). \tag{2.47}$$

Therefore, we can write $G|x'\rangle = C|mx'\rangle$ and the constant C is determined by

$$\langle x_1 | x_2 \rangle = \delta(x_1 - x_2) = \langle x_1 | G^\dagger G | x_2 \rangle,$$

$$= C^2 \langle mx_1 | mx_2 \rangle,$$

$$= C^2 \delta(mx_1 - mx_2)$$

$$= \frac{C^2}{m}\delta(x_1 - x_2), \tag{2.48}$$

$$\Rightarrow C = \sqrt{m}, \tag{2.49}$$

where we have used the property of the Dirac δ-function, $\delta(a(x - x')) = \delta(x - x')/a$. In other words, the field $|x'\rangle$ of a light source at a point x' in the object plane is mapped to a point-like light field $|x''\rangle = G|x'\rangle = \sqrt{m}|mx'\rangle$, which is located at $x'' = mx$ on the x-axis of the image plane. Thus, we identify m as the magnification of the lens as a function of object and image distances:

$$m = \frac{x''}{x'} = -\frac{s_2}{s_1}. \tag{2.50}$$

With these results, the sought kernel $g(x, x')$ reads

$$\langle x|G|x'\rangle = g(x, x') = \sqrt{m}\langle x| mx'\rangle,$$

$$= \sqrt{m}\delta(x - mx'). \tag{2.51}$$

Hence, we can express the scalar part u_2 of the electric field in the image plane given a general input field u_1 by

$$u_2(x) = \int_{x'} G(x, x')u_1(x')dx',$$

$$= \int_{x'} \sqrt{m}\delta(x - mx')u_1(x')dx',$$

$$= \frac{1}{\sqrt{m}}u_1\left(\frac{x}{m}\right). \tag{2.52}$$

This equation shows that the intensity of the field and the size of the image are increased by the factor m compared to the intensity and size of the object. If m is negative (e.g., for the real object ($s_1 > 0$) and real image ($s_2 > 0$)), then the image is inverted, otherwise it is upright.

If we choose the object and image plane to be positioned in the focal planes in front and behind the lens respectively, that is, $s_1 = s_2 = f$, the X-term on the right-hand side of Equation 2.43 vanishes and we obtain

$$GXG^\dagger = -\frac{f}{k}P. \tag{2.53}$$

As a result, one can show, similar to the calculation above, that $G\,|\,x'\rangle$ is an eigenvector $|\,k'\rangle$ of the momentum operator P with eigenvalue $k' = -kx'/f$. Therefore, we can write $G\,|\,x'\rangle = \gamma\,|\,k'\rangle$, where $\gamma = \sqrt{k/f}$ is the normalization constant. Hence, the scalar part of the electric field in the focal plane behind the lens reads

$$u_2(x) = \int_{x'} \langle x\,|G\,|\,x'\rangle\langle x'\,|\,u_1\rangle dx',$$

$$= \sqrt{\frac{k}{f}} \int_{x'} \langle x\,|\,k' = -\frac{kx'}{f}\rangle\langle x'\,|\,u_1\rangle dx',$$

$$= \sqrt{\frac{k}{2\pi f}} \int_{x'} \exp\left(-i\frac{kx'x}{f}\right) u_1(x')dx'. \tag{2.54}$$

The last equation represents the Fourier transform property of the thin lens, namely, the image in the focal plane behind the lens is proportional to the Fourier transform of the object in the focal plane in front of the lens: $u_2(x) = \gamma\tilde{u}_1(k'')$ with $k'' = kx/f$. Here, we have used $\langle x\,|\,k'\rangle = \exp(ik'x)/\sqrt{2\pi}$.

2.5 Polarization Optics

In the beginning of Section 2.2.1, we made the assumption that the electric field $\vec{E}$ can be written as the product of a scalar part, which carries the spatial dependence of $\vec{E}$ and a vectorial part representing the polarization:

$$\vec{E}(x, y, z) = g(x, y, z)\vec{n}. \tag{2.55}$$

In the following sections, we discuss the bases representation, measurement, and propagation of the scalar part g using methods from quantum mechanics. In this section, the polarization $\vec{n}$ of light is studied under the assumption that it is spatially uniform as in Equation 2.55.* This will lead to the use of the tensor product in the Dirac formulation of optics explained in the next section. In Equation 2.55, polarization vectors $\vec{n}$ are two-dimensional complex vectors and as such form a Hilbert space $\mathcal{H}_p = \mathbb{C}^2$. In Dirac notation, we can thus represent the polarization vector $\vec{n}$ by $|\vec{n}\rangle \in \mathcal{H}_p$, which can be written as a linear combination of the basis vectors $|h\rangle$ and $|v\rangle$ representing horizontal and vertical polarization, respectively.

$$|\vec{n}\rangle = \alpha|h\rangle + \beta|v\rangle, \tag{2.56}$$

with complex numbers α and β and $|\alpha|^2 + |\beta|^2$ proportional to the intensity of light in the x–y-plane. The component vector of $|\vec{n}\rangle$ with respect to this basis thus reads:

$$\vec{n} = \begin{pmatrix} \alpha \\ \beta \end{pmatrix}, \tag{2.57}$$

which is known as *Jones vector*.

The action of linear optical elements, such as wave plates, phase plates, mirrors, and beam splitters (BSs), is denoted by a 2×2 matrix called *Jones matrix* [7,8,10]. For example, the Jones matrix corresponding to a horizontal polarizer is given by

$$\tilde{J}_h = \begin{pmatrix} 1 & 0 \\ 0 & 0 \end{pmatrix}. \tag{2.58}$$

The corresponding operator in Dirac notation reads

$$J_h = |h\rangle\langle h|. \tag{2.59}$$

Hence, the action of the Jones matrix J_p on the Jones vector $\vec{n}$ is given by

$$\tilde{J}_h \vec{n} = \begin{pmatrix} 1 & 0 \\ 0 & 0 \end{pmatrix} \begin{pmatrix} \alpha \\ \beta \end{pmatrix} = \begin{pmatrix} \alpha \\ 0 \end{pmatrix} \tag{2.60}$$

$$\Leftrightarrow \quad J_h|\vec{n}\rangle = |h\rangle\langle h|(\alpha|h\rangle + \beta|v\rangle) = \alpha|h\rangle. \tag{2.61}$$

---

* The opposite case does not allow to write the electric field as a product of the scalar part and a polarization vector. This is known as classical entanglement and is discussed in Section 2.8.

In general, the Jones matrix can be written as

$$\tilde{J} = \tilde{T} \begin{pmatrix} \gamma_1 & 0 \\ 0 & \gamma_2 \end{pmatrix} \tilde{T}^\dagger, \tag{2.62}$$

where γ_1, γ_2 are complex numbers and $\tilde{T}$ is a unitary matrix.

The Jones-matrix formalism can be applied only if light is completely polarized. In other words, the polarization state is pure as opposed to an incoherent (or statistical) mixture of different polarizations. For incoherent mixtures, one can instead use the *Müller-matrix* formalism. In the latter, the polarization state of light is given by a Stokes vector that is defined as

$$\vec{S} = \begin{pmatrix} S_0 \\ S_1 \\ S_2 \\ S_3 \end{pmatrix} \tag{2.63}$$

with

$$S_0 = |E_x|^2 + |E_y|^2, \tag{2.64}$$

$$S_1 = |E_x|^2 - |E_y|^2, \tag{2.65}$$

$$S_2 = 2\text{Re}(E_x E_y^*), \tag{2.66}$$

$$S_3 = 2\text{Im}(E_x E_y^*), \tag{2.67}$$

where E_x and E_y are the electric field amplitudes in x- and y-direction, respectively, whereas Re and Im denote the real and imaginary part. The degree of polarization of light is defined by

$$p = \frac{\sqrt{S_1^2 + S_2^2 + S_3^2}}{S_0}. \tag{2.68}$$

Completely polarized light possesses $p = 1$. For a given Jones vector $\vec{n}$, one obtains the corresponding Stokes vector $\vec{S}$ by means of the tensor product $\otimes$:

$$\vec{S} = \tilde{U}(\vec{n} \otimes \vec{n}^*), \tag{2.69}$$

with

$$\tilde{U} = \begin{pmatrix} 1 & 0 & 0 & 1 \\ 1 & 0 & 0 & -1 \\ 0 & 1 & 1 & 0 \\ 0 & i & -i & 0 \end{pmatrix}. \tag{2.70}$$

The transformation due to linear optical elements is represented by Müller matrices $\tilde{M}$ that transform the vector $\vec{S}$ as

$$\vec{S}' = \tilde{M}\vec{S}. \tag{2.71}$$

For a Jones matrix $\tilde{J}$, the Müller matrix $\tilde{M}$ can be written as

$$\tilde{M} = \frac{1}{2}\tilde{U}(\tilde{J} \otimes \tilde{J}^*)\tilde{U}^\dagger. \tag{2.72}$$

Therefore, the Müller matrix for the horizontal polarizer amounts to (cf. $\tilde{J}_h$ in Equation 2.58),

$$\tilde{M}_p = \frac{1}{2}\begin{pmatrix} 1 & 1 & 0 & 0 \\ 1 & 1 & 0 & 0 \\ 0 & 0 & 0 & 0 \\ 0 & 0 & 0 & 0 \end{pmatrix}. \tag{2.73}$$

In general, any Müller matrix $\tilde{M}$ can be decomposed as

$$\tilde{M} = \frac{1}{2}\tilde{U}\left(\sum_n p_n \tilde{J}_n \otimes \tilde{J}_n^*\right)\tilde{U}^\dagger, \tag{2.74}$$

where p_n are real and positive numbers such that $\sum_n p_n = 1$ and $\tilde{J}_n$ are Jones matrices.

2.6 Tensor Product Structure in Optics

We are now ready to discuss the representation of the entire electric field in Dirac notation. Let us start with the case where the scalar part and the polarization separate as reflected by condition Equation 2.55. To describe a pure state of such a field, we are looking for a vector that incorporates the state $|g(z)\rangle \in \mathcal{H}_s$ of the scalar part and the polarization represented by $|\vec{n}\rangle \in \mathcal{H}_p$. Moreover, the electric field in conventional notation is linear in both parts, that is, we can superpose electric fields with different spatial modes as well as polarizations and write the resulting E-field as

$$\vec{E} = g_1\vec{n}_1 + g_1\vec{n}_2 + g_2\vec{n}_1 + g_2\vec{n}_2 = (g_1 + g_2)(\vec{n}_1 + \vec{n}_2). \tag{2.75}$$

This property is reflected by the tensor product of vectors $|g(z)\rangle \otimes |\vec{n}\rangle$, which is linear in each argument:

$$\begin{aligned}
\big(a_1\,|g_1(z)\rangle &+ a_2\,|g_2(z)\rangle\big) \otimes \big(b_1\,|\vec{n}_1\rangle + b_2\,|\vec{n}_2\rangle\big) \\
&= a_1 b_1\,|g_1(z)\rangle \otimes |\vec{n}_1\rangle + a_1 b_2\,|g_1(z)\rangle \otimes |\vec{n}_2\rangle \\
&\quad + a_2 b_1\,|g_2(z)\rangle \otimes |\vec{n}_1\rangle + a_2 b_2\,|g_2(z)\rangle \otimes |\vec{n}_2\rangle,
\end{aligned} \tag{2.76}$$

where a_i, b_j are complex coefficients. The tensor product of $|g(z)\rangle \in \mathcal{H}_s$ and $|\vec{n}\rangle \in \mathcal{H}_p$ is a vector in a new Hilbert space, that is, $|g(z)\rangle \otimes |\vec{n}\rangle \in \mathcal{H}_s \otimes \mathcal{H}_p$, which is spanned by the images of all combinations of basis vectors of the two original Hilbert spaces:

$$\mathcal{H}_s \otimes \mathcal{H}_p = \text{span}\{\,|l,p\rangle \otimes |s\rangle \mid (|l,p\rangle)_{l,p} \text{ basis of } \mathcal{H}_s\,;\,(|s\rangle)_{s=h,v} \text{ basis of } \mathcal{H}_p\}.$$

Here, we have chosen the LG modes (cf. Chapter 1, Equation 1.48) and horizontal/vertical polarization (cf. Equation 2.56) as bases that are as good as any other choice. The inner product of $\mathcal{H}_s \otimes \mathcal{H}_p$ is defined by multiplying the inner products of the factor spaces. That is, for general vectors $|v\rangle, |v'\rangle \in \mathcal{H}_s$ and $|w\rangle, |w'\rangle \in \mathcal{H}_p$:

$$\big(\langle v' | \otimes \langle w' |\big)\big(| v \rangle \otimes | w \rangle\big) = \langle v' | v \rangle \langle w' | w \rangle. \tag{2.77}$$

This implies, for example, that three orthonormal vectors from $\mathcal{H}_s$, say LG beams $|l, p = 0\rangle$ with different orbital angular momenta $l = -1, 0, 1$ together with two orthonormal polarization vectors $|h\rangle, |v\rangle \in \mathcal{H}_p$ result in six orthonormal tensor product vectors $|l, p = 0\rangle \otimes |s\rangle$ with $l = \pm 1, 0$ and $s = h, v$ that reflects the possibility to combine the two degrees of freedom of light *independently*. More importantly, two light modes of the form (2.55) are orthogonal if either their spatial parts or their polarization vectors (or both) are orthogonal. This is faithfully represented by the inner product in Equation 2.77.

Hence, we can postulate:

Postulate 4

If light beams possess different spatial modes and polarization as two independent degrees of freedom with corresponding Hilbert spaces $\mathcal{H}_s$ and $\mathcal{H}_p$, then the corresponding Hilbert space for the beam is the tensor product space $\mathcal{H}_s \otimes \mathcal{H}_p$.

Thus, the most general state $|\eta\rangle \in \mathcal{H}_s \otimes \mathcal{H}_p$ of a paraxial beam with arbitrary polarization can be written as

$$|\eta\rangle = \sum_{l,p} a_{l,p}\,|l,p\rangle \otimes |h\rangle + \sum_{l',p'} b_{l',p'}\,|l',p'\rangle \otimes |v\rangle, \tag{2.78}$$

where $a_{l,p}$ and $b_{l',p'}$ are expansion coefficients. The basis expansion (2.78) points out that not all states of the electromagnetic field can be expressed in product form: $|\eta\rangle \neq |g(z)\rangle \, |n\rangle$. For example, this is not possible for the state

$$|\Phi^+\rangle = (|\vec{k}\rangle \otimes |h\rangle + |\vec{k}_\perp\rangle \, |v\rangle)/\sqrt{2},$$

which can be prepared by means of a polarizing BS from the state $|\vec{k}\rangle (|h\rangle + |v\rangle)/\sqrt{2}$. This nonseparability of spatial modes and polarization is reminiscent of entanglement in quantum mechanics. The relation between this type of "classical" entanglement and quantum entanglement will be discussed after we have introduced the quantum theory of light in the following sections.

This similarity is due to the fact that also in quantum mechanics, the state of particles in terms of position and motion in space on one hand and spin on the other hand are described by a tensor product of Hilbert spaces. However, quantum mechanics also comprises a tensor product structure to represent the state of composite systems. A classical analog of the latter is apparently not available within optics.

2.7 Quantum Optics

2.7.1 Introduction

QO is the quantum theory of light and its interaction with matter. While quantum electrodynamics (QED) studies the interaction of light with elementary particles such as leptons and quarks, QO deals with matter typically in the form of composite, systems, such as atoms, molecules, and even solid-state objects. In this sense, it is rather an effective theory as opposed to the elementary theory of QED. In addition, in QO, matter is usually dealt with in the nonrelativistic regime of velocities that are small compared to the speed of light. In this regime, antiparticles (e.g., positrons, antineutrinos, and antiquarks) do not need to be taken into account.

The formulation of a new theory of light, which deviates from classical optics, was initiated by Max Planck in the year 1900. Planck [13] was able to explain the energy spectrum of the radiation of a black body, by assuming that light of a given frequency v can only be absorbed or emitted in multiples of energy

$$E = hv, \tag{2.79}$$

where $h = 6.626 \times 10^{-34}$ Js is Planck's constant. This is considered as the birth of quantum mechanics. In 1905, Albert Einstein [4] postulated that light itself consists of particles carrying this energy called *light quanta* and later on *photons*, and thus obtained a model for the photoelectric effect. The idea that some physical quantities do not assume continuous but only discrete—quantized—values agreed with the observation of discrete emission spectra of atoms and molecules. Eventually it led to the development of quantum mechanics and modern quantum field theories that form the leading paradigms to explain the foundations of physics.

2.7.2 Quantization of the Electromagnetic Field in Vacuum

It is well known that the vector potential $\vec{A}$ can be chosen in the coulomb (or transversal gauge) given by the condition

$$\nabla \vec{A} = 0, \tag{2.80}$$

such that the electric and magnetic field can be expressed as

$$\vec{E} = -\frac{\partial}{\partial t}\vec{A} \quad \text{and} \quad \vec{B} = \nabla \times \vec{A}. \tag{2.81}$$

It is obvious that such a choice satisfies Maxwell's equations in vacuum in the form of both the Gauss' laws and Faraday's law, whereas Ampére's law leads to the wave equations for all three components of the vector potential:

$$\Delta \vec{A} - \frac{1}{c^2}\frac{\partial^2}{\partial t^2}\vec{A} = 0, \tag{2.82}$$

where the vacuum velocity of light $c = 1/\sqrt{\epsilon_0 \mu_0}$ and Δ is the Laplacian. Let us consider the electromagnetic field in a large box with sides of length L and periodic boundary conditions, that is, $\vec{A}(\vec{x}, t) = \vec{A}(\vec{x} + L\vec{e}_i, t)$ for $i = 1, 2, 3$. Then any solution of the wave equations 2.82 can be expressed in terms of $\exp(i\vec{k}\vec{x})/\sqrt{V}$ that form a basis (a complete orthonormal system) with varying $\vec{k}$ for all square-integrable complex-valued functions defined in the volume of the box. Here and in the following sections, the components of the wave vector $\vec{k}$ vary, such that the vector potential satisfies the periodic boundary conditions, that is, k_i is an integer multiple of $2\pi/L$ or 0. Thus

$$\vec{A}(\vec{x}, t) = \frac{1}{\sqrt{V}}\sum_{\vec{k}}\sum_{s}\vec{\epsilon}_{\vec{k},s}\left(c_{\vec{k},s}e^{-i\omega_k t}e^{i\vec{k}\vec{x}} + c.c.\right), \tag{2.83}$$

where c.c. denotes the complex conjugate, $c_{\vec{k},s}$ denotes complex coefficients, $\omega_k = ck$, and $\vec{\epsilon}_{\vec{k},s}$ denotes polarization vectors that are orthogonal to the wave vector $\vec{k}$ because of condition (2.80). Because of the latter, the summation index $s = 1, 2$ and only two orthonormal polarization vectors suffice. It is worth noting for the quantization procedure that follows we could have expanded the vector potential in any set of basis functions that obey the boundary conditions. Here, we used plane waves but the procedure is the same for the quantization of all modes. In the context of plane waves, a specific wave vector $\vec{k}$ and polarization direction given by index s defines a particular mode. In general, the choice of modes depends on the boundary conditions.

2.7.2.1 Hamiltonian of the Electromagnetic Field

The Hamiltonian of the electromagnetic field in vacuum inside the box reads

$$H = \frac{1}{2}\int_V \epsilon_0 \vec{E}\vec{E} + \frac{1}{\mu_0}\vec{B}\vec{B}d^3\vec{x}. \tag{2.84}$$

Inserting the expressions for the electric and the magnetic field (2.81) in terms of the vector potential (2.83) we obtain

$$H = 2\epsilon_0 \sum_{\vec{k},s} \omega_k c^*_{\vec{k},s} c_{\vec{k},s}. \tag{2.85}$$

It is worth noting that the Hamiltonian of the electromagnetic field is related to that of a collection of uncoupled harmonic oscillators. To make this relation obvious, let us absorb the time dependence of $\vec{A}$ in the coefficients of its expansion

$$c_{\vec{k},s}(t) := c_{\vec{k},s}(0)e^{-i\omega t}, \tag{2.86}$$

and introduce the generalized coordinates $q_{\vec{k},s}$ and the conjugate momenta $p_{\vec{k},s}$:

$$c_{\vec{k},s}(t) = \frac{1}{2\omega_k \sqrt{\epsilon_0}} \left(\omega_{\vec{k},s} q_{\vec{k},s}(t) + i p_{\vec{k},s}(t) \right), \tag{2.87}$$

$$c^*_{\vec{k},s}(t) = \frac{1}{2\omega_k \sqrt{\epsilon_0}} \left(\omega_{\vec{k},s} q_{\vec{k},s}(t) - i p_{\vec{k},s}(t) \right). \tag{2.88}$$

Replacing the coefficients $c_{\vec{k},s}$, the Hamiltonian now reads in terms of the new variables:

$$H = \frac{1}{2} \sum_{\vec{k},s} \left(p^2_{\vec{k},s}(t) + \omega^2_k q^2_{\vec{k},s}(t) \right), \tag{2.89}$$

which resembles the Hamiltonian of many uncoupled harmonic oscillators in one spatial dimension with unit mass, momenta $p_{\vec{k},s}(t)$, and position coordinates $q_{\vec{k},s}(t)$. It is a simple matter to check that these variables are canonically conjugate—by verifying that they obey Hamitonian equations:

$$\frac{\partial H}{\partial q_{\vec{k},s}} = -\dot{p}_{\vec{k},s}, \quad \frac{\partial H}{\partial p_{\vec{k},s}} = \dot{q}_{\vec{k},s}. \tag{2.90}$$

2.7.2.2 Canonical Quantization

Once the generalized positions and the corresponding canonically conjugate momenta of a classical system are identified, the quantization procedure is given to find the quantum analog of the system. In our case, this is the quantized electromagnetic field. The quantization procedure is to replace the classical coordinates by operators that satisfy the commutation relation of positions and conjugate momenta given by

$$\left[\hat{q}_{\vec{k},s}, \hat{p}_{\vec{k}',s'} \right] := \hat{q}_{\vec{k},s}\hat{p}_{\vec{k}',s'} - \hat{p}_{\vec{k}',s'}\hat{q}_{\vec{k},s} = i\hbar \delta_{\vec{k},\vec{k}'} \delta_{s,s'}. \tag{2.91}$$

This is reminiscent of the algebra of the corresponding Poisson brackets. The eigenstates and eigenvalues of the operator analog of Hamiltonian of a sequence of simple harmonic oscillators—for each mode of the electromagnetic field if there is one oscillator—can be found by introducing the creation and annihilation operators for excitations of each mode[*]

$$\hat{a}_{\vec{k},s}(t) = \frac{1}{\sqrt{2\hbar\omega_k}} \left(\omega_{\vec{k},s} \hat{q}_{\vec{k},s}(t) + i\hat{p}_{\vec{k},s}(t) \right), \tag{2.92}$$

$$\hat{a}^{\dagger}_{\vec{k},s}(t) = \frac{1}{\sqrt{2\hbar\omega_k}} \left(\omega_{\vec{k},s} \hat{q}_{\vec{k},s}(t) - i\hat{p}_{\vec{k},s}(t) \right). \tag{2.93}$$

These operators correspond to the right shift and left shift on the Hilbert space of energy eigenstates of a single mode (a harmonic oscillator) and are characterized by the following commutation relations:

$$\left[\hat{a}_{\vec{k},s}, \hat{a}_{\vec{k}',s'} \right] = \left[\hat{a}^{\dagger}_{\vec{k},s}, \hat{a}^{\dagger}_{\vec{k}',s'} \right] = 0 \quad \text{and} \quad \left[\hat{a}_{\vec{k},s}, \hat{a}^{\dagger}_{\vec{k}',s'} \right] = \delta_{\vec{k},\vec{k}'} \delta_{s,s'}, \tag{2.94}$$

which follow from relation (2.91) for generalized positions and momenta. By means of these operators, the Hamiltonian of the quantized electromagnetic field can be expressed as

$$\hat{H} = \sum_{\vec{k},s} \hbar\omega_k \left(\hat{a}^{\dagger}_{\vec{k},s} \hat{a}_{\vec{k},s} + \frac{1}{2} \right), \tag{2.95}$$

or with the so-called "number operator" $\hat{n}_{\vec{k},s} := \hat{a}^{\dagger}_{\vec{k},s} \hat{a}_{\vec{k},s}$

$$\hat{H} = \sum_{\vec{k},s} \hbar\omega_k \left(\hat{n}_{\vec{k},s} + \frac{1}{2} \right). \tag{2.96}$$

Analogous to the algebraic diagonalization of the Hamiltonian operator of a simple harmonic oscillator (cf. e.g., Ref. [16]), the commutation relation in Equation 2.94 implies that the number operators $\hat{n}_{\vec{k},s}$ only have nonnegative, integer eigenvalues $n_{\vec{k},s} \geq 0$. Let us call the corresponding simultaneous eigenstates of the number operators of all modes

$$|\{n\}\rangle \equiv |n_{\vec{k}_1,s_1}, n_{\vec{k}_2,s_2}, n_{\vec{k}_3,s_3}, \ldots\rangle \tag{2.97}$$

$$\text{with} \quad \hat{n}_{\vec{k}_i,s_i} |\{n\}\rangle = n_{\vec{k}_i,s_i} |\{n\}\rangle \quad \text{for } i = 1, 2 \ldots. \tag{2.98}$$

---

[*]   Note that we indicate the operators related to the quantum properties of light by a "hat" in order to distinguish them from operators that are used to describe classical optics in the first part and are then denoted by capital letters without "hat."

The state $|\{n\}\rangle$ for an arbitrary sequence $\{n\} = (n_{\vec{k}_1,s_1}, n_{\vec{k}_2,s_2}, n_{\vec{k}_3,s_3}, \ldots)$ of many finitely nonnegative integers is also an eigenstate of the electromagnetic Hamiltonian:

$$\hat{H}|\{n\}\rangle = \sum_{\vec{k},s} \hbar\omega_k \left(n_{\vec{k},s} + \frac{1}{2} \right) |\{n\}\rangle. \tag{2.99}$$

The annihilation and creation operators act on the number states just as in the case of a simple harmonic oscillator and annihilate or create one elementary excitation of the associated mode:

$$\hat{a}_{\vec{k},s}|n_{\vec{k},s}\rangle = \sqrt{n}|(n-1)_{\vec{k},s}\rangle \quad \text{and} \quad \hat{a}_{\vec{k},s}^{\dagger}|n_{\vec{k},s}\rangle = \sqrt{n+1}|(n+1)_{\vec{k},s}\rangle. \tag{2.100}$$

The energy eigenvalue $E_{\{n\}} = \sum_{\vec{k},s} \hbar\omega_k \left(n_{\vec{k},s} + \frac{1}{2} \right)$ corresponds to the energy of the electromagnetic field in state $|\{n\}\rangle$. Unfortunately, this value as represented here is infinite and, therefore, the eigenvalue equation (2.99), is mathematically not sound. To see this, let us calculate the energy E_0 of the ground state

$$|\{0\}\rangle = |0,0,0\ldots\rangle =: |0\rangle, \tag{2.101}$$

also called the vacuum state since it contains no photons.

$$\hat{H}|0\rangle = \sum_{\vec{k},s} \frac{\hbar\omega_k}{2}|0\rangle. \tag{2.102}$$

Thus,

$$E_0 = \frac{\hbar c}{\pi L} \sum_{n_1=-\infty}^{\infty} \sum_{n_2=-\infty}^{\infty} \sum_{n_3=-\infty}^{\infty} \sqrt{n_1^2 + n_2^2 + n_3^2} \tag{2.103}$$

is infinitely large and needs to be renormalized. Since only energy differences can be detected, the energy can be renormalized by subtracting the ground-state (vacuum) energy from the Hamiltonian. The "renormalized" Hamiltonian thus reads

$$\hat{H} = \sum_{\vec{k},s} \hbar\omega_k \hat{n}_{\vec{k},s}. \tag{2.104}$$

In this representation, the ground-state energy itself vanishes:

$$\hat{H}|0\rangle = 0|0\rangle, \quad \text{hence } E_0 = 0.$$

Any state of the electromagnetic field can be represented as a linear combination of eigenstates $|\{n\}\rangle$ of the renormalized Hamiltonian $\hat{H}$ in Equation 2.104. The renormalization did not change the eigenstates but only the eigenvalues of $\hat{H}$.

2.7.2.3 Quantized Electric and Magnetic Fields

Equation 2.83 denotes an expansion of the vector potential in the Coulomb gauge in terms of plane waves. The expansion coefficients correspond to the amplitudes of the modes and are commuting (classical) numbers. What does the vector potential of the quantized electromagnetic field look like? A comparison between the expansion coefficients $c_{\vec{k},s}(t)$ (2.87) and the annihilation operators $\hat{a}_{\vec{k},s}(t)$ (2.92) yields the answer. Apart from the fact that the latter are operators (q-numbers), both only differ by a constant. Taking this constant into account, we obtain the following rule for the transition from classical to quantum light:

$$c_{\vec{k},s}(t) \rightarrow \hat{c}_{\vec{k},s}(t) = \frac{\sqrt{2\hbar\omega_k}}{2\omega_k\sqrt{\epsilon_0}}\hat{a}_{\vec{k},s}(t) = \left(\frac{\hbar}{2\omega_k\epsilon_0}\right)^{\frac{1}{2}}\hat{a}_{\vec{k},s}(t). \tag{2.105}$$

Replacing the coefficients in the expansion of the vector potential $\vec{A}$ with respect to plane waves, we obtain

$$\hat{\vec{A}}(\vec{x},t) = \frac{1}{\sqrt{V}}\sum_{\vec{k}}\sum_s \left(\frac{\hbar}{2\omega_k\epsilon_0}\right)^{\frac{1}{2}}\vec{\epsilon}_{\vec{k},s}\left(\hat{a}_{\vec{k},s}e^{-i\omega_k t}e^{i\vec{k}\vec{x}} + \hat{a}^\dagger_{\vec{k},s}e^{i\omega_k t}e^{-i\vec{k}\vec{x}}\right), \tag{2.106}$$

where we use $c^*_{\vec{k},s}(t) \rightarrow \hat{c}^*_{\vec{k},s}(t) = \hat{a}^\dagger_{\vec{k},s}(t)(\hbar/2\omega_k\epsilon_0)^{\frac{1}{2}}$. The quantized electric field $\hat{\vec{E}}$ and magnetic field $\hat{\vec{B}}$ can be readily calculated from the quantized vector potential $\hat{\vec{A}}$:

$$\hat{\vec{E}}(\vec{x},t) = -\frac{\partial\hat{\vec{A}}(\vec{x},t)}{\partial t}$$

$$= \frac{i}{\sqrt{V}}\sum_{\vec{k}}\sum_s \left(\frac{\hbar\omega_k}{2\epsilon_0}\right)^{\frac{1}{2}}\vec{\epsilon}_{\vec{k},s}\left(\hat{a}_{\vec{k},s}e^{-i\omega_k t}e^{i\vec{k}\vec{x}} - \hat{a}^\dagger_{\vec{k},s}e^{i\omega_k t}e^{-i\vec{k}\vec{x}}\right), \tag{2.107}$$

$$\hat{\vec{B}}(\vec{x},t) = \nabla \times \hat{\vec{A}}(\vec{x},t)$$

$$= \frac{i}{c\sqrt{V}}\sum_{\vec{k}}\sum_s \left(\frac{\hbar\omega_k}{2\epsilon_0}\right)^{\frac{1}{2}}\frac{\vec{k}}{k}\times\vec{\epsilon}_{\vec{k},s}\left(\hat{a}_{\vec{k},s}e^{-i\omega_k t}e^{i\vec{k}\vec{x}} - \hat{a}^\dagger_{\vec{k},s}e^{i\omega_k t}e^{-i\vec{k}\vec{x}}\right). \tag{2.108}$$

We note that the quantized fields $\hat{\vec{A}}$, $\hat{\vec{E}}$, and $\hat{\vec{B}}$ are represented by self-adjoint operators (e.g., $\hat{\vec{A}}^\dagger = \hat{\vec{A}}$). But, only $\hat{\vec{E}}$ and $\hat{\vec{B}}$ are directly measurable and thus correspond to observables. Here, these observables are time dependent corresponding to the Heisenberg picture. In the Schrödinger picture, the time dependence is carried by the states of the electromagnetic field and the vector potential is expressed by Equation 2.106 simply without the time-dependent phase factors $\exp(\pm i\omega_k t)$.

2.7.2.4 Transition to the Continuum

So far, we have considered the electromagnetic field in a box with sides of length L and periodic boundary conditions that lead to discrete wavelengths and a lattice of wave vectors. A continuum of wavelengths and wave vectors for the electromagnetic field in free space is obtained by taking the limit of infinite side lengths $L \to \infty$. This amounts to replacing all expressions of the electromagnetic quantities derived above, the volume $V = L^3$ by the normalization factor $(2\pi)^{3/2}$, and sums over countably many modes by continuous integrals:

$$\frac{1}{\sqrt{L^3}} \sum_{\vec{k},s} (\ldots) \xrightarrow{L \to \infty} \frac{1}{\sqrt{(2\pi)^3}} \int d^3\vec{k} \sum_s (\ldots) \tag{2.109}$$

2.7.3 Classical and Nonclassical States

The state space of a general multimode electromagnetic field is a bosonic Fock space* $\mathcal{F}$ and is given by an infinite tensor product of copies of the state space $\mathcal{H}$ of the harmonic oscillator; one copy for each mode of the electromagnetic field.

$$\mathcal{F} = \mathcal{H} \otimes \mathcal{H} \otimes \mathcal{H} \ldots \tag{2.110}$$

$$= \left\{ \sum_{n_1,n_2,n_3,\ldots} b_{n_1,n_2,n_3,\ldots} | n_1, n_2, n_3 \ldots \rangle : \right.$$

$$\left. b_{n_1,n_2,n_3,\ldots} \in \mathbb{C} \text{ and } \sum_{n_1,n_2,n_3,\ldots} |b_{n_1,n_2,n_3,\ldots}|^2 < \infty \right\},$$

where n_i represents the number of photons in mode i and thus, the states

$$| n_1, n_2, n_3 \ldots \rangle = \frac{\left(\hat{a}_1^\dagger\right)^{n_1}}{\sqrt{n_1}} \frac{\left(\hat{a}_1^\dagger\right)^{n_2}}{\sqrt{n_2}} \frac{\left(\hat{a}_1^\dagger\right)^{n_3}}{\sqrt{n_3}} \ldots | 0 \rangle, \tag{2.111}$$

are simultaneous eigenstates of the number operator $\hat{n} = \sum_i \hat{n}_i = \sum_i \hat{a}_i^\dagger \hat{a}_i$ and the Hamiltonian $\hat{H}$ of the electromagnetic field. Here, $| 0 \rangle$ symbolizes the vacuum state (cf. Equation 2.101). Such eigenstates are called Fock states and form a complete orthonormal

---

* The symmetrization of the bosonic states of the electromagnetic field is intrinsically contained in our description based on the occupation number basis. A detailed mathematical discussion of Fock space can be found in Ref. [14].

system (or Hilbert space basis) of $\mathcal{F}$

$$\langle n_1, n_2, n_3 \ldots \mid n_1', n_2', n_3' \ldots \rangle = \delta_{n_1,n_1'} \delta_{n_2,n_2'} \delta_{n_3,n_3'} \cdots, \qquad (2.112)$$

$$\sum_{n_1,n_2,n_3,\ldots} \mid n_1, n_2, n_3 \ldots \rangle \langle n_1, n_2, n_3 \ldots \mid = \mathbb{I}. \qquad (2.113)$$

In the completeness relation (2.113), the symbol $\mathbb{I}$ represents the identity operator on the state space $\mathcal{F}$.

Since in many physical situations, not all basis modes of the electromagnetic field are occupied, only those photon numbers of excited modes ($n \neq 0$) are denoted in the eigenstates of the Hamiltonian. For example, a state with a single photon in mode $(\vec{k}, s)$ can be written as $\mid 1_{\vec{k},s} \rangle$ or simply $\mid 1 \rangle$ if the mode is obvious. Hence, a general pure single-mode state and a two-mode state with obvious modes read

$$\mid \psi \rangle = \sum_{n=0}^{\infty} b_n \mid n \rangle \qquad (2.114)$$

$$\mid \phi \rangle = \sum_{n}^{\infty} \sum_{n'}^{\infty} b_{n,n'} \mid n, n' \rangle \qquad (2.115)$$

2.7.3.1 Single-Photon States

Eigenstates of the number operator $\hat{n}$ with eigenvalue 1, that is, $\hat{n} \mid \varphi \rangle = \mid \varphi \rangle$, are called single-photon states. These can be any superposition of a single elementary excitation of modes. For example, a continuous superposition of excitations of plane waves with varying wave number

$$\mid \varphi \rangle = \frac{1}{(2\pi(\Delta k)^2)^{1/4}} \int dk\, e^{-\frac{(\vec{k}-\vec{k}_0)^2}{4(\Delta k)^2}} \mid 1_{\vec{k},s} \rangle$$

leads to an excitation of a localized Gaussian wave packet. The width of the wave packet is given by $\Delta z = 1/(2\Delta k)$, where we assume a superposition of plane waves propagating in the z-direction. Similarly, n photon states are eigenstates of the number operator $\hat{n}$ with eigenvalue n. The expectation value of the electric field $\hat{\vec{E}}$ vanishes for all Fock states (cf. Equations 2.100)

$$\langle \{n\} \mid \hat{\vec{E}} \mid \{n\} \rangle \quad \text{since } \langle n_i \mid a_i \mid n_i \rangle = \langle n_i \mid a_i^\dagger \mid n_i \rangle = 0 \quad \text{for } i = 0, 1, 2, \ldots \qquad (2.116)$$

For the same reason, the expectation value of the magnetic field vanishes as well. However, the expectation value of $\hat{\vec{E}}^2$ is proportional to the average detected intensity of the electromagnetic field and does not vanish. Fock states show nonclassical behavior (cf. Hong–Ou–Mandel effect below) and are considered as characteristic quantum states of the electromagnetic field. With the current technology, Fock states are difficult to generate in particular with a large number of photons.

2.7.3.2 Coherent States

Coherent states [2] are the eigenstates of mode operator a and are usually represented by $|\alpha\rangle$, where α is a complex number. We know that the ground state of a quantum harmonic oscillator $|0\rangle$ is an eigenstate of the operator a with eigenvalue 0, that is,

$$a|0\rangle = 0. \tag{2.117}$$

If we can find a unitary operator Z that commutes with a, then $Z|0\rangle$ will also be an eigenstate of a. In fact, any unitary operator with the property

$$Z^\dagger a Z = a + \alpha, \tag{2.118}$$

for any complex number α gives rise to a state $Z|0\rangle$, which is an eigenstate of a with eigenvalue α. One such operator is

$$Z = D(\alpha) = \exp(\alpha a^\dagger - \alpha^* a). \tag{2.119}$$

The operator $D(\alpha)$ is called the displacement operator and results in the state

$$|\alpha\rangle = D(\alpha)|0\rangle = e^{-\frac{|\alpha|^2}{2}} \sum_{n=0}^{\infty} \frac{\alpha^n (a^\dagger)^n}{n!} |0\rangle$$

$$= e^{-\frac{|\alpha|^2}{2}} \sum_{n=0}^{\infty} \frac{\alpha^n}{\sqrt{n!}} |n\rangle, \tag{2.120}$$

such that

$$a|\alpha\rangle = \alpha|\alpha\rangle. \tag{2.121}$$

Expression (2.120) implies that a coherent state is a superposition of vacuum, a single excitation, and multiple excitations of the corresponding light mode. Since they are eigenstates of the annihilation operator, the expectation value of the electric field with respect to a coherent state $|\alpha\rangle$ yields the electric field of classical light with a complex amplitude proportional to $\alpha = |\alpha| \exp i\phi$:

$$\langle \alpha | \hat{\vec{E}} | \alpha \rangle = -i \sqrt{\frac{\hbar\omega}{2\epsilon_0 V}} \vec{\epsilon} \left(\alpha e^{i(\vec{k}\vec{x} - \omega_k t)} - \alpha^* e^{-i(\vec{k}\vec{x} - \omega_k t)} \right) \tag{2.122}$$

$$= -2\sqrt{\frac{\hbar\omega}{2\epsilon_0 V}} \vec{\epsilon} |\alpha| \sin(\vec{k}\vec{x} - \omega_k t + \phi). \tag{2.123}$$

The standard deviation of the electric field amounts to

$$\Delta E = (\langle \alpha | \hat{\vec{E}}^2 | \alpha \rangle - \langle \alpha | \hat{\vec{E}} | \alpha \rangle^2)^{1/2} = \sqrt{\frac{\hbar\omega}{2\epsilon_0 V}}, \tag{2.124}$$

which is independent of the amplitude α and, therefore, leads to a vanishing relative uncertainty $\Delta E / \langle |\hat{E}| \rangle$ in the limit of high average photon number given by $\langle \alpha \, | \hat{n} \, | \alpha \rangle = |\alpha|^2$. For these reasons, coherent states $|\alpha\rangle$ are referred to as "classical" states since they reproduce the properties of light as described by classical optics. The light emitted by high-quality lasers can be described as a coherent state. However, laser modes are not plane waves but rather Hermite–Gaussian or LG modes. Therefore, laser light is described better as a coherent state of one of these modes.

2.7.4 Quantization of the Electromagnetic Field in the Paraxial Approximation and Orbital Angular Momentum

In this section, the quantization of the electromagnetic field in the paraxial limit is studied according to Ref. [3]. In this context, we focus on LG modes. For this purpose, we consider the quantized vector potential in terms of plane waves, (cf. Equation 2.106). In the paraxial limit, the z-component of the wave vector $\vec{k}$ varies slowly as a function of the divergence of the beam. This results in the dispersion relation

$$\vec{k} = \vec{q} + k_0 (1 - \theta^2) \vec{e}_z, \tag{2.125}$$

where $\vec{q}$ is the transverse wave vector, $k_0 = |\vec{k}|$, and $\theta = q/(\sqrt{2}k_0)$ governs the order of paraxiality of the beam and approximately equals the divergence angle of the beam. $\vec{e}_z$ denotes a unit vector in the z-direction. Since Equation 2.125 captures the paraxial approximation, substituting Equation 2.125 in Equation 2.106 will amount to the quantization of the electromagnetic field in the paraxial limit

$$\hat{\vec{A}}(\vec{x}, t) = \frac{1}{\sqrt{V}} \sum_{\vec{k}} \sum_{s} \left(\frac{\hbar}{2\omega_k \epsilon_0} \right)^{\frac{1}{2}} \vec{\epsilon}_{\vec{k},s} \left(\hat{a}_{\vec{k},s}(t) e^{ik_0 z} e^{i\vec{q}\cdot\vec{x}_\perp - i\theta^2 z} + h.c. \right). \tag{2.126}$$

In the continuum limit, the vector potential reads:

$$\hat{\vec{A}}(\vec{x}, t) = \int_{k_0} dk_0 \int_{\vec{q}} d^2q \sum_{s} \left(\frac{\hbar}{16\pi^3 \omega_k \epsilon_0} \right)^{\frac{1}{2}} \vec{\epsilon}_{\vec{k},s} \left(\hat{a}_{\vec{k},s}(t) e^{ik_0 z} e^{i\vec{q}\cdot\vec{x}_\perp - i\theta^2 z} + h.c. \right). \tag{2.127}$$

To express the vector potential $\hat{\vec{A}}$ with respect to LG modes, one can expand the factor $e^{i\vec{q}\cdot\vec{x}_\perp - i\theta^2 z}$ as follows:

$$e^{i\vec{q}\cdot\vec{x}_\perp - i\theta^2 z} = \int_{\vec{q}'} d^2 q' e^{i\vec{q}'\cdot\vec{x}_\perp - i\theta^2 z} \delta^{(2)}(\vec{q} - \vec{q}'),$$

$$= \int_{\vec{q}'} d^2 q' e^{i\vec{q}'\cdot\vec{x}_\perp - i\theta^2 z} \sum_{l,p} \mathcal{LG}^*_{l,p}(\vec{q}) \mathcal{LG}_{lp}(\vec{q}'),$$

$$= \sum_{l,p} \mathcal{LG}^*_{l,p}(\vec{q}) \int_{\vec{q}'} d^2 q' e^{i\vec{q}'\cdot\vec{x}_\perp - i\theta^2 z} \mathcal{LG}_{lp}(\vec{q}'),$$

$$= \sum_{l,p} \mathcal{LG}^*_{l,p}(\vec{q}) LG_{l,p}(\vec{x}_\perp, z; k_0). \tag{2.128}$$

where $\mathcal{LG}_{l,p}(\vec{q})$ is the LG mode in the Fourier basis. The last part of the equation uses the propagation of LG modes in the z-direction.

Substituting Equation 2.128 in Equation 2.127 results in

$$\hat{\vec{A}}(\vec{x}, t) = \int_{k_0} dk_0 \int_{\vec{q}} d^2 q \sum_s \left(\frac{\hbar}{16\pi^3 \omega_k \epsilon_0} \right)^{\frac{1}{2}} \vec{\epsilon}_{\vec{k},s}$$

$$\times \left(\hat{a}_{\vec{k},s}(t) e^{ik_0 z} \sum_{l,p} \mathcal{LG}^*_{l,p}(\vec{q}) LG_{l,p}(\vec{x}_\perp, z; k_0) + h.c. \right), \tag{2.129}$$

$$= \int_{k_0} dk_0 \sum_{s,l,p} \left(\frac{\hbar}{16\pi^3 \omega_k \epsilon_0} \right)^{\frac{1}{2}} \vec{\epsilon}_{\vec{k}} \left(\hat{a}_{k_0,s,l,p}(t) e^{ik_0 z} LG_{l,p}(\vec{x}_\perp, z; k_0) + h.c. \right). \tag{2.130}$$

where

$$\hat{a}_{k_0,s,l,p}(t) \equiv \int_{\vec{q}} d^2 q \, \hat{a}_{\vec{k},s}(t) \mathcal{LG}^*_{l,p}(\vec{q}) \tag{2.131}$$

is the LG mode annihilation operator that satisfies the necessary commutation relations:

$$[\hat{a}_{k_0,s,l,p}(t), \hat{a}_{k'_0,s',l',p'}(t)] = 0, \tag{2.132}$$

$$[\hat{a}^\dagger_{k_0,s,l,p}(t), \hat{a}^\dagger_{k'_0,s',l',p'}(t)] = 0, \tag{2.133}$$

$$[\hat{a}_{k_0,s,l,p}(t), \hat{a}^\dagger_{k'_0,s',l',p'}(t)] = \delta_{ss'} \delta_{ll'} \delta_{pp'} \delta(k_0 - k'_0). \tag{2.134}$$

To arrive at expressions (2.129) and (2.130), corresponding to the paraxial approxima-tion, the polarization vectors $\vec{\epsilon}_{\vec{k},s}$ are approximated to lie in a plane transversal to the

z-direction. The coherent state in LG modes can be expressed as

$$|\alpha_{LG}\rangle = \prod_{s,l,p}\left(e^{-\frac{|\alpha_{s,l,p}|^2}{2}} \sum_n \frac{\alpha_{s,l,p}^n (\hat{a}_{k_0,s,l,p}^\dagger)^n}{n!} \right) |0\rangle, \tag{2.135}$$

$$= \prod_{s,l,p} |\alpha_{s,l,p}\rangle. \tag{2.136}$$

2.7.5 Passive Linear Optics

As an application of the quantum formalism introduced above, we now discuss the action of loss-less BSs and phase shifters on the states of the quantized electromagnetic field. This is a part of passive linear optics *passive* because these devices do not change the number of photons and *linear* because the state evolution can be described in an in–out formalism where the final creation operators for any mode *l* are superpositions of the initial creation operators:

$$(a_l^\dagger)' = \sum_m u_{ml} a_m^\dagger, \quad \text{the numbers } u_{ml} \text{ form a unitary matrix.} \tag{2.137}$$

For the sake of simplicity, we omit to specify operators by a "hat" in this section. There is of course a microscopic description of the BSs and phase shifters in terms of the interaction between the incident light and the atoms constituting these optical devices. However, here, we seek a phenomenological model. We will also neglect polarization and consider only two-mode fields for the sake of simplicity. In the Schrödinger picture, the time dependence is incorporated in the states and not in the observables of the system. In the absence of noise, the passive linear optical devices transform pure states into pure states. Probability conservation requires a unitary evolution from the initial state $|\psi\rangle$, in front of the optical device, to the state $|\psi'\rangle$ behind the device

$$|\psi\rangle \xrightarrow{\text{Optical device}} |\psi'\rangle = U|\psi\rangle \quad \text{with} \quad U^\dagger U = \mathbb{I}. \tag{2.138}$$

Since the photon number is preserved, the vacuum state $|00\rangle$ (zero photons in both modes) does not change:

$$|00\rangle \xrightarrow{\text{Optical device}} U|00\rangle = |00\rangle. \tag{2.139}$$

A general state of a two-mode field in Fock state representation is given by

$$|\psi\rangle = \sum_{n,m}^{\infty} c_{nm} |nm\rangle \quad \text{with} \quad \sum_{n,m}^{\infty} |c_{nm}|^2 = 1. \tag{2.140}$$

The annihilation and creation operators of the first and second mode (denoted by letters a and b, respectively) act on the number states as follows:

$$a\,|\,nm\rangle = \sqrt{n}\,|\,n-1,m\rangle, \quad a^{\dagger}\,|\,nm\rangle = \sqrt{n+1}\,|\,n+1,m\rangle, \tag{2.141}$$

$$b\,|\,nm\rangle = \sqrt{n}\,|\,n-1,m\rangle, \quad b^{\dagger}\,|\,nm\rangle = \sqrt{n+1}\,|\,n+1,m\rangle. \tag{2.142}$$

To see the working principle of our description, consider a single-photon state of the two-mode system:

$$|\,\varphi\rangle = \alpha\,|\,10\rangle + \beta\,|\,01\rangle = \alpha a^{\dagger}\,|\,00\rangle + \beta b^{\dagger}\,|\,00\rangle. \tag{2.143}$$

The new state after a unitary operation can be expressed by its action on the creation operators:

$$|\,\varphi'\rangle = U\,|\,\varphi\rangle = U(\alpha a^{\dagger} + \beta b^{\dagger})\,|\,00\rangle, \tag{2.144}$$

$$= U(\alpha a^{\dagger} + \beta b^{\dagger})U^{\dagger}U\,|\,00\rangle, \tag{2.145}$$

$$= (\alpha \underbrace{Ua^{\dagger}U^{\dagger}}_{=:(a^{\dagger})'} + \beta \underbrace{Ub^{\dagger}U^{\dagger}}_{=:(b^{\dagger})'})\,|\,00\rangle, \tag{2.146}$$

where we used $U^{\dagger}U = \mathbb{I}$ to obtain Equation 2.145 and the invariance of the vacuum (2.139) to arrive at Equation 2.146. The expressions for the new creation operators are related to and can be obtained* from the changes of the single-photon basis states $|\,10\rangle$ and $|\,01\rangle$

$$a^{\dagger} \rightarrow (a^{\dagger})' = u_{aa}a^{\dagger} + u_{ba}b^{\dagger} \quad \Leftrightarrow \quad |\,10\rangle \rightarrow u_{aa}\,|\,10\rangle + u_{ba}\,|\,01\rangle, \tag{2.147}$$

$$b^{\dagger} \rightarrow (b^{\dagger})' = u_{ab}a^{\dagger} + u_{bb}b^{\dagger} \quad \Leftrightarrow \quad |\,01\rangle \rightarrow u_{ab}\,|\,10\rangle + u_{bb}\,|\,01\rangle. \tag{2.148}$$

Remarkably, the linear optical change of any state of the two-mode electromagnetic field with infinite-dimensional Hilbert space $\mathcal{H} \otimes \mathcal{H}$ of two uncoupled oscillators can be determined from the change of the two single-photon states above, that is, from the coefficients of the unitary 2×2 matrix $(U_{ij}) \equiv (u_{ij})$. The outgoing state is obtained from the ingoing state by replacing the corresponding creation operators:

$$|\,\psi'\rangle = U\,|\,\psi\rangle = U\sum_{n,m}^{\infty} c_{nm}\frac{(a^{\dagger})^{n}}{\sqrt{n!}}\frac{(b^{\dagger})^{m}}{\sqrt{m!}}\,|\,00\rangle = \sum_{n,m}^{\infty} c_{nm}U\frac{(a^{\dagger})^{n}}{\sqrt{n!}}\frac{(b^{\dagger})^{m}}{\sqrt{m!}}U^{\dagger}U\,|\,00\rangle$$

$$= \sum_{n,m}^{\infty} c_{nm}\frac{((a^{\dagger})')^{n}}{\sqrt{n!}}\frac{((b^{\dagger})')^{m}}{\sqrt{m!}}\,|\,00\rangle. \tag{2.149}$$

---

* The transformations on the right-hand side follow from applying the creation operators on the left-hand side onto the vacuum state and the left-hand side is derived by expressing the states in terms of creation operators.

Here, the last line follows by inserting the identity operator $\mathbb{I} = U^\dagger U$ between any two creation operators in the first line.

2.7.5.1 Phase Shifters, Mirrors, and BSs

Phase shifters are usually glass plates with an antireflective coating. The light is transmitted but suffers a phase shift relative to other light modes that are propagating through an optically less dense medium. Here, we assume two modes propagating in orthogonal directions (wave vectors $\vec{k}$ and $\vec{k}'$) where only the first mode traverses a phase shifter. A single-photon state (cf. Equation 2.143) evolves according to

$$\alpha \,|\, 10 \rangle + \beta \,|\, 01 \rangle \rightarrow \alpha e^{i\phi} \,|\, 10 \rangle + \beta \,|\, 01 \rangle, \tag{2.150}$$

which is equivalent to the transformation of creation operators

$$a^\dagger \rightarrow (a^\dagger)' = e^{i\phi} a^\dagger \text{ and } b^\dagger \rightarrow (b^\dagger)' = b^\dagger. \tag{2.151}$$

Therefore, a general state 2.140 of a two-mode light field transforms according to

$$\sum_{n,m}^{\infty} c_{nm} \,|\, nm \rangle \rightarrow \sum_{n,m}^{\infty} c_{nm} \frac{(e^{i\phi} a^\dagger)^n}{\sqrt{n!}} \frac{(b^\dagger)^m}{\sqrt{m!}} \,|\, 00 \rangle = \sum_{n,m}^{\infty} c_{nm} e^{in\phi} \,|\, nm \rangle. \tag{2.152}$$

A planar mirror swaps modes, that is, a single-photon state transforms according to

$$\alpha \,|\, 10 \rangle + \beta \,|\, 01 \rangle \longrightarrow \alpha \,|\, 01 \rangle + \beta \,|\, 10 \rangle, \tag{2.153}$$

Or in terms of creation operators

$$a^\dagger \rightarrow (a^\dagger)' = b^\dagger \text{ and } b^\dagger \rightarrow (b^\dagger)' = a^\dagger. \tag{2.154}$$

The transformation of a general two-mode state due to the mirror thus leads to a transposition of the coefficient matrix:

$$\sum_{n,m}^{\infty} c_{nm} \,|\, nm \rangle \longrightarrow \sum_{n,m}^{\infty} c_{mn} \,|\, nm \rangle. \tag{2.155}$$

Eventually, the action of a BS (cf. Figure 2.2) can be expressed by means of a general 2×2 unitary matrix:

$$\begin{pmatrix} (a^\dagger)' \\ (b^\dagger)' \end{pmatrix} = \begin{pmatrix} te^{i\delta_t} & -re^{-i\delta_r} \\ re^{i\delta_r} & te^{-i\delta_t} \end{pmatrix} \begin{pmatrix} a^\dagger \\ b^\dagger \end{pmatrix} \quad \text{with } r^2 + t^2 = 1. \tag{2.156}$$

At the same time, this is the most general linear transformation of the creation operators of a two-mode field (cf. Equations 2.147 and 2.148). Thus, any linear optics

FIGURE 2.2 Each of the input modes a and b is mapped by a BS onto a superposition of output modes a and b.

transformation of an appropriate two-mode field can be realized by choosing a BS with the appropriate transmitivity $t^2 = 1 - r^2$ ($t, r > 0$) and phase shifts δ_t and δ_r of the transmitted and reflected beam. In this sense, phase shifters ($t = 1$) and mirrors ($t = 0, \phi = 0$) are just extreme cases of BSs. BSs can be combined with mirrors to arrays of the so-called multiports, that is, linear optic devices with a number of n input ports and n output ports. By means of these, an arbitrary $n \times n$ unitary transformation of n optical modes corresponding to n different beams can be realized [19]. Two BS classes that are often used are represented by the following unitaries:

$$U_1 = \begin{pmatrix} \cos\theta & i\sin\theta \\ i\sin\theta & \cos\theta \end{pmatrix} \quad \text{and} \quad U_2 = \begin{pmatrix} \cos\theta & -\sin\theta \\ \sin\theta & \cos\theta \end{pmatrix}. \tag{2.157}$$

2.7.5.2 Hong–Ou–Mandel Effect

Single photons incident on an array of BSs show detection characteristics similar to laser beams with very small intensity [11]. This can even be used to simulate quantum walks— a quantum analog of random walks—for single photons by means of classical light. A different picture is obtained for multiple photons. Here, the results can differ considerably, from the classical analog. One example of a genuine quantum phenomenon in linear optics is the Hong–Ou–Mandel effect. For one photon in each input mode of a 50:50 BS of the type U_2, that is, $\theta = \pi/4$, we obtain:

$$|11\rangle = a^\dagger b^\dagger |00\rangle \longrightarrow \frac{1}{2}(a^\dagger + b^\dagger)(-a^\dagger + b^\dagger)|00\rangle \equiv |\psi'\rangle,$$

$$|\psi'\rangle = \frac{1}{2}\left((a^\dagger)^2 + a^\dagger b^\dagger - b^\dagger a^\dagger - (b^\dagger)^2\right)|00\rangle = \frac{1}{2}\left((a^\dagger)^2 - (b^\dagger)^2\right)|00\rangle,$$

$$|\psi'\rangle = \frac{1}{\sqrt{2}}(|20\rangle - |02\rangle) = \frac{1}{\sqrt{2}}(|20\rangle - |02\rangle) \tag{2.158}$$

This means that there is no coincidence count in the two detectors that are located one at each output port. This so-called Hong–Ou–Mandel dip has been experimentally confirmed for the first time by Hong, Ou and Mandel [6] in 1987 and it is used to verify single-photon sources. The Hong–Ou–Mandel dip does also appear for a 50:50 BS of the type U_1 in Equation 2.157. It is not there if single photons (elementary excitations of wave pulses) arrive at different times at the input ports of the BS. Therefore, the Hong–Ou–Mandel effect cannot be explained by assuming that each photon only interferes with itself. Thus, it is a result of many particle interferences.

2.8 Classical Entanglement

The tensor product structure of light with independent spatial and polarization degree of freedom (cf. Postulate 4 in Section 2.6) reveals an interesting feature of classical optics, namely, "classical" entanglement. For example, consider the following state of a laser beam described by means of Dirac notation in classical optics:

$$|\eta\rangle = \frac{1}{\sqrt{2}} \left(|\ell\rangle \otimes |h\rangle + |-\ell\rangle \otimes |v\rangle \right), \qquad (2.159)$$

where $|h\rangle$ and $|v\rangle$ represent horizontal/vertical polarization and $|\pm\ell\rangle$ is the LG modes with $\pm\ell$ and $p = 0$. Note that the state $|\eta\rangle$ cannot be written as a tensor product of any particular spatial mode and polarization state. In fact, the beam does not carry a particular polarization or orbital angular momentum. Instead, both properties are entangled. In quantum mechanics, pure states of composite systems that cannot be written in product form are called entangled. In classical optics, they are termed "nonquantum entangled" or "classical entangled" states.

The formal analogy between both forms of entanglement has been used to show that in classical optics, those Muller matrices that represent positive, but not completely positive operations, are not physically realizable [17]. Another application of classical entanglement is a coined quantum walk with LG modes in Ref. [5].

Here we demonstrate that classically entangled optical fields such as the one given in Equation 2.159, do not contain quantum entanglement. To this end, we consider the state of the laser beam in its quantum description (cf. Section 2.7.3) as a pure coherent state $|\alpha_{\ell,s}\rangle$, where the indices ℓ and s specify the mode and polarization as above:

$$|\alpha_{\ell,s}\rangle = \exp(\alpha a_{\ell,s}^{\dagger} - \alpha^{*} a_{\ell,s}) |0\rangle. \qquad (2.160)$$

Here, $a_{\ell,s}^{\dagger}$ $(a_{\ell,s})$ is the creation (annihilation) operator of the corresponding mode. The state $|\eta\rangle$ in Equation 2.159 can be prepared by applying a unitary operation U on the state $|\chi\rangle$ with

$$U = \sum_{\ell=-\infty}^{\infty} \left(|\ell\rangle\langle\ell| \otimes |h\rangle\langle h| + |-\ell\rangle\langle\ell| \otimes |v\rangle\langle v| \right), \qquad (2.161)$$

$$|\chi\rangle = |\ell\rangle \otimes \frac{1}{\sqrt{2}} (|h\rangle + |v\rangle). \qquad (2.162)$$

The quantum optical equivalent of state $|\chi\rangle$ reads:

$$|\chi\rangle_Q = \exp\left(\alpha \frac{1}{\sqrt{2}} \left(a_{\ell,h}^{\dagger} + a_{\ell,v}^{\dagger} \right) - \alpha^{*} \frac{1}{\sqrt{2}} \left(a_{\ell,h} + a_{\ell,v} \right) \right) |0\rangle, \qquad (2.163)$$

$$= |\frac{\alpha_{\ell,h}}{\sqrt{2}}\rangle \otimes |\frac{\alpha_{\ell,v}}{\sqrt{2}}\rangle, \qquad (2.164)$$

while U is represented by its action on the creation and annihilation operators:

$$U : a_{\ell,h} \to a_{\ell,h}, \quad U : a_{\ell,v} \to a_{-\ell,v}. \tag{2.165}$$

Thus, we find the quantum description of $|\eta\rangle$:

$$U|\chi\rangle_Q = |\frac{\alpha_{\ell,h}}{\sqrt{2}}\rangle \otimes |\frac{\alpha_{-\ell,v}}{\sqrt{2}}\rangle, \tag{2.166}$$

$$= |\eta\rangle_Q, \tag{2.167}$$

which is clearly a product state. Hence, although the so-called classically entangled states in classical optics do not factorize, they do in the corresponding QO description and, therefore, they are not entangled in terms of quantum theory.

References

1. Allen, L., Beijersbergen, M. W., Spreeuw, R. J. C., and Woerdmann, J. P. 1992. Orbital angular momentum of light and the transformation of Laguerre–Gaussian laser modes, *Phys. Rev. A* **45**, 8185.
2. Cahill, K. E. and Glauber, R. J. 1969. Ordered expansion in boson amplitude operators, *Phys. Rev.* **177**, 1857.
3. Calvo, G. F., Picn, A., and Bagan, E. 2006. Quantum field theory of photons with orbital angular momentum, *Phys. Rev. A* **73**, 013805.
4. Einstein, A. 1905. Über einen die Erzeugung und Verwandlung des Lichtes betreffenden heuristischen Gesichtspunkt, *Ann. Phys.* **17** (6), 132–148.
5. Goyal, S. K., Roux, F. S., Forbes, A., and Konrad, T. 2013. Implementing quantum walks using orbital angular momentum of classical light, *Phys. Rev. Lett.* **110**, 263602.
6. Hong, C. K., Ou, Z. Y., and Mandel, L. 1987. Measurement of subpicosecond time intervals between two photons by interference, *Phys. Rev. Lett.* **59**, 2044.
7. Hurwitz, J. R. H. and Jones, R. C. 1941. A new calculus for the treatment of optical systems. II, *J. Opt. Soc. Am.* **31**, 493.
8. Jones, R. C. 1941. A new calculus for the treatment of optical systems. III, *J. Opt. Soc. Am.* **31**, 500.
9. Jones, R. C. 1941. A new calculus for the treatment of optical systems. I, *J. Opt. Soc. Am.* **31**, 488.
10. Jones, R. C. 1942. A new calculus for the treatment of optical systems. IV, *J. Opt. Soc. Am.* **32**, 486.
11. Mayer, K., Tichy, M. C., Mintert, F., Konrad, T., and Buchleitner, A. 2011. Counting statistics of many-particle quantum walks, *Phys. Rev. A* **83**, 062307.
12. Pederson, G. K. 1989. *Analysis Now*, Springer-Verlag, New York.
13. Planck, M. 1901. Über das Gesetz der Energieverteilung im Normalspectrum, *Ann. Phys.* **309** (3), 553–563.

14. Reed, M. and Simon, B. 1972. *Methods of modern mathematical physics*, Academic Press, San Diego, CA.
15. Ryder, L. H. 1985. *Quantum Field Theory*, Vol. 155, Cambridge University Press, Cambridge, UK.
16. Sakurai, J. J. 1985. *Modern Quantum Mechanics*, Addison-Wesley, Reading, MA.
17. Simon, B. N., Simon, S., Gori, F., Santarsiero, M., Borghi, R., Mukunda, N., and Simon, R. 2010. Nonquantum entanglement resolves a basic issue in polarization optics, *Phys. Rev. Lett.* **104**, 023901.
18. Stoler, D. 1981. Operator methods in physical optics, *J. Opt. Soc. Am.* **71**, 334–341.
19. Zukowski, M., Zeilinger, A., and Horne, M. A. 1997. Realizable higher-dimensional two-particle entanglements via multiport beam splitters, *Phys. Rev. A* **55**, 2564.

3

Time Domain Laser Beam Propagation

Thomas Feurer
University of Bern

This chapter reviews the propagation of time domain pulses through optical media, such as glasses, optical fibers, or nonlinear crystals, and special optical systems, such as stretchers or compressors, which are designed to modify time domain pulses in a very specific way. The chapter starts with the wave equation as it is commonly used in optics and as introduced in Chapter 1. The material equations are derived from a classical model of a driven oscillator in an anharmonic potential. Despite the simplicity of the model, it is sufficient to qualitatively explain a great number of effects in linear as well as in nonlinear optics. The entire chapter will be based on a one-dimensional propagation equation, that is, the results are applicable whenever the transverse beam size remains more or less constant as the pulse propagates through the dielectric material. We then proceed with more complex linear systems, which are typically composed of various optical elements and act on the temporal or spectral part of the electric field. To further illustrate the different effects, we show simulation results for 100-fs-long and 10-fs-long Gaussian-shaped time domain pulses, respectively; their center wavelength is

800 nm. All the simulations in this chapter have been performed with the Lab2 simulation software [1].

3.1 Wave Equation

From Maxwell's equations in coordinate space and time domain as defined in Chapter 1 and assuming dielectric media, the wave equation

$$\nabla^2 E(x, t) - \frac{1}{c^2} \partial_t^2 E(x, t) - \mu_0 \partial_t^2 P(x, t) = 0 \tag{3.1}$$

can be derived. Assuming that the electric field can be written as a product $E(x, t) = f(x, y) E(z, t)$ and that the transverse beam support is such that the transverse derivatives of $f(x, y)$ may be safely neglected, the wave equation in time domain becomes

$$\partial_z^2 E(z, t) - \frac{1}{c^2} \partial_t^2 E(z, t) - \mu_0 \partial_t^2 P(z, t) = 0 \tag{3.2}$$

and in frequency domain

$$\partial_z^2 E(z, \omega) + \frac{\omega^2}{c^2} E(z, \omega) + \omega^2 \mu_0 P(z, \omega) = 0. \tag{3.3}$$

To proceed, we employ a simple classical picture to find a constituent relationship between the polarization and the electric field in dielectric media, that is, $P(E)$. We assume that the dielectric material consists of atomic-scale subsystems that can couple to an external electric field through their dipole moment. Once accelerated, they start to radiate and act as sources of radiation. Further, we assume that the subsystems are independent, that is do not couple to each other, and are sufficiently well described by driven oscillators in a generally anharmonic potential. For example, such subsystems may be electrons bound to ions or optical phonons. When the driving electric field is strong enough, the oscillators will leave the mostly harmonic part and start to explore the increasingly anharmonic parts of the potential surface. To employ a perturbation approach to solve the oscillator's equation of motion, it is convenient to expand the potential energy surface in a Taylor series [2]. The material's polarization is related to the oscillator position vector s through $P_i(z, t) = N_i q_i s_i(t)$ with the number of oscillators N_i and their charge q_i and we find

$$\partial_t^2 P_i(z, t) + \gamma_i \partial_t P_i(z, t) + \omega_{i0}^2 P_i(z, t) + \sum_{j,k} a_{ijk} P_j(z, t) P_k(z, t)$$

$$+ \sum_{jkl} b_{ijkl} P_j(z, t) P_k(z, t) P_l(z, t) + \cdots = \varepsilon_0 f_i E_i(z, t), \tag{3.4}$$

Equation 3.4 is solved by inserting a power series expansion for the polarization, that is

$$P_i(z,t) = P_i^{(0)}(z,t) + \lambda P_i^{(1)}(z,t) + \lambda^2 P_i^{(2)}(z,t) + \lambda^3 P_i^{(3)}(z,t) + \cdots, \qquad (3.5)$$

where λ is related to the strength of the perturbation. The original perturbation $\varepsilon_0 f_i E_i(z,t)$ is replaced by $\lambda \varepsilon_0 f_i E_i(z,t)$. The power series Ansatz is a solution to Equation 3.4 if all terms proportional to λ^n satisfy the equation separately. The resulting system of equations is now solved in order until reaching some n after which the remaining terms may be safely neglected. In the following, we assume that the oscillator starts at rest, which means that $P_i^{(0)}(z,t)$ must be zero for all times. For molecules, this is not always true as they frequently have a static dipole moment. Collecting orders up to $n = 3$ and transforming to the frequency domain yields

$$P_i^{(1)}(z,\omega) = \varepsilon_0 \chi_{ii}^{(1)}(\omega;\omega) E_i(z,\omega) \qquad (3.6a)$$

$$P_i^{(2)}(z,\omega) = \frac{\varepsilon_0}{2\pi} \sum_{j,k} \int d\omega_2 \, \chi_{ijk}^{(2)}(\omega;\omega_2,\omega-\omega_2) E_j(z,\omega_2) E_k(z,\omega-\omega_2) \qquad (3.6b)$$

$$P_i^{(3)}(z,\omega) = \frac{\varepsilon_0}{4\pi^2} \sum_{j,k,l} \iint d\omega_2 d\omega_3 \, \chi_{ijkl}^{(3)}(\omega;\omega_2,\omega_3,\omega-\omega_2-\omega_3)$$

$$\times E_j(z,\omega_2) E_k(z,\omega_3) E_l(z,\omega-\omega_2-\omega_3), \qquad (3.6c)$$

where we have used the definitions

$$\chi_{ii}^{(1)}(\omega;\omega) \doteq \frac{f_i}{\omega_{i0}^2 - \omega^2 + i\omega\gamma_i} \qquad (3.7a)$$

$$\chi_{ijk}^{(2)}(\omega;\omega_2,\omega-\omega_2) \doteq -\frac{a_{ijk}\varepsilon_0}{f_i} \chi_{ii}^{(1)}(\omega;\omega) \, \chi_{jj}^{(1)}(\omega_2;\omega_2)$$

$$\times \chi_{kk}^{(1)}(\omega-\omega_2;\omega-\omega_2) \qquad (3.7b)$$

$$\chi_{ijkl}^{(3)}(\omega;\omega_2,\omega_3,\omega-\omega_2-\omega_3) \doteq \frac{\varepsilon_0^2}{f_i} \left[\sum_m \frac{2a_{ijm}a_{mkl}}{f_m} \chi_{mm}^{(1)}(\omega-\omega_2;\omega-\omega_2) - b_{ijkl} \right]$$

$$\times \chi_{ii}^{(1)}(\omega;\omega) \chi_{jj}^{(1)}(\omega_2;\omega_2) \chi_{kk}^{(1)}(\omega_3;\omega_3)$$

$$\times \chi_{ll}^{(1)}(\omega-\omega_2-\omega_3;\omega-\omega_2-\omega_3). \qquad (3.7c)$$

Because of energy conservation, the first frequency argument in $\chi^{(n)}$ must be equal to the sum of all the following frequencies. Note that in this classical picture, all higher-order susceptibilities can be related to a combination of first-order susceptibilities. Most optical materials possess more than just a single type of oscillator and the first-order susceptibility (Equation 3.7a) turns into a sum over several resonances. While phonon resonances typically appear in the THz to IR part of the spectrum, electronic resonances are found in the blue to UV part. Glasses and crystals are probably the most important dielectric materials in optics and all their resonances are typically far outside the visible spectral region and, consequently, the imaginary part of the susceptibility may be safely neglected. For not-too-short pulses and for frequencies far away from the resonance frequencies ω_{i0}, the dispersive properties of the second- and third-order susceptibilities may be safely neglected. With this, $\chi^{(2)}$ and $\chi^{(3)}$ become constant and their Fourier transforms are $R^{(2)}(t_1, t_2) = \chi^{(2)}\delta(t_1)\delta(t_2)$ and $R^{(3)}(t_1, t_2, t_3) = \chi^{(3)}\delta(t_1)\delta(t_2)\delta(t_3)$, respectively. Then the solutions (3.6a) simplify to

$$P_i^{(1)}(z, t) = \varepsilon_0 \int dt_1 \, R_{ii}^{(1)}(t_1)E_i(z, t - t_1) \tag{3.8a}$$

$$P_i^{(2)}(z, t) = \varepsilon_0 \sum_{j,k} \chi_{ijk}^{(2)} E_j(z, t)E_k(z, t) \tag{3.8b}$$

$$P_i^{(3)}(z, t) = \varepsilon_0 \sum_{j,k,l} \chi_{ijkl}^{(3)} E_j(z, t)E_k(z, t)E_l(z, t). \tag{3.8c}$$

Inserting Equations 3.8 for the polarization yields the wave equation in the time domain

$$\partial_z^2 E(z, t) = \frac{1}{c^2} \, \partial_t^2$$
$$\times \left[\int dt_1 \, \epsilon_r(t_1) \cdot E(z, t - t_1) + \chi^{(2)} : E(z, t) \cdot E(z, t) + \chi^{(3)} \vdots E(z, t) \cdot E(z, t) \cdot E(z, t) \right], \tag{3.9}$$

with the relative permittivity $\epsilon_{ii}(t) = \delta(t) + R_{ii}^{(1)}(t)$. Similarly, in the frequency domain, we obtain

$$\partial_z^2 E(z, \omega) = -\frac{\omega^2}{c^2} \left[\epsilon_r(\omega) \cdot E(z, \omega) + \frac{1}{2\pi} \int d\omega_1 \, \chi^{(2)} : E(z, \omega_1) \cdot E(z, \omega - \omega_1) \right.$$
$$\left. + \frac{1}{4\pi^2} \iint d\omega_1 d\omega_2 \, \chi^{(3)} \vdots E(z, \omega_2) \cdot E(z, \omega_1 - \omega_2) \cdot E(z, \omega - \omega_1) \right], \tag{3.10}$$

with $\epsilon_{ii}(\omega) = 1 + \chi_{ii}^{(1)}(\omega)$. There is no clear preference whether to solve the nonlinear wave equation in the time or the frequency domain, as in both cases, convolutions appear. In practice, we mostly encounter second- or third-order effects separately, which is why we treat them independently.

3.2 Solution to the Linear Wave Equation

Neglecting all nonlinear contributions to the polarization brings us to the realms of linear optics. The wave equation simplifies to

$$\partial_z^2 E(z,t) = \frac{1}{c^2} \partial_t^2 \int dt_1 \; \epsilon_r(t_1) \cdot E(z, t - t_1). \tag{3.11}$$

Generally, the relative permittivity ϵ_r is a tensor of rank two and simplifies to a scalar for isotropic materials, such as glasses or liquids [3]. Moreover, we can always rotate the coordinate system such that the tensor becomes diagonal. The index of refraction of an optical material is defined through

$$n_{ii}(\omega) \doteq \sqrt{\epsilon_{ii}(\omega)}. \tag{3.12}$$

That is, a material's linear response to an external electric field may either be described through its susceptibility, its dielectric function, or its index of refraction. For frequencies where $\partial_\omega \Re\{\epsilon_r(\omega)\} > 0$, the material is said to have normal dispersion and for $\partial_\omega \Re\{\epsilon_r(\omega)\} < 0$, the material shows anomalous dispersion. The linear wave equation is readily solved and the most general solution presents itself as a superposition of plane waves propagating in the positive z-direction

$$E_i(z,t) = \frac{1}{2\pi} \int d\omega \; E_i(z,\omega) \; e^{i\omega t - i k_i(\omega) z}. \tag{3.13}$$

Each plane wave component is characterized by an amplitude, a frequency, and a wave number k_i that is related to the plane wave's frequency ω through the dispersion relation

$$k_i(\omega) = \frac{\omega}{c} \; n_{ii}(\omega). \tag{3.14}$$

In the following, we use the phasor field description as introduced in Chapter 1, that is

$$E(z,t) = E^+(z,t) + E^-(z,t), \tag{3.15}$$

with $E^-(z,t) = [E^+(z,t)]^*$. Frequently, the spectral support of a pulse is limited to a narrow region around the carrier frequency ω_0. Then, we often employ the slowly varying envelope form

$$E^+(z,t) = \frac{1}{2} \; \mathcal{E}^+(z,t) \; e^{i\omega_0 t - i k_0 z}. \tag{3.16}$$

With Equations 3.13 and 3.16, it is obvious that the slowly varying field at some distance z is related to the field at $z = 0$ through

$$\mathcal{E}_i^+(z,t) = \frac{1}{2\pi} \int d\Omega \; \mathcal{E}_i^+(\Omega) \; e^{i\Omega t - i k_i(\omega) z}, \tag{3.17}$$

with $\mathcal{E}_i^+(\Omega) \doteq \mathcal{E}_i^+(z = 0, \Omega)$ being the Fourier transform of the pulse $\mathcal{E}_i^+(z = 0, t)$ and the relative frequency $\Omega \doteq \omega - \omega_0$. If the index of refraction is a relatively smooth function within the region of interest, the wave number may be expanded in a Taylor series and it is sufficient to consider only low-order Taylor coefficients, that is

$$k(\omega) = \sum_j \frac{1}{j!} \frac{d^j k(\omega)}{d\omega^j}\bigg|_{\omega_0} (\omega - \omega_0)^j. \tag{3.18}$$

As a rule of thumb, if

$$\frac{1}{j!} \frac{d^j k(\omega)}{d\omega^j}\bigg|_{\omega_0} \Delta\omega^j z \ll 1 \tag{3.19}$$

holds, with $\Delta\omega$ being the spectral width of the pulse, then the corresponding orders may be safely neglected. The phase velocity is related to the zero-order term through

$$\frac{1}{v_p} \doteq \frac{k_0}{\omega_0} = \frac{n(\omega_0)}{c}, \tag{3.20}$$

where we have used $k_0 \doteq k(\omega_0)$. The group velocity is defined through

$$\frac{1}{v_g} = \frac{dk}{d\omega}\bigg|_{\omega_0} = \frac{1}{c}\left[n(\omega_0) + \omega_0 \frac{dn}{d\omega}\bigg|_{\omega_0}\right] = \frac{1}{c}\left[n(\lambda_0) - \lambda_c \frac{dn}{d\lambda}\bigg|_{\lambda_0}\right] = \frac{n_g(\omega_0)}{c}, \tag{3.21}$$

with n_g being the group index. The group velocity dispersion (GVD) is defined through

$$D_\omega \doteq \frac{d^2 k}{d\omega^2}\bigg|_{\omega_0}. \tag{3.22}$$

In optical communication, the dispersion parameter D_λ is more frequently used

$$D_\lambda \doteq -\frac{2\pi c}{\lambda_0^2} D_\omega = -\frac{2\pi c}{\lambda_0^2} \frac{d^2 k}{d\omega^2}\bigg|_{\omega_0}. \tag{3.23}$$

Terminating the series expansion (3.18) at $j = 2$ yields

$$\mathcal{E}^+(z, t) = \frac{1}{2\pi} \int d\Omega\, E_0^+(\Omega)\, \exp\left[i\left(t - \frac{z}{v_g}\right)\Omega - i\frac{D_\omega z}{2}\Omega^2\right]. \tag{3.24}$$

Often, we are interested in changes of the shape of the pulse and not the propagation itself. Then it is useful to transform to a coordinate system, which moves with the group velocity of the pulse. In this coordinate system, the new coordinates are $\zeta = z$

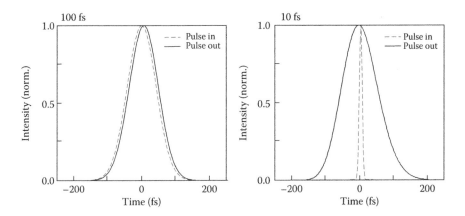

FIGURE 3.1 A 100-fs and a 10-fs pulse before and after propagation through a 10-mm-thick BK7 glass window.

and $\tau = t - z/v_{\mathrm{g}}$ and we obtain

$$\mathcal{E}^+(\zeta,\tau) = \frac{1}{2\pi} \int d\Omega \; \mathcal{E}_0^+(\Omega) \, \exp\left[i\tau\Omega - i\frac{D_\omega\zeta}{2}\Omega^2\right]. \tag{3.25}$$

Figure 3.1 shows two pulses with a duration of 100 and 10-fs, respectively, before and after propagation through a 10-mm-thick piece of BK7 glass (e.g., a vacuum window). While the longer pulse is almost unaffected, the 10-fs pulse shows a considerable broadening, which is mostly due to GVD, but also an asymmetry, which is due to higher-order Taylor coefficients not considered in Equation 3.25. Generally, dispersion effects become more of an issue with shorter pulses.

3.3 Solution to the Nonlinear Wave Equation

Nonlinear effects in ultrafast optics are often explored to change the spectral content of the original laser pulse, for example, in second harmonic or super continuum generation. Nonlinear effects are also prerequisites for soliton formation in optical fibers and are at the heart of all short-pulse characterization methods. In the following, we restrict the review on nonlinear effects to those relevant for ultrafast optics. We address second- and third-order effect separately, because they mostly appear separately.

3.3.1 Second-Order Nonlinearities: Three Wave Mixing

In the first section, we consider second-order nonlinear effects only, which reduces the wave equation in the frequency domain to

$$\partial_z^2 E(z,\omega) + \frac{\omega^2}{c^2}\epsilon_r(\omega)\cdot E(z,\omega) = -\frac{\omega^2}{2\pi c^2}\int d\omega_1 \; \chi^{(2)} : E(z,\omega_1)\cdot E(z,\omega-\omega_1). \tag{3.26}$$

The nonlinearity is such that two fields within the convolution may mix to produce a third, which is why second-order nonlinear effects are also labeled three wave mixing. That is, the total field can be assumed to be a superposition of three fields, which are centered around frequencies $\omega_{0\alpha}$ and which may be polarized in different directions $\hat{e}_\alpha$. In order to simplify, we invoke the following Ansatz:

$$E(z,\omega) = \frac{1}{2} \sum_{\alpha=1}^{3} \left[\mathcal{E}_\alpha^+(z, \omega - \omega_{0\alpha}) + \mathcal{E}_\alpha^-(z, \omega + \omega_{0\alpha}) \right] e^{-ik_\alpha z} \, \hat{e}_\alpha. \tag{3.27}$$

To arrive at the next expression, we first insert Equation 3.27 in Equation 3.26, treat all fields $\mathcal{E}_\alpha^\pm$ on the left-hand side of the equation separately assuming that they are all well separated on the frequency axis, invoke the slowly varying approximation $\partial_z^2 \mathcal{E}_\alpha^+ \ll k_\alpha \partial_z \mathcal{E}_\alpha^+$, and multiply the resulting equation from the left with $\hat{e}_\alpha$.

$$\partial_z \mathcal{E}_\alpha^+(z, \omega - \omega_{0\alpha}) = -i \frac{\omega}{8\pi \, cn_\alpha} \sum_{\beta,\gamma} \hat{e}_\alpha \cdot \chi^{(2)} : \hat{e}_\beta \cdot \hat{e}_\gamma \int d\omega_1$$

$$\left\{ \mathcal{E}_\beta^+(z, \omega_1 - \omega_{0\beta}) \, \mathcal{E}_\gamma^+(z, \omega - \omega_1 - \omega_{0\gamma}) + \mathcal{E}_\beta^+(z, \omega_1 - \omega_{0\beta}) \, \mathcal{E}_\gamma^-(z, \omega - \omega_1 \right.$$

$$+ \omega_{0\gamma}) + \mathcal{E}_\beta^-(z, \omega_1 + \omega_{0\beta}) \, \mathcal{E}_\gamma^+(z, \omega - \omega_1 - \omega_{0\gamma}) + \mathcal{E}_\beta^-(z, \omega_1 + \omega_{0\beta})$$

$$\left. \times \mathcal{E}_\gamma^-(z, \omega - \omega_1 + \omega_{0\gamma}) \right\} e^{ik_\alpha(\omega)z - ik_\beta(\omega_1)z - ik_\gamma(\omega - \omega_2)z}. \tag{3.28}$$

All slowly varying field envelopes have nonzero contributions only when the frequency argument is close to zero. Therefore, the four terms on the right-hand side of Equation 3.28 do contribute only if certain frequency relations hold. From the left-side, we deduce that $\omega \approx \omega_{0\alpha}$ must hold, and for the right-hand side, we find similar arguments for the four terms, that is

Field β	Field γ		
$\omega_1 \approx \omega_{0\beta}$	$\omega_{0\alpha} - \omega_{0\beta} - \omega_{0\gamma} \approx 0$	$\omega_{0\alpha} \approx \omega_{0\beta} + \omega_{0\gamma}$	SFG
$\omega_1 \approx \omega_{0\beta}$	$\omega_{0\alpha} - \omega_{0\beta} + \omega_{0\gamma} \approx 0$	$\omega_{0\alpha} \approx \omega_{0\beta} - \omega_{0\gamma}$	DFG
$\omega_1 \approx -\omega_{0\beta}$	$\omega_{0\alpha} + \omega_{0\beta} - \omega_{0\gamma} \approx 0$	$\omega_{0\alpha} \approx -\omega_{0\beta} + \omega_{0\gamma}$	DFG
$\omega_1 \approx -\omega_{0\beta}$	$\omega_{0\alpha} + \omega_{0\beta} + \omega_{0\gamma} \approx 0$	$\omega_{0\alpha} \approx -\omega_{0\beta} - \omega_{0\gamma}$	

The first term contributes if the sum of the two center frequencies of the two fields on the right-hand side is equal to the center frequency of the field on the left-hand side. Such a process is called sum-frequency generation (SFG). Similarly, the two following terms result in a difference of frequencies, which is why these contributions are labeled difference-frequency generation (DFG). The fourth term apparently does not contribute at all. These conditions are somewhat reminiscent of energy conservation. If all three center frequencies are identical and nonzero, none of these conditions can be fulfilled. Therefore, in second-order nonlinear processes, or three wave mixing, we usually have at least two different fields, typically three, with different center frequencies involved. In order to proceed and to arrive at somewhat more practical expressions, we need to specify in detail the process of interest. In the following, we select SFG as a practical example.

In SFG, two low-frequency input fields generate a third, high-frequency field. Energy conservation requires $\omega_{03} \approx \omega_{01} + \omega_{02}$. We consider a negative uniaxial crystal and the phase matching is type I (ooe) [4]. That is, $\mathcal{E}_1^{\pm}$ and $\mathcal{E}_2^{\pm}$ are polarized in the x direction (ordinary) and $\mathcal{E}_3^{\pm}$ is polarized along the y direction (extraordinary). Note that for every field on the left-hand side of Equation 3.28, there are two contributions on the right-hand side corresponding to different combinations of β and γ. By use Kleinann's conjecture, introducing relative frequencies, and assuming a loss-less medium, we arrive at the final expressions

$$\partial_z \mathcal{E}_1^+ (z, \Omega_1) = -i\kappa_1 \int d\Omega_3 \; \mathcal{E}_2^- (z, \Omega_1 - \Omega_3) \; \mathcal{E}_3^+ (z, \Omega_3) \; e^{-i\Delta k_1 z} \tag{3.29a}$$

$$\partial_z \mathcal{E}_2^+ (z, \Omega_2) = -i\kappa_2 \int d\Omega_3 \; \mathcal{E}_1^- (z, \Omega_2 - \Omega_3) \; \mathcal{E}_3^+ (z, \Omega_3) \; e^{-i\Delta k_2 z} \tag{3.29b}$$

$$\partial_z \mathcal{E}_3^+ (z, \Omega_3) = -i\kappa_3 \int d\Omega_1 \; \mathcal{E}_1^+ (z, \Omega_1) \; \mathcal{E}_2^+ (z, \Omega_3 - \Omega_1) \; e^{i\Delta k_3 z}, \tag{3.29c}$$

with

$$\kappa_\alpha \doteq \frac{\omega_{0\alpha} + \Omega_\alpha}{4\pi c n_\alpha} \chi_{xxy}^{(2)}$$

$$\Delta k_1 \doteq k_3(\omega_{03} + \Omega_3) - k_2(\omega_{02} + \Omega_3 - \Omega_1) - k_1(\omega_{01} + \Omega_1)$$

$$\Delta k_2 \doteq k_3(\omega_{03} + \Omega_3) - k_2(\omega_{02} + \Omega_2) - k_1(\omega_{01} + \Omega_3 - \Omega_2)$$

$$\Delta k_3 \doteq k_3(\omega_{03} + \Omega_3) - k_2(\omega_{02} + \Omega_3 - \Omega_1) - k_1(\omega_{01} + \Omega_1).$$

The same set of equations is found for type II (eoe) phase matching in negative uniaxial crystals. Here, the polarizations of the incoming fields are perpendicular to each other and the generated field is polarized along the extraordinary axis.

In order for the individual integrals to contribute to the result, the differences of the three wave vectors should also vanish, a condition that is called phase matching and is related to momentum conservation. As all three interacting fields have a spectral distribution and thus experience dispersion, phase matching is usually only fulfilled for the center frequencies. The phase matching condition is sometimes written as a Taylor series, for example, in the case of the first integral

$$\Delta k_1 = k_3(\omega_{03} + \Omega_3) - k_2(\omega_{02} + \Omega_3 - \Omega_1) - k_1(\omega_{01} + \Omega_1)$$

$$= [k_{03} - k_{02} - k_{01}] + \left[\frac{dk_3}{d\omega} \bigg|_{\omega_{03}} \Omega_3 - \frac{dk_2}{d\omega} \bigg|_{\omega_{02}} (\Omega_3 - \Omega_1) - \frac{dk_1}{d\omega} \bigg|_{\omega_{01}} \Omega_1 \right] + \cdots$$

If the lowest-order terms add to zero, the process is labeled phase-matched to first order. If, in addition, the next three terms add to zero, the process is phase-matched to second order, and so on. Note that phase matching to first order is identical to matching the phase velocities at the center frequencies. Phase matching to second order requires matching the group velocities. It has to be emphasized at this point that phase matching is a condition that is fulfilled only if special care is taken. Usually, appropriate crystals have to be employed and, moreover, the crystals have to be aligned correctly with respect to the propagation direction and the polarization of the fields. That is to say, three wave mixing processes are observed to be extremely weak unless the optical setup is deliberately designed. In addition, we have only considered cases where all three fields propagate in the same direction, and consequently, we have to match wave numbers rather than wave vectors. There are situations where this is not the case, and the angles between the wave vectors are nonzero. For example, such geometries are used when both phase velocity matching and group velocity matching must be fulfilled.

A frequently used approximation is the nondepleted fundamental wave approximation. Here, it is assumed that the overall conversion efficiency is low and the two input fields remain constant in amplitude as they propagate through the medium, that is $\mathcal{E}_{1,2}^{\pm}(z, \Omega) \equiv \mathcal{E}_{1,2}^{\pm}(\Omega)$. Therefore, only one differential equation remains to be solved

$$\partial_z \mathcal{E}_3^+(z, \Omega_3) = -i\kappa_3 \int d\Omega_1 \, \mathcal{E}_1^+(\Omega_1) \, \mathcal{E}_2^+(\Omega_3 - \Omega_1) \, e^{i\Delta kz}. \tag{3.30}$$

This equation can be integrated from $z = 0$ to $z = L$ and we find

$$\mathcal{E}_3^+(L, \Omega_3) = -i\kappa_3 L \int d\Omega_1 \, \mathcal{E}_1^+(\Omega_1) \, \mathcal{E}_2^+(\Omega_3 - \Omega_1) \, \text{sinc} \frac{\Delta kL}{2} \, e^{i\Delta kL/2}, \tag{3.31}$$

where L is the length of the crystal. A special case of SFG is second harmonic generation. Here, the two input fields are identical and we obtain

$$\mathcal{E}_3^+(L, \Omega_3) = -i\kappa_3 L \int d\Omega_1 \, \mathcal{E}_1^+(\Omega_1) \, \mathcal{E}_1^+(\Omega_3 - \Omega_1) \, \text{sinc} \frac{\Delta kL}{2} \, e^{i\Delta kL/2}. \tag{3.32}$$

Figure 3.2 shows the spectral intensity $|\mathcal{E}_3^+(z, \Omega_3)|^2$ as the 100-fs-long and 10-fs-long pulses propagate through a suitable nonlinear crystal. The second harmonic field is initially zero and grows through frequency conversion. While the 100-fs-long pulse is frequency doubled almost without any spectral distortions, the 10-fs-long pulse is modified and one can see the effect of the sinc function, that is the effect of phase matching, through the appearance of side bands.

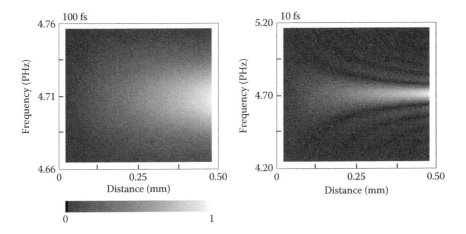

FIGURE 3.2 A 100-fs and a 10-fs pulse (800 nm) frequency doubled in a 500-μm-thick BBO crystal. The intensity plots show the evolution of the spectral intensity as the pulses propagate through the crystal.

3.3.1.1 Limit of a Thin Crystal

In the limit of a thin crystal, the phase mismatch ΔkL is close to zero for all frequency combinations within the spectra of the two input pulses

$$\mathcal{E}_3^+(L, \Omega_3) = -i\kappa_3 L \int d\Omega_1 \, \mathcal{E}_1^+(\Omega_1) \, \mathcal{E}_2^+(\Omega_3 - \Omega_1).$$

This situation is desirable for converting broad fundamental spectra, especially in pulse characterization techniques. However, thin crystals necessarily imply a low conversion efficiency and there is always a trade-off between bandwidth and efficiency, which has to be optimized for a given situation.

3.3.1.2 Limit of a Thick Crystal

For very thick crystals, ΔkL is large except for a single frequency Ω_3. If we assume that the crystal is oriented such that this condition is fulfilled for $\Omega_3 = 0$, then

$$\mathcal{E}_3^+(L, 0) = -i\kappa_3 L \int d\Omega_1 \, \mathcal{E}_1^+(\Omega_1) \, \mathcal{E}_2^+(-\Omega_1).$$

If we consider second harmonic generation and split the two input fields into amplitudes and phases, we obtain

$$\mathcal{E}_3^+(L, 0) = -i\kappa_3 L \int d\Omega_1 \, |\mathcal{E}_1^+(\Omega_1)| \, |\mathcal{E}_1^+(-\Omega_1)| \, e^{i\phi(\Omega_1) + i\phi(-\Omega_1)}.$$

Thus, two very distinct situations arise depending on whether the slowly varying phases are odd or even functions with respect to frequency

$$\mathcal{E}_3^+(L,0) = -i\kappa_3 L \int d\Omega_1 \; |\mathcal{E}_1^+(\Omega_1)| \; |\mathcal{E}_1^+(-\Omega_1)| \begin{cases} e^{i2\phi(\Omega_1)} & \text{even} \\ 1 & \text{odd} \end{cases}.$$

for odd phase functions, the SHG process is completely independent of the exact nature of the phase modulation. On the contrary, for even phase functions, the detailed shape may even cause the second harmonic signal to completely disappear.

3.3.2 Third-Order Nonlinearities: Four Wave Mixing

In materials with inversion symmetry, such as glasses or liquids, all even-order suscepti-bilities must be zero. Then, third-order nonlinearities become important as they are the lowest-order contribution to the nonlinear material response. For third-order nonlinear-ities, also labeled four wave mixing processes, the wave equation is almost always solved the in time domain

$$\partial_z^2 E(z,t) = \frac{1}{c^2} \partial_t^2 \left[\int dt_1 \; \epsilon_r(t_1) E(z, t-t_1) + \chi^{(3)} \vdots E(z,t) \cdot E(z,t) \cdot E(z,t) \right]. \quad (3.33)$$

Here, the total field may be composed of four individual fields

$$E(z,t) = \frac{1}{2} \sum_{\alpha=1}^{4} \left[\mathcal{E}_\alpha^+(z,t) e^{i\omega_{0\alpha}t - ik_{0\alpha}z} + \mathcal{E}_\alpha^-(z,t) e^{-i\omega_{0\alpha}t + ik_{0\alpha}^* z} \right] \hat{e}_\alpha. \quad (3.34)$$

Inserting the Ansatz (3.34) in the nonlinear wave equation (3.33) yields a relatively complex expression. With the following approximations, we try to simplify this expres-sion to a level that permits either an analytic solution or leads to a form that can be easily solved by numerical means. First, we replace the linear dispersion term by Fourier transforming it to the frequency domain, expressing the squared wave vector through its Taylor expansion, and Fourier transforming the resulting series expansion back to the time domain. For each of the four fields, we then obtain

$$\frac{1}{c^2} \partial_t^2 \int dt_1 \; \varepsilon(t_1) \; \mathcal{E}_\alpha^+(z, t-t_1) \; e^{i\omega_{0\alpha}(t-t_1) - ik_{0\alpha}z}$$

$$= - \sum_{n,m=0}^{\infty} \frac{k_{n\alpha} k_{m\alpha}}{n! \, m! \, i^{n+m}} \partial_t^{n+m} \mathcal{E}_\alpha^+(z,t) \; e^{i\omega_{0\alpha}t - ik_{0\alpha}z}, \quad (3.35)$$

with $k_{n\alpha} \doteq d^n k / d\omega^n |_{\omega_{0\alpha}}$. Explicitly evaluating the three lowest-order terms on the right-hand side of Equation 3.35, that is $[n,m] = [0,0], [0,1]$ and $[1,0]$, employing the slowly varying envelope approximation, writing a separate equation for each field, and multiplying the resulting equation from the left with $\hat{e}_\alpha$ yields

$$\left[\partial_z \mathcal{E}_\alpha^+(z,t) + \frac{1}{v_{g\alpha}} \partial_t \mathcal{E}_\alpha^+(z,t) + \frac{i}{2k_{0\alpha}} \sum_{n,m=1}^{\infty} \frac{k_{n\alpha} k_{m\alpha}}{n!\, m!\, i^{n+m}} \partial_t^{n+m} \mathcal{E}_\alpha^+(z,t) \right] e^{i\omega_{0\alpha}t - ik_{0\alpha}z}$$

$$= -\frac{i}{8k_{0\alpha} c^2} \sum_{\beta,\gamma,\delta} \hat{e}_\alpha \cdot \chi^{(3)} \vdots \hat{e}_\beta \cdot \hat{e}_\gamma \cdot \hat{e}_\delta \, \partial_t^2$$

$$\times \Big\{ \mathcal{E}_\beta^+(z,t)\mathcal{E}_\gamma^+(z,t)\mathcal{E}_\delta^+(z,t)\, e^{i(\omega_{0\beta}+\omega_{0\gamma}+\omega_{0\delta})t - i(k_{0\beta}+k_{0\gamma}+k_{0\delta})z}$$

$$+ \mathcal{E}_\beta^+(z,t)\mathcal{E}_\gamma^+(z,t)\mathcal{E}_\delta^-(z,t)\, e^{i(\omega_{0\beta}+\omega_{0\gamma}-\omega_{0\delta})t - i(k_{0\beta}+k_{0\gamma}-k_{0\delta})z}$$

$$+ \mathcal{E}_\beta^+(z,t)\mathcal{E}_\gamma^-(z,t)\mathcal{E}_\delta^+(z,t)\, e^{i(\omega_{0\beta}-\omega_{0\gamma}+\omega_{0\delta})t - i(k_{0\beta}-k_{0\gamma}+k_{0\delta})z}$$

$$+ \mathcal{E}_\beta^+(z,t)\mathcal{E}_\gamma^-(z,t)\mathcal{E}_\delta^-(z,t)\, e^{i(\omega_{0\beta}-\omega_{0\gamma}-\omega_{0\delta})t - i(k_{0\beta}-k_{0\gamma}-k_{0\delta})z}$$

$$+ \mathcal{E}_\beta^-(z,t)\mathcal{E}_\gamma^+(z,t)\mathcal{E}_\delta^+(z,t)\, e^{i(-\omega_{0\beta}+\omega_{0\gamma}+\omega_{0\delta})t - i(-k_{0\beta}+k_{0\gamma}+k_{0\delta})z}$$

$$+ \mathcal{E}_\beta^-(z,t)\mathcal{E}_\gamma^+(z,t)\mathcal{E}_\delta^-(z,t)\, e^{i(-\omega_{0\beta}+\omega_{0\gamma}-\omega_{0\delta})t - i(-k_{0\beta}+k_{0\gamma}-k_{0\delta})z}$$

$$+ \mathcal{E}_\beta^-(z,t)\mathcal{E}_\gamma^-(z,t)\mathcal{E}_\delta^+(z,t)\, e^{i(-\omega_{0\beta}-\omega_{0\gamma}+\omega_{0\delta})t - i(-k_{0\beta}-k_{0\gamma}+k_{0\delta})z}$$

$$+ \mathcal{E}_\beta^-(z,t)\mathcal{E}_\gamma^-(z,t)\mathcal{E}_\delta^-(z,t)\, e^{i(-\omega_{0\beta}-\omega_{0\gamma}-\omega_{0\delta})t - i(-k_{0\beta}-k_{0\gamma}-k_{0\delta})z} \Big\}. \tag{3.36}$$

Evaluating the second-order time derivative of each term on the right-hand side of Equation 3.36 yields three terms with decreasing magnitude. Sometimes, it will be sufficient to consider the largest contribution only, as we will see in the next section. Also, we change the coordinate system and transform it to a system that propagates at the group velocity of the field α.

3.3.2.1 Third Harmonic Generation

The first practical example of Equation 3.36 is third harmonic generation, where we assume that the three incoming fields are identical and polarized along x and the resulting third harmonic is polarized along y. Energy conservation requires $\omega_{0\alpha} = \omega_{0\beta} + \omega_{0\gamma} + \omega_{0\delta}$ and only the first term in Equation 3.36 yields a nonzero contribution, that is

$$\partial_\xi \mathcal{E}_4^+(\xi,\tau) + \frac{i}{2k_{04}} \sum_{n,m=1}^{\infty} \frac{k_{n4} k_{m4}}{n!\, m!\, i^{n+m}} \partial_\tau^{n+m} \mathcal{E}_4^+(\xi,\tau) \approx -i \frac{k_{04} \chi_{yxxx}^{(3)}}{8} \mathcal{E}_1^{+3}(\xi,\tau)\, e^{i\Delta k\xi},$$

$$\tag{3.37}$$

with $\Delta k = k_{04} - 3k_{01}$. The third-order nonlinear susceptibilities are usually low and phase matching is not easy to achieve, that is, efficient third harmonic generation is difficult to realize. This is why third harmonic generation in solid materials has found only a few practical applications.

3.3.2.2 Degenerate Four Wave Mixing

If all interacting fields have the same center frequency, the nonlinear interaction is called degenerate four wave mixing (DFWM). It plays an important role in ultrafast optics as it is responsible for an intensity-dependent change of the index of refraction, for self-steepening, it facilitates white light generation, and it is important for soliton formation and pulse propagation in optical fibers. Although beyond the scope of this chapter, it is also responsible for spatial effects, such as self-focusing. The four interacting fields may or may not have the same polarization. If they have the same polarization, then for some terms in Equation 3.36, phase matching is automatically fulfilled, that is for those terms with two positive and two negative wave number contributions. Without the necessity of extra phase matching measures, these processes will be strong in almost any material if only the intensity is high enough.

In the following, we assume that all interacting fields have the same center frequency and are polarized along the same axis. Then, three out of the eight terms on the right-hand side of Equation 3.36 contribute and considering the two largest contributions on the right-hand side of Equation 3.37 yields

$$\partial_\xi \mathcal{E}^+(\xi, \tau) + \frac{i}{2k_0} \sum_{n,m=1}^{\infty} \frac{k_n k_m}{n!\, m!\, i^{n+m}} \partial_\tau^{n+m} \mathcal{E}^+(\xi, \tau)$$

$$\approx -i\gamma \left[1 - \frac{2i}{\omega_0} \partial_\tau \right] \mathcal{E}^+(\xi, \tau) |\mathcal{E}^+(\xi, \tau)|^2 \qquad (3.38)$$

where we have used $\gamma \doteq 3\omega_0 \chi_{xxxx}^{(3)}/(8cn_0)$. In the following sections, we will analyze different aspects of this equation. That is, we will assume different approximations with the goal to find analytic solutions for specific situations.

3.3.2.3 Self-Phase Modulation

If the dispersion of a material is vanishingly small, the material is very thin, or if the pulses have relatively small spectral bandwidth, we may neglect dispersion and consider only the strongest third-order contribution

$$\partial_\xi \mathcal{E}^+(\xi, \tau) = -i\gamma\, \mathcal{E}^+(\xi, \tau) |\mathcal{E}^+(\xi, \tau)|^2. \qquad (3.39)$$

To solve this equation, we separate the field in its amplitude and phase part and split the resulting differential equation in two, one for the real and one for the imaginary part, that is

$$\partial_\xi |\mathcal{E}^+(\xi, \tau)| = 0 \tag{3.40a}$$

$$\partial_\xi \Phi(\xi, \tau) = -\gamma \, |\mathcal{E}^+(\xi, \tau)|^2. \tag{3.40b}$$

The solution is found to be

$$\mathcal{E}^+(\xi, \tau) = |\mathcal{E}^+(0, \tau)| \, e^{-i\gamma|\mathcal{E}^+(0,\tau)|^2 \xi}. \tag{3.41}$$

Apparently, the slowly varying amplitude is not changing as the pulse propagates through the medium. The temporal phase, however, grows linearly with ξ and causes a modulation of the spectral amplitude. This is why this process has been named self-phase modulation. Sometimes Equation (3.41) is cast into the following form:

$$\mathcal{E}^+(\xi, \tau) = |\mathcal{E}^+(0, \tau)| \, \exp\left[-i\frac{\omega_0}{c} n_2 I(0, \tau)\xi\right], \tag{3.42}$$

with the nonlinear index of refraction

$$n_2 = \frac{2\gamma}{\varepsilon_0 n_0 \omega_0} = \frac{3\chi_{xxxx}^{(3)}}{4\varepsilon_0 c n_0^2}. \tag{3.43}$$

The intensity-induced refractive index change causes an instantaneous phase modulation and the instantaneous frequency can deviate considerably from ω_0, that is

$$\omega(t) - \omega_0 = -\frac{\omega_0}{c} n_2 \partial_\tau I(0, \tau)\xi. \tag{3.44}$$

Figure 3.3 shows the spectrum of a 100-fs-long and a 10-fs-long pulse after undergoing self-phase modulation in a 0.1-mm-thick fused silica plate. Both pulses have the same peak intensity.

What looks like spectral interference is a result of the fact that self-phase modulation produces pairs of instantaneous frequencies at different times. The number of such interference peaks M in the spectrum (see Figure 3.3) increases with intensity and propagation length ξ and is approximately given by

$$M = \frac{1}{2} + \frac{\gamma|\mathcal{E}^+(0,0)|^2 \xi}{\pi} = \frac{1}{2} + \frac{\omega_0 n_2 I_0 \xi}{c\pi}, \tag{3.45}$$

where I_0 is the peak intensity of the pulse. Note that self-phase modulation does not always cause spectral broadening. If the pulse has an initial negative chirp, even spectral narrowing may be observed.

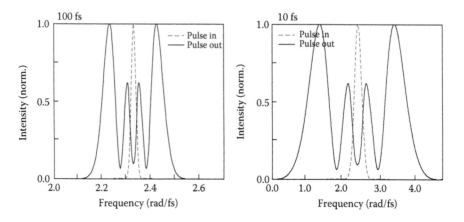

FIGURE 3.3 Self-phase-modulated spectrum of a 100-fs and a 10-fs laser pulse after propagation through a 0.1-mm-thick fused silica plate.

3.3.2.4 Two-Photon Absorption

Here, we investigate the case when $\chi^{(3)}$ is not purely real, but has a small imaginary contribution, that is $\chi^{(3)} = \chi^{(3)'} + i\chi^{(3)''}$. Again, we neglect dispersion and the wave equation reads

$$\partial_\xi \mathcal{E}^+(\xi,\tau) = -i(\gamma' + i\gamma'')\mathcal{E}^+(\xi,\tau)|\mathcal{E}^+(\xi,\tau)|^2. \tag{3.46}$$

To find a solution, we separate the field in an amplitude and a phase part, that is $\mathcal{E}^+(\xi,\tau) = \sqrt{2I(\xi,\tau)/\varepsilon_0 c n_0}\ \exp[i\Phi(\xi,\tau)]$. Inserting this Ansatz in Equation 3.46 yields

$$\partial_\xi I(\xi,\tau) = \frac{8\gamma''}{(\varepsilon_0 c n_0)^2} I^2(\xi,\tau) \tag{3.47a}$$

$$\partial_\xi \Phi(\xi,\tau) = -\frac{2\gamma'}{\varepsilon_0 c n_0} I(\xi,\tau). \tag{3.47b}$$

The solution is

$$I(\xi,\tau) = \frac{I(0,\tau)}{1 - \frac{4\gamma''}{\varepsilon_0 c n_0} I(0,\tau)\xi} \tag{3.48a}$$

$$\Phi(\xi,\tau) = \frac{\gamma'}{2\gamma''} \ln\left[1 - \frac{4\gamma''}{\varepsilon_0 c n_0} I(0,\tau)\xi\right]. \tag{3.48b}$$

For $\gamma'' < 0$, the intensity decreases as the pulse propagates through the medium and the infinitesimal change is proportional to the square of the intensity; this situation is labeled two-photon absorption.

3.3.2.5 Self-Steepening

Neglecting dispersion but considering the two strongest third-order contributions yields

$$\partial_\xi \mathcal{E}^+(\xi, \tau) = -i\gamma \left(1 - \frac{2i}{\omega_0} \partial_\tau \right) \mathcal{E}^+(\xi, \tau) |\mathcal{E}^+(\xi, \tau)|^2. \tag{3.49}$$

To find an analytic solution, we separate the field in amplitude and phase, $\mathcal{E}^+(\xi, \tau) = \sqrt{2I(\xi, \tau)/\varepsilon_0 c n_0} \, \exp[i\Phi(\xi, \tau)]$ and solve the resulting differential equation for the real and the imaginary part separately, that is

$$\partial_\xi I(\xi, \tau) = -\frac{12\gamma}{\varepsilon_0 c n_0 \omega_0} I(\xi, \tau) \, \partial_\tau I(\xi, \tau) \tag{3.50a}$$

$$\partial_\xi \Phi(\xi, \tau) = -\frac{2\gamma}{\varepsilon_0 c n_0} I(\xi, \tau) - \frac{4\gamma}{\varepsilon_0 c n_0 \omega_0} I(\xi, \tau) \partial_\tau \Phi(\xi, \tau). \tag{3.50b}$$

The solution to Equation 3.49a is of the form

$$I(\xi, \tau) = I\left(\tau - \frac{12\gamma I}{\varepsilon_0 c n_0 \omega_0} \xi \right). \tag{3.51}$$

Even in the absence of dispersion, the pulse shape changes slightly upon propagation. It appears as if the group velocity becomes intensity dependent. Those parts of the pulse that have a higher intensity propagate slower than those with a lower intensity and, consequently, the pulse shape will steepen toward the trailing edge. It is because of this behavior that this contribution has been labeled self-steepening.

3.3.2.6 Nonlinear Schrödinger Equation

In the following, we allow for dispersion but consider only second-order dispersion. In this case, the so-called nonlinear Schrödinger equation results

$$\partial_\xi \mathcal{E}^+(\xi, \tau) - \frac{ik_2}{2} \partial_\tau^2 \mathcal{E}^+(\xi, \tau) = -i\gamma \mathcal{E}^+(\xi, \tau) |\mathcal{E}^+(\xi, \tau)|^2. \tag{3.52}$$

With a change of variables, that is $\zeta = \xi |k_2|/\Delta t_0^2$ and $\Theta = \tau/\Delta t_0$, where Δt_0 is some pulse duration, and the substitution $u(\zeta, \Theta) = \sqrt{\Delta t_0^2 \gamma / |k_2|} \, \mathcal{E}^+(\xi, \tau)$, we find

$$i\partial_\zeta u(\zeta, \Theta) - \frac{1}{2}\partial_\Theta^2 u(\zeta, \Theta) - u(\zeta, \Theta) \, |u(\zeta, \Theta)|^2 = 0, \tag{3.53}$$

given that $\gamma > 0$ and $k_2 < 0$. This equation has the solution

$$u(\zeta, \Theta) = \text{sech}\Theta \, e^{-i\zeta/2}, \tag{3.54}$$

as can be easily shown. That is, the solution is a pulse with a temporal amplitude invariant with propagation. Such a pulse is called an optical soliton [5]. In terms of the original variables, we find

$$\mathcal{E}^+(\xi, \tau) = \sqrt{\frac{|k_2|}{\gamma \Delta t_0^2}} \operatorname{sech} \frac{\tau}{\Delta t_0} \exp\left(-i\xi \frac{|k_2|}{2\Delta t_0^2}\right). \tag{3.55}$$

Inspecting this solution shows that an optical soliton is only found if a number of requirements are fulfilled. First, the medium must have a positive nonlinear constant and a negative GVD (or vice versa). Second, given the material parameters, the amplitude and the temporal width of the soliton are linked to each other and cannot be chosen arbitrarily. Third, the soliton acquires no temporal phase upon propagation, that is, not only the temporal but also the spectral amplitude remains unaltered. Besides the fundamental soliton, there exist higher-order solitons of order N whenever the amplitude satisfies $N\sqrt{|k_2|/(\gamma \Delta t_0^2)}$, with $N = 1, 2, \ldots$. Figure 3.4 shows the temporal and spectral intensity of a pulse propagating through a suitable medium. If the nonlinearity is set to zero (top row), the pulse undergoes dispersion and as a consequence broadens. If the dispersion is set to zero but the nonlinearity is nonzero, the spectrum broadens and shows the characteristic sequence of peaks. If both dispersion and nonlinearity are nonzero, neither the temporal nor the spectral shape of the pulse seems to change upon propagation, indicating the existence of a soliton.

3.3.2.7 Intrapulse Raman Scattering

In order to formulate an appropriate equation describing the propagation of pulses through very long media, such as optical fibers, we need to include the Raman effect. Laser pulses may couple to molecular vibrations or optical phonons, which typically have much lower eigenfrequencies, through an induced polarizability. To gain some insight, we again treat the system classically and assume that the material is composed of independent oscillators. The driving force is proportional to the derivative of the time-averaged electric energy density with respect to the direction of motion s, that is

$$F(z, t) = \frac{1}{2}\varepsilon_0 \left.\partial_s \alpha\right|_{s=0} \langle E^2(z, t)\rangle, \tag{3.56}$$

with the polarizability α. The time average reflects the fact that the resonance frequencies of molecules or phonons are much lower than the center frequency of the laser pulse. The solution may be found in the frequency domain

$$s(z, \omega) = \frac{\varepsilon_0}{2m} \left.\partial_s \alpha\right|_{s=0} \frac{1}{\omega_0^2 - \omega^2 + i\gamma\omega} \int dt \langle E^2(z, t)\rangle e^{-i\omega t}. \tag{3.57}$$

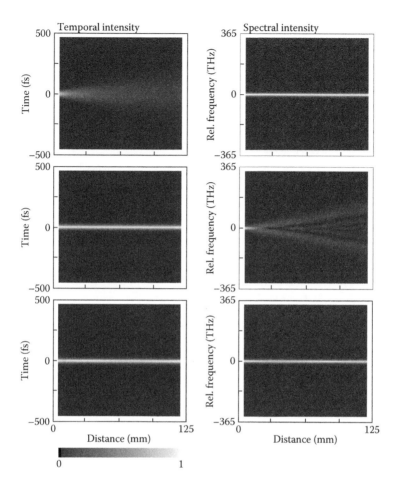

FIGURE 3.4 Temporal and spectral intensity of a sech-square pulse propagating through a 125-mm-long material, which has a quadratic dispersion $k_2 = -2.6710^{-26}$ s^2/m and a nonlinear index of refraction of $n_2 = 2.510^{-16}$ cm^2/W. Top: $n_2 \equiv 0$, center: $k_2 \equiv 0$, and bottom $n_2 \neq 0$ and $k_2 \neq 0$.

The macroscopic polarization is $P(z,t) = N\alpha(z,t)E(z,t)$ and we are only interested in the field-induced part, that is $P(z,t) = N\partial_s\alpha(z,t)|_{s=0} \, s(z,t)E(z,t)$. With this and using $\langle E^2(z,t) \rangle = |\mathcal{E}^+(z,t)|^2/2$, we find for the slowly varying polarization

$$\mathcal{P}^+(z,t) = \varepsilon_0 \int dt_1 R_R(t_1)|\mathcal{E}^+(z,t-t_1)|^2\mathcal{E}^+(z,t), \tag{3.58}$$

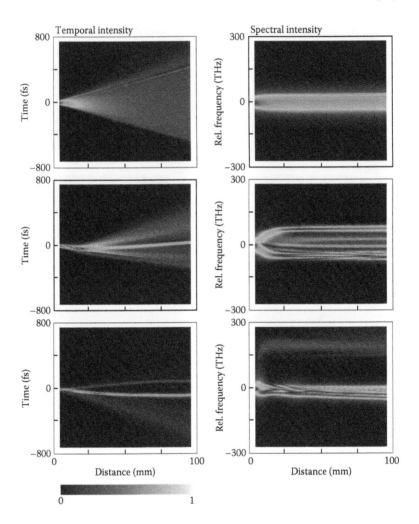

FIGURE 3.5 Temporal and spectral intensity as a function of propagation distance through a 100-mm-long optical fiber. The center wavelength of the pulse is in the normal dispersion region (top), coincides with the zero dispersion wavelength (center), or is within the anomalous dispersion region (bottom).

with $R_R(t)$ being the Fourier transform of

$$\frac{N}{2m}(\partial_s\alpha|_{s=0})^2 \frac{1}{\omega_0^2 - \omega^2 + i\gamma\omega}. \tag{3.59}$$

Inspecting the imaginary part of the Raman susceptibility indicates absorption for positive relative frequencies and gain for negative relative frequencies. That is, the Raman effect shifts pulses toward lower frequencies as they propagate through the medium.

3.3.2.8 Optical Fibers

Adding the Raman contribution to Equation 3.38 results in an equation that is commonly used to describe the propagation of time domain pulses in optical fibers [6]

$$\partial_\xi \mathcal{E}^+(\xi,\tau) + i \sum_{n,m=1}^{\infty} \frac{k_n k_m}{n!\, m!\, i^{n+m}} \partial_\tau^{n+m} \mathcal{E}^+(\xi,\tau)$$

$$= -i\gamma \left[1 - \frac{2i}{\omega_0} \partial_\tau \right] \mathcal{E}^+(\xi,\tau) \int d\tau_1\, R(\tau_1) |\mathcal{E}^+(\xi,\tau-\tau_1)|^2, \qquad (3.60)$$

where we have combined the instantaneous electronic response $\delta(t)$ and the Raman response $R_R(t)$ of the medium to $R(t) = (1-f)\,\delta(t) + fR_R(t)$. The relative weight is governed by the constant f, which for fused silica is approximately 0.18.

Figure 3.5 shows the evolution of the temporal and the spectral intensity as a function of propagation distance through a 100-mm-long fiber. The zero GVD wavelength of the optical fiber is around 754 nm. If we inject a 20-fs-long laser pulse with a center wavelength of 654 nm, we are well within the normal dispersion regime and we observe spectral broadening due to self-phase modulation and substantial broadening due to dispersion (top row). After a few millimeters, the pulse broadens so much that the intensity drops below a level where it can cause self-phase modulation and the spectral broadening stops. Conversely, if the center wavelength is well within the anomalous dispersion regime, the temporal intensity profile shows the formation of several solitons, which due to their different spectral content propagate with different group velocities (bottom row). The spectrum is structured, also because of interference effects. Injecting a pulse with a center wavelength that matches the zero dispersion wavelength shows a mixture of the two scenarios discussed (center row).

3.4 Time Domain Pulses in Linear Optical Systems

In practically every ultrafast experiment, we find optical units that are designed to change the properties of a time domain pulse in a very specific way. Such systems can often be described by means of a linear transfer function thereby neglecting the detailed layout of the unit. They manipulate the spectral amplitude $|\mathcal{E}_{in}^+(\omega)|$ or phase $\phi(\omega)$ of the incident pulse through a linear operation, that is

$$\mathcal{E}_{out}^+(\omega) = \mathcal{E}_{in}^+(\omega)\, M(\omega) = |\mathcal{E}_{in}^+(\omega)|\, |M(\omega)|\, e^{i\phi(\omega)+i\Phi(\omega)}. \qquad (3.61)$$

If $|M(\omega)| \neq 1$, we find amplitude modulation, and if $\Phi(\omega) \neq 0$, phase modulation. In the following section, we start by introducing systems that are based on angular dispersion and mostly aim at stretching and compressing pulses in a well-controlled way. The next class of systems is based on resonances and generally modifies both amplitude and phase of the incident pulse. We conclude this section with the most flexible device, a so-called pulse shaper, which can manipulate the spectral amplitude and phase in an almost arbitrary way. More details can be found in a number of textbooks [7].

3.4.1 Systems Based on Angular Dispersion

The phase modulation in optical systems based on angular dispersion is mostly a result of the fact that different colors within the spectrum travel along different paths. Loosely speaking, the phase is frequency dependent because the geometrical path length is a function of the frequency. In the following, we neglect all amplitude modulations that may arise due to spectral clipping, wavelength-dependent reflectivities at interfaces, and so on. Furthermore, at the exit of the optical system, the pulse must be recovered, that is, all angular dispersion present at some point within the system must be compensated for by the time the pulse has passed through the entire system. Our task is to find the phase modulation for a given sequence of optical elements, that is the optical path length as a function of frequency. There are different routes to proceed from here.

The first is based on a general argument, which is valid for any system possessing angular dispersion. Suppose an incoming pulse impinging on a dispersive element after which the spectrum is dispersed and different colors travel in different directions. We select two diffracted rays, namely, the reference ray, which corresponds to the center frequency, and any other ray. The distance between the dispersive element and the plane of interest measured along the reference ray is L. It is easy to see that the phase front of the arbitrary ray intersects with the phase front of the reference ray at the plane of interest, if the length of the corresponding ray is $L \cos \Delta\beta$, where $\Delta\beta = \beta(\omega) - \beta_0$ is the relative diffraction angle. With this, the phase modulation becomes

$$\Phi(\omega) = \frac{\omega}{c} L \cos \Delta\beta. \tag{3.62}$$

Equation 3.62 produces correct results if no dispersive material is involved. From Equation 3.62, analytic expressions for the Taylor coefficients of the phase expansion can be obtained. The quadratic phase coefficient amounts to

$$\Phi_2 = -\frac{L\omega_0}{c} \left(d_\omega \beta \big|_{\omega_0} \right)^2, \tag{3.63}$$

where we have assumed that $\Delta\beta \ll 1$. Note that Equation 3.63 is always negative irrespective of the sign of the angular dispersion. Similarly, we find for the cubic phase

$$\Phi_3 = -\frac{3L}{c} \left[\left(d_\omega \beta \big|_{\omega_0} \right)^2 + \omega_0 \, d_\omega \beta \big|_{\omega_0} \, d_\omega^2 \beta \big|_{\omega_0} \right], \tag{3.64}$$

which may be positive or negative depending on the angular dispersion.

A different approach to calculate the total phase modulation of an optical system is to treat the system as a sequence of homogeneous materials separated by interfaces with specific properties. The total phase then is

$$\Phi(\omega) = \sum_j \left[\frac{\omega}{c} P_j(\omega) + R_j(\omega) \right]. \tag{3.65}$$

The optical path length $P_j(\omega) = n_j(\omega) z_j(\omega)$ is the product of the refractive index $n_j(\omega)$ and the frequency-dependent geometrical path length $z_j(\omega)$ between two interfaces. At the interfaces, an additional phase jump $R_j(\omega)$ may have to be considered.

The last approach uses the fact that the first derivative of the spectral phase with respect to frequency can be interpreted as the "time" each frequency requires to travel through the optical system

$$t(\omega) = d_\omega \Phi(\omega). \tag{3.66}$$

Brorson and Haus have shown that for an optical system without dispersive material

$$t(\omega) = \frac{P(\omega)}{c} \tag{3.67}$$

is a consequence of Fermat's principle. Therefore, if the phase modulation of the complete optical system is written as

$$\Phi(\omega) = \frac{\omega}{c} P(\omega) + R(\omega), \tag{3.68}$$

we find with Equations 3.66 and 3.67 by comparison that

$$d_\omega R(\omega) = -\frac{\omega}{c} d_\omega P(\omega) \quad \text{and} \quad R(\omega) = -\frac{1}{c} \int d\omega' \omega' \, d_{\omega'} P(\omega'). \tag{3.69}$$

Next, we will apply the different approaches to three relatively simple, but widely used, systems, namely, the grating compressor, the grating stretcher, and the prism compressor.

3.4.1.1 Grating Compressor

Figure 3.6 shows a grating compressor consisting of two parallel gratings and an end mirror. The incoming beam is spectrally dispersed on the first grating, impinges onto the second grating, and is sent back through the system by the end mirror. To calculate the optical path $P(\omega)$ as a function of frequency, the full path is split into two parts. The first part is the distance between the two gratings $\overline{AB} = L/\cos\beta$, and the second part is the distance between the second grating and the mirror (see Figure 3.6) $\overline{BC} = r - \overline{B_0 B} \cos(90 - \alpha)$, where r is the distance between the second grating and the end mirror. The total path is the sum of the two contributions

$$\overline{ABC} = \frac{L}{\cos\beta} + r - L \tan\beta_0 \sin\alpha + L \tan\beta \sin\alpha, \tag{3.70}$$

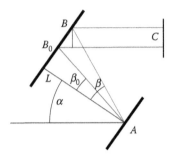

FIGURE 3.6 The total path is the sum of two contributions. The distance between the two gratings $\overline{AB}$ and the distance between the second grating and end mirror $\overline{BC}$. B_0 is the point where the center frequency impinges on the second grating.

where β_0 is the diffraction angle at center frequency. From Equation 3.70, we calculate $P(\omega)$, $R(\omega)$, and the total phase modulation $\Phi(\omega)$. Note that in a full compressor the pulse travels to the end mirror and back again through the whole system, yielding

$$\Phi(\omega) = \frac{2\omega}{c}\left[\frac{L}{\cos\beta} + r + L(\tan\beta - \tan\beta_0)\sin\alpha\right] - \frac{4\pi L}{G}\tan\beta. \tag{3.71}$$

From the phase, we derive the second- and third-order Taylor coefficients

$$\Phi_2 = -\frac{8\pi^2 cL}{G^2\omega_0^3}\frac{1}{\cos^3\beta_0} = -\frac{\lambda_0}{\pi c^2}\left(\frac{\lambda_0}{G}\right)^2\frac{L}{\cos^3\beta_0}, \tag{3.72}$$

$$\Phi_3 = \frac{3\lambda_0^2}{2\pi^2 c^3}\left(\frac{\lambda_0}{G}\right)^2\frac{L}{\cos^3\beta_0}\left[1 + \frac{\lambda_0}{G}\frac{\sin\beta_0}{\cos^2\beta_0}\right]. \tag{3.73}$$

The characteristic length is the normal distance between the two gratings L. Table 3.1 lists the second- and third-order dispersion coefficients for different center wavelengths and grating constants. The angle of incidence is arbitrarily fixed to 30°. Note that the second- and third-order dispersions of a grating compressor typically have opposite signs.

3.4.1.2 Grating Stretcher

As we have seen in the previous section, a grating compressor consists of only two gratings and has a negative second-order dispersion. Grating stretchers are used to compensate for this negative dispersion or, to put it in the right perspective, to generate a positive dispersion in the first place. Obviously, these devices require more than just two gratings. Let us start with a classic stretcher design, as shown in Figure 3.7. The incident beam is dispersed on the first grating, passes through a series of two identical lenses with focal length f, and is recombined at the second grating. Then it is sent all the way back through the optical system by the end mirror. If all optical elements are separated by one

TABLE 3.1 Second- and Third-Order Dispersion for a Grating
Compressor with Different Gratings

λ_c (nm)	Lines/mm	Φ_2/L (fs^2/cm)	Φ_3/L (fs^3/cm)
400	600	−882	527
400	1200	−3292	2086
400	1800	−7727	4603
800	300	−1754	2091
800	600	−6545	8290
800	1200	−32,723	32,140
1064	300	−3971	6372
1064	600	−15,663	29,000
1064	1200	−153,724	520,604

Note: First column: wavelength; second column: grating constant; third
column: second-order dispersion; and fourth column: third-order dispersion.
The angle of incidence on the first grating in all cases is 30°.

focal length, the net phase modulation is zero (4f setup). This configuration is labeled
"zero dispersion compressor" (Figure 3.7a).

 If both gratings are translated simultaneously toward the lenses by a distance g, the
optical setup shows positive dispersion. If the gratings are moved further away, the dispersion becomes negative. What is commonly labeled a "stretcher" is actually able to act
as a stretcher or a compressor depending on the position of the gratings with respect to
the lenses. Chromatic dispersion due to the two lenses can be avoided if the lenses are
replaced by reflective optics, as shown in Figure 3.7d. The number of elements may be
further reduced if the setup is folded along the line of symmetry. Figure 3.7e displays the
grating stretcher geometry that is used in most standard laser systems. We now calculate
the optical path of each spectral component in order to obtain a quantitative measure of
the phase modulation introduced. With all beams propagating in vacuum, $P(\omega)$, $R(\omega)$,

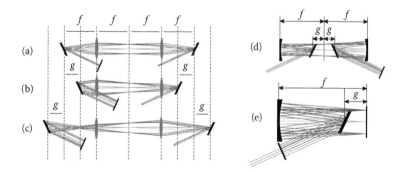

FIGURE 3.7 Classic stretcher design. Depending on the value of g the device has (a) zero, (b)
positive, or (c) negative dispersion. (d) The lenses have been replaced by reflective mirrors in order
to avoid additional material dispersion. (e) Folding the setup once again leads to an even simpler
design with less optical components to adjust.

and, therefore, $\Phi(\omega)$ are readily obtained:

$$\Phi(\omega) = \frac{\omega}{c}\left[8f + 2r - 4g\frac{\cos\beta_0 - \sin(\beta_0 - \beta)\sin\alpha}{\cos\beta}\right] + \frac{8\pi g\cos\beta_0}{G}\tan\beta. \quad (3.74)$$

Before comparing the results to those of the grating compressor, we calculate the expressions for second- and third-order dispersion:

$$\Phi_2 = \frac{\lambda_0}{\pi c^2}\left(\frac{\lambda_0}{G}\right)^2\frac{2g}{\cos^2\beta_0} \quad (3.75)$$

$$\Phi_3 = -\frac{3\lambda_0^2}{2\pi^2 c^3}\left(\frac{\lambda_0}{G}\right)^2\frac{2g}{\cos^2\beta_0}\left[1 + \frac{\lambda_0}{G}\frac{\sin\beta_0}{\cos^2\beta_0}\right]. \quad (3.76)$$

Clearly, the results are identical to Equations 3.72 and 3.73 if we replace $2g\cos\beta_0$ by $(-L)$. Therefore, it is possible to design a stretcher–compressor pair with opposite phase modulation properties by adjusting L, g, and β_c accordingly. This is at the heart of all chirped pulse amplification (CPA) schemes, which stretch the pulse prior to amplification and then compress it again and rely on the fact that stretching and compressing preserves the temporal shape of the input pulse as much as possible.

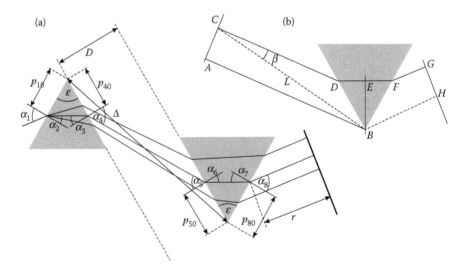

FIGURE 3.8 (a) Geometry of a prism compressor consisting of two prisms and an end mirror. The pulse enters the prism sequence from the left, is reflected at the end mirror, and passes the prism sequence a second time. (b) Beam path from the tip of the first prism (point C) through the second prism for the center frequency (solid curve) and some other frequency (dashed curve).

3.4.1.3 Prism Compressor

The geometry of a typical prism compressor is depicted in Figure 3.8a. Two identical isosceles prisms are arranged opposite to each other with their inner sides parallel. The laser pulse comes from the left, is dispersed by the first prism, and propagates to the second prism, after which all frequencies are parallel again. The end mirror reverses the direction and the pulse travels back through the entire system. The prism geometry, that is the apex angle, and the angle of incidence α_1 are usually, but not necessarily, determined by two constrains. First, in order to minimize reflection losses, the electric field is p-polarized with respect to the prism faces and the angle of incidence is identical to Brewster's angle. Second, the apex angle is such that the center frequency experiences minimum deviation, that is it travels through the prism parallel to its base to avoid any astigmatism to the beam. In order to derive analytic expressions for the second- and the third-order phase, it is instructive to evaluate the situation depicted in Figure 3.8b, which shows the second prism only. We assume that the pulse is refracted at the tip of the first prism (C in Figure 3.8b). Then possible wave fronts at center frequency are $\overline{AC}$ and $\overline{BE}$. From this, we deduce that $\overline{CDE} = \overline{AB}$ and we may apply the formalism derived in the beginning of this chapter with L being the distance between the two prism tips. With the angle of incidence being equal to Brewster's angle and taking into account that the pulse travels twice through the two prisms, one finds

$$\Phi_2 = -\frac{4\lambda_0^3 L}{\pi c^2} \left(d_\lambda n \big|_{\lambda_0} \right)^2 \tag{3.77}$$

$$\Phi_3 = -\frac{6\lambda_0^4 L}{\pi^2 c^3} \left\{ \left(d_\lambda n \big|_{\lambda_0} \right)^2 \left[1 - \lambda_0 \, d_\lambda n \big|_{\lambda_0} \left(\frac{1}{n^3} - n^2 \right) \right] \right.$$
$$\left. + \lambda_0 \left(d_\lambda n \big|_{\lambda_0} \, d_\lambda^2 n \big|_{\lambda_0} \right) \right\}. \tag{3.78}$$

As stated before, these equations do not account for the chromatic dispersion introduced by the two prisms themselves. That is why the second-order dispersion is negative as soon as the distance L is greater than zero, which in practice is not the case. The dashed curve in Figure 3.9 shows Φ_2 as a function of prism separation as calculated from Equation 3.77. Taking into account the material dispersion of the prisms themselves shows that for zero prism separation the second-order phase is positive and the prisms need to have a minimal separation before Φ_2 becomes negative.

3.4.2 Systems Based on Resonances

In optical systems, such as etalons, interferometers, or multilayer systems, the output pulse may be interpreted as a result of interferences between an infinite number of time-delayed replicas. Interferences lead to resonant behavior and it is expected that such systems display a complex modulation behavior, which generally cannot be described by a pure phase modulation. We start with the most general case, namely, a sequence of homogeneous media separated by parallel plane interfaces. From the general case, we derive, as an example, the transfer function of a Fabry–Perot etalon. After this, we

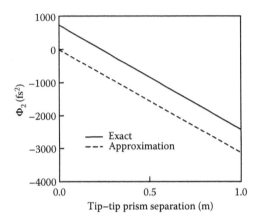

FIGURE 3.9   Second-order phase Φ_2 as a function of prism separation. The dashed curve shows the approximation according to Equation 3.77 and the solid curve the exact values incorporating the chromatic dispersion of the prisms themselves.

introduce the Fabry–Perot interferometer and a close relative, the Gires–Tournois interferometer. Note that some of these devices may have two rather than one output channels, for example, the Fabry–Perot interferometer has a reflected and a transmitted pulse, which are calculated from the incident pulse by applying two different transfer functions.

3.4.2.1 Multilayer System

Multilayer systems are composed of a stack of layers with different relative permittivities and thicknesses and may be considered as a piecewise homogeneous medium. All interfaces are perpendicular to the z-axis and without any loss of generality we assume that the wave vector is within the yz plane. In optics, the angle of incidence is usually the angle between the wave vector and the normal to the interface. If the incidence angle is nonzero, two cases need to be distinguished. The electric field vector is either perpendicular or parallel to the yz plane. While the first case is referred to as s- or TE-polarization, the latter case is called p- or TM-polarization. Any other state of polarization can be projected into the s/p basis system. From electromagnetic field theory, we know that all tangential field components plus the tangential wave vector component are preserved when passing from one medium to the next. With this knowledge, it is possible to find the transfer functions of a multilayer system for the two states of polarization. In the following, we will use the notion $M_{ab}(\omega)$ with $a = $ r,t for reflection or transmission, and, $b = $ s,p for TE(s) or TM(p) polarization. Thus, a multilayer system in general has four transfer functions. A multilayer system can be designed to have almost any desired reflection or transmission as a function of wavelength. Limitations are mostly technological in nature, for example, the maximum number of layers is currently restricted to a few hundred. Often, multilayer systems are designed to be high reflectivity mirrors, beam splitters, or harmonic separators to name but a few. It should be mentioned that an ideal mirror for beam steering must have a vanishingly small phase modulation as any phase distortion has to be avoided. There are, however, cases where mirrors are used to introduce or compensate for a specific phase modulation. Those mirrors are commonly referred to as chirped mirrors.

3.4.2.1.1 TE (s) Polarization

For TE polarization, the only nonzero component of the electric field is $E_x(y, z, \omega)$ and the magnetic field has components parallel to y and z. Boundary conditions require E_x, H_y, and k_y to be continuous. With the Ansatz $E_x(y, z, \omega) = E_x(y, \omega) \exp(-ik_z z)$, we find with the help of Maxwell's equations and the Helmholtz equation

$$\left(\partial_z^2 + k_{jz}^2\right) E_x(z, \omega) = 0 \tag{3.79a}$$

$$H_y(z, \omega) = \frac{i}{\omega \mu_0} \partial_z E_x(z, \omega) \tag{3.79b}$$

$$H_z(z, \omega) = -\frac{k_z}{\omega \mu_0} E_x(z, \omega), \tag{3.79c}$$

with k_{jz} being the z component of the wave vector in layer j. After some algebra, it can be shown that the continuous fields at the exit side of each layer are obtained from the fields at the input side of each layer through

$$\begin{pmatrix} E_x(d_j, \omega) \\ i\omega\mu_0 H_y(d_j, \omega) \end{pmatrix} = m_j^{\text{TE}}(d_j) \begin{pmatrix} E_x(0, \omega) \\ i\omega\mu_0 H_y(0, \omega) \end{pmatrix}, \tag{3.80}$$

with d_j being the thickness of layer j and the matrix $m_j^{\text{TE}}(d_j)$ is found to be

$$m_j^{\text{TE}}(d_j) = \begin{pmatrix} \cos(k_{jz}d_j) & -\sin(k_{jz}d_j)/k_{jz} \\ k_{jz}\sin(k_{jz}d_j) & \cos(k_{jz}d_j) \end{pmatrix}, \tag{3.81}$$

where $k_{jz} = \sqrt{k_j^2 - k_z^2}$. For N subsequent layers, the matrix of the total system is obtained through successive matrix multiplication

$$M^{\text{TE}} = \prod_{j=1}^{N} m_j^{\text{TE}}(d_j). \tag{3.82}$$

From the boundary conditions at the first and the last layer, we obtain the reflection and transmission coefficients of the whole system:

$$M_{rs}(\omega) = \frac{(k_{Iz}M_{22} - k_{Tz}M_{11}) - i(M_{21} + k_{Iz}k_{Tz}M_{12})}{(k_{Iz}M_{22} + k_{Tz}M_{11}) + i(M_{21} - k_{Iz}k_{Tz}M_{12})} \tag{3.83}$$

$$M_{ts}(\omega) = \frac{2k_{Iz}}{(k_{Iz}M_{22} + k_{Tz}M_{11}) + i(M_{21} - k_{Iz}k_{Tz}M_{12})}, \tag{3.84}$$

where the M_{ij} are the four matrix elements of $\mathbf{M}^{\mathrm{TE}}$. The z components of the incident and the transmitted wave vector are derived from the angle of incidence through

$$k_{Iz} = \frac{\omega}{c}\, n_I \cos\theta_I \tag{3.85a}$$

$$k_{Tz} = \frac{\omega}{c}\sqrt{n_T^2 - n_I^2 \sin^2\theta_I}. \tag{3.85b}$$

3.4.2.1.2 TM (p) Polarization

For TM polarization, the magnetic field is parallel to the x-direction and the electric field has components along y and z. Boundary conditions require H_x, E_y, and k_y to be continuous. Similarly to the previous section, we find

$$M_{\mathrm{rp}}(\omega) = \frac{(\epsilon_T k_{Iz} M_{22} - \epsilon_I k_{Tz} M_{11}) - \mathrm{i}(\epsilon_I \epsilon_T M_{21} + k_{Iz} k_{Tz} M_{12})}{(\epsilon_T k_{Iz} M_{22} + \epsilon_I k_{Tz} M_{11}) + \mathrm{i}(\epsilon_I \epsilon_T M_{21} - k_{Iz} k_{Tz} M_{12})} \tag{3.86}$$

$$M_{\mathrm{tp}}(\omega) = \sqrt{\frac{\epsilon_I}{\epsilon_T}}\, \frac{2\epsilon_T k_{Iz}}{(\epsilon_T k_{Iz} M_{22} + \epsilon_I k_{Tz} M_{11}) + \mathrm{i}(\epsilon_I \epsilon_T M_{21} - k_{Iz} k_{Tz} M_{12})}, \tag{3.87}$$

with ϵ_I being the relative permittivity in the first layer and ϵ_T the relative permittivity of the last layer, respectively.

3.4.2.2 Fresnel Equations

First, we apply the formalism to the most simple system, namely, a single interface in which case we have

$$\mathbf{M}^{\mathrm{TE}} = \mathbf{M}^{\mathrm{TM}} = \begin{pmatrix} 1 & 0 \\ 0 & 1 \end{pmatrix}. \tag{3.88}$$

With the help of Equations 3.83 and 3.84, we find for the TE polarized light

$$M_{\mathrm{rs}}(\omega) = \frac{k_{Iz} - k_{Tz}}{k_{Iz} + k_{Tz}} = \frac{n_I \cos\theta_I - n_T \cos\theta_T}{n_I \cos\theta_I + n_T \cos\theta_T} \tag{3.89}$$

$$M_{\mathrm{ts}}(\omega) = \frac{2k_{Iz}}{k_{Iz} + k_{Tz}} = \frac{2n_I \cos\theta_I}{n_I \cos\theta_I + n_T \cos\theta_T} \tag{3.90}$$

and with Equations 3.86 and 3.87

$$M_{\mathrm{rp}}(\omega) = \frac{\epsilon_T k_{Iz} - \epsilon_I k_{Tz}}{\epsilon_T k_{Iz} + \epsilon_I k_{Tz}} = \frac{n_T \cos\theta_I - n_I \cos\theta_T}{n_T \cos\theta_I + n_I \cos\theta_T} \tag{3.91}$$

$$M_{\mathrm{tp}}(\omega) = \sqrt{\frac{\epsilon_I}{\epsilon_T}}\, \frac{2\epsilon_T k_{Iz}}{\epsilon_T k_{Iz} + \epsilon_I k_{Tz}} = \frac{n_I}{n_T}\, \frac{2n_I \cos\theta_I}{n_T \cos\theta_I + n_I \cos\theta_T}. \tag{3.92}$$

Equations 3.89 through 3.92 are known as Fresnel equations for a beam passing from a medium with a refractive index of n_I to a medium with a refractive index of n_T.

3.4.2.3 Fabry–Perot Etalon

The next example is that of a Fabry–Perot etalon. It is composed of a medium (2) with thickness d sandwiched in between air (1), that is, we have two interfaces. Its characteristic matrices for TE(s) and TM(p) polarization are

$$M^{\text{TE}}(d) = \begin{pmatrix} \cos(k_{1z}d) & -\sin(k_{1z}d)/k_{1z} \\ k_{1z}\sin(k_{1z}d) & \cos(k_{1z}d) \end{pmatrix} \quad (3.93)$$

and

$$M^{\text{TM}}(d) = \begin{pmatrix} \cos(k_{1z}d) & -\frac{\epsilon_1}{k_{1z}}\sin(k_{1z}d) \\ \frac{k_{1z}}{\epsilon_1}\sin(k_{1z}d) & \cos(k_{1z}d) \end{pmatrix}. \quad (3.94)$$

The incident wave vector equals the transmitted wave vector, that is $k_{Iz} = k_{Tz}$ and with this and Equations 3.83 and 3.84, we find for the TE polarized light

$$M_{rs}(\omega) = \frac{-i\sin\delta\left(\frac{k_{1z}}{k_{Iz}} - \frac{k_{Iz}}{k_{1z}}\right)}{2\cos\delta + i\sin\delta\left(\frac{k_{1z}}{k_{Iz}} + \frac{k_{Iz}}{k_{1z}}\right)}, \quad (3.95)$$

$$M_{ts}(\omega) = \frac{2}{2\cos\delta + i\sin\delta\left(\frac{k_{1z}}{k_{Iz}} + \frac{k_{Iz}}{k_{1z}}\right)}, \quad (3.96)$$

where we have used $\delta \doteq k_{1z}d$. With Equations 3.86 and 3.87, one obtains

$$M_{rp}(\omega) = \frac{-i\sin\delta\left(\frac{\epsilon_I k_{1z}}{\epsilon_1 k_{Iz}} - \frac{\epsilon_1 k_{Iz}}{\epsilon_I k_{1z}}\right)}{2\cos\delta + i\sin\delta\left(\frac{\epsilon_I k_{1z}}{\epsilon_1 k_{Iz}} + \frac{\epsilon_1 k_{Iz}}{\epsilon_I k_{1z}}\right)}, \quad (3.97)$$

$$M_{tp}(\omega) = \frac{2}{2\cos\delta + i\sin\delta\left(\frac{\epsilon_I k_{1z}}{\epsilon_1 k_{Iz}} + \frac{\epsilon_1 k_{Iz}}{\epsilon_I k_{1z}}\right)}. \quad (3.98)$$

Figure 3.10 shows the two transfer functions for a Fabry–Perot etalon, which is a 10-μm-thick fused silica plate in air, and the angle of incidence is 30°. The etalon introduces a wavelength-dependent loss to the transmitted beam. By rotating the plate, the positions of maximum loss can be fine-tuned to different wavelengths.

3.4.2.4 Fabry–Perot Interferometer

A Fabry–Perot interferometer is almost identical to a Fabry–Perot etalon; however, the reflectivity at the two interfaces is dominated by separate coatings (metallic or dielectric) rather than by Fresnel equations. The medium has a thickness of d and is sandwiched in between two partially reflective mirrors. The two interfaces, 1 and 2, are not necessarily identical and may have different reflectivities. The multilayer approach is somewhat

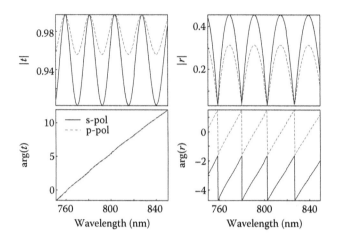

FIGURE 3.10 Amplitude and phase of the transfer functions for the reflected and the transmitted field. The Fabry–Perot etalon is a 10-μm-thick fused silica plate and the angle of incidence is 30°.

impractical to apply here, because of the different types of mirrors that could potentially be used. That is why one usually refers to an alternative method, which is based on accounting for all possible transmission or reflection pathways. The reflected and the transmitted fields are obtained as

$$M_r(\omega) = \frac{r_{12} + (t_{11}t_{12} - r_{11}r_{12})r_{22}\,e^{-2ikL}}{1 - r_{11}r_{22}\,e^{-2ikL}}, \tag{3.99}$$

$$M_t(\omega) = \frac{t_{12}t_{22}\,e^{-ikL}}{1 - r_{11}r_{22}\,e^{-2ikL}}. \tag{3.100}$$

To proceed, we concentrate on a single interface. By reversing the beam path of the reflected and the transmitted beam and sending the complex conjugate of both through the interface in reverse direction, we find the following relation:

$$t_{11}t_{12} - r_{11}r_{12} = \frac{t_{12}}{t_{12}^*} = -\frac{r_{12}}{r_{11}^*}. \tag{3.101}$$

With this relation, we obtain for the reflected part

$$M_r(\omega) = \frac{r_{12} - \frac{r_{12}}{r_{11}^*}r_{22}\,e^{-2ikL}}{1 - r_{11}r_{22}\,e^{-2ikL}}. \tag{3.102}$$

If we assume that the interferometer has dielectric mirrors on both sides, for which

$$\arg(r_{11}) = \pi \quad \arg(r_{21}) = 0$$
$$\arg(r_{12}) = 0 \quad \arg(r_{22}) = \pi$$

$$|r_{11}| = |r_{12}| \qquad\qquad |r_{21}| = |r_{22}|$$
$$\Rightarrow r_{11} = -r_{12} = r_{11}^*$$

is approximately true, and if we further assume that the interferometer is symmetric, that is both mirrors are identical, we may derive

$$M_r(\omega) = \frac{\sqrt{R}\,(1 - e^{-2ikL})}{1 - R\,e^{-2ikL}} \tag{3.103}$$

$$M_t(\omega) = \frac{(1 - R)\,e^{-ikL}}{1 - R\,e^{-2ikL}}, \tag{3.104}$$

with $r_{12} = \sqrt{R}$ and $t_{12} = \sqrt{1 - R}$. Of course R may still be dependent on the polarization if the angle of incidence is nonzero. Generally, such interferometers are operated in two different regimes. First, if the round trip time through the interferometer is larger than the pulse duration, the Fabry–Perot interferometer will produce a train of pulses with decreasing amplitude. Second, if the round trip time is smaller than the pulse duration, the pulse will appear stretched in time and thus the interferometer may be used as a stretcher or a compressor.

3.4.2.5 Gires–Tournois Interferometer

The Gires–Tournois interferometer is a special case of the Fabry–Perot interferometer with one of the two mirrors being 100% reflective, that is $|r_{22}| = 1 \Rightarrow r_{22} = -1$ and $t_{22} = 0$. Thus, there is no transmitted beam and the amplitude of the reflected beam must be equal to one and

$$M_r(\omega) = \frac{\sqrt{R} - e^{-2ikL}}{1 - \sqrt{R}\,e^{-2ikL}}. \tag{3.105}$$

The Gires–Tournois interferometer introduces a pure phase modulation

$$\Phi(\omega) = \arctan\left[\frac{(1 - R)\,\sin(2kL)}{2\sqrt{R} - (1 + R)\,\cos(2kL)}\right]. \tag{3.106}$$

The phase is almost sinusoidal however with a pronounced asymmetry. As for the Fabry–Perot interferometer, we may distinguish between two scenarios: first, the round trip time being larger and, second, being smaller than the pulse duration. While the first case leads to a train of well-separated pulses, the second produces a single-phase-modulated pulse.

3.4.3 Pulse Shaping

All optical systems considered so far impose an amplitude and/or phase modulation on the pulse that is determined by either the dispersive properties of the material or the geometry of the optical system. In contrast, the so-called pulse shaping devices allow to generate almost arbitrary amplitude and/or phase modulations [8]. The most commonly used experimental setup is based on a zero dispersion compressor as shown in

Figure 3.7a. Its 4f arrangement may be thought of as a sequence of two 2f imaging setups and each of those performs a perfect optical Fourier transformation of the spatial wavepacket in the object plane. That is, the angular dispersion from the first grating is mapped to different locations at the so-called Fourier plane of the first lens. The second 2f arrangement annuls the first one and the original pulse is recovered. The actual phase modulator is a one-dimensional device and allows for an amplitude and/or phase modulation of each single-frequency component at the Fourier plane. Through computer programming the spatial light modulator, almost any linear transfer function can be realized.

3.5 Summary

In this chapter, we derived the one-dimensional propagation equation of time domain pulses through a dielectric material starting from Maxwell's equations and a simple classical model describing the dielectric properties of the material. Going to a quantum description of the dielectric material often only requires to replace the classical oscillator strength by a quantum mechanical dipole transition element. The one-dimensional description of pulse propagation is accurate as long as the transverse beam profile is more or less constant from the point where the pulse enters the material to the point where it leaves the material. That is to say, either the Rayleigh length of the beam is larger than the thickness of the material or the transverse profile remains constant because the pulse is confined in a waveguide. We then discussed different aspects of the propagation equation, such as self-phase modulation, and so on by making educated simplifications. For very long propagation distances, for example, in optical fibers, we expanded the propagation equation by including the Raman process again through a simple classical model. The chapter was concluded with more complex optical systems, which are all designed to modify the temporal or spectral properties of a time domain pulse while leaving the transverse spatial beam profile unaffected.

References

1. Schmidt, B., Hacker, M., Stobrawa, and G., Feurer, T. LAB2—A virtual femtosecond laser lab, http://www.lab2.de.
2. Boyd, R.W. 1992. *Nonlinear Optics.* San Diego: Academic Press.
3. Huard, S. 1996. *Polarization of Light.* Chichester: Wiley.
4. Dmitriev, V.G., Gurzadyan, G.G., and Nikogosyan, D.N. 1997. *Handbook of Nonlinear Crystals.* Berlin: Springer.
5. Remoissenet, M. 1996. *Waves Called Solitons.* Berlin: Springer.
6. Agrawal G.P. 2001. *Nonlinear Fiber Optics.* San Diego: Academic Press.
7. Akhmanov, S.A., Vysloukh, V.A., and Chirkin, A.S. 1992. *Optics of Femtosecond Laser Pulses*, New York: AIP; Diels, J.-C. and Rudolph, W. 1996. *Ultrashort Laser Pulse Phenomena.* San Diego: Academic Press; Weiner, A.M. 2009. *Ultrafast Optics.* Hoboken: Wiley.
8. Weiner, A.M. 2000. Femtosecond pulse shaping using spatial light modulators, *Rev.Sci. Instrum.* 71: 1929.

Generation
and
Characterization
of Laser Beams

4

Spatial Laser Beam Characterization

Christian Schulze
Friedrich-Schiller University Jena

Daniel Flamm
Friedrich-Schiller University Jena

4.1 Introduction

There are several approaches to the spatial characterization of laser beams. One of the simplest characterization techniques is probably the measurement of beam power, which is often recorded over time to reveal potential instabilities. However, spatial resolution is lost here and is only achieved using a camera to record the power density in a transverse plane instead of integration. Although this approach gives a more detailed picture, there is still a crucial beam characteristic missing: its propagation. Two beams can exhibit exactly the same power density distribution but completely differ in propagation. A simple and commonly used measure to evaluate laser beam propagation is the beam propagation ratio M^2, which compares the propagation properties of the beam to those of a perfect diffraction-limited Gaussian beam. Moreover, laser light has a certain phase and wave front distribution, as well as a certain polarization. While the former is commonly measured with interferometers or Shack–Hartmann sensors (SHS) the latter can be obtained from recording the Stokes parameters of the beam. Amplitude, phase, and polarization completely describe a laser beam and are traditionally measured by various experiments. A modern and comprehensive approach that enables to gather the complete information on an optical field at once is given by modal decomposition using an inner-product measurement. Since intensity, phase, wave front, polarization distribution, and even orbital angular momentum density can be inferred from that approach, we will focus on this technique in the next sections and show how it can be applied to measure the above-mentioned fundamental beam quantities.

4.2 Concept

The description of optical fields in terms of eigenmodes is an intuitive approach for beam characterization since useful insights into the physics of the laser light during the generation and propagation can be extracted from the actual mode composition. In the following sections, the basic principles are illustrated to perform a complete modal decomposition of laser beams using optical correlation filters. Such filters are based on knowledge about the spatial field properties of modes in the respective optical system and can be realized as digital holograms implemented into a simple Fourier setup. The basic modal properties in different optical systems (resonators as well as optical fibers) are outlined in Section 4.2.1 and are used in Section 4.2.2 to explain the working principle of a correlation filter. Here, particular attention is paid to the performance of calculating an inner product optically. Finally, the realization of such filters as digital holograms, including the discussion on different coding techniques and the required measurement process, is discussed in Section 4.2.3.

4.2.1 Fundamental Properties of Modes

As has been discussed in Chapter 1, a stationary, monochromatic, and linearly polarized (scalar) optical field U that is propagating in an isotropic and linear medium has to be a solution of the Helmholtz equation

$$\left[\Delta + k_0^2 n^2 \left(\mathbf{r}\right)\right] U \left(\mathbf{r}\right) = 0, \tag{4.1}$$

with the Laplace operator Δ, the vacuum wave number k_0, and the refractive index distribution $n\left(\mathbf{r}\right)$. Searching for beam-like (paraxial) and self-reproducible fundamental field solutions [1] from which every possible field distribution can be constructed, Equation 4.1 is conveniently transformed into an eigenvalue problem

$$\mathcal{P}\psi_n \left(\mathbf{r}\right) = \gamma_n \psi_n \left(\mathbf{r}\right). \tag{4.2}$$

Here, the linear propagation operator $\mathcal{P}$ contains geometric information about the respective optical system (e.g., resonator geometry, fiber refractive index distribution) and determines the propagation properties and the transverse field distribution of the eigenfunctions $\psi_n \left(\mathbf{r}\right)$ with corresponding eigenvalues γ_n. Since $\mathcal{P}$ is a Hermitian operator, the eigensolutions of Equation 4.2 $\{\psi_n\}$ build an orthonormal set

$$\langle \psi_m | \psi_n \rangle = \iint_{\mathbb{R}^2} d^2 r \, \psi_m^*(\mathbf{r}) \psi_n(\mathbf{r}) = \delta_{mn} \tag{4.3}$$

and enable one to express an arbitrary propagating field as a modal composition

$$U(\mathbf{r}) = \sum_{n=1}^{n_{max}} c_n \psi_n(\mathbf{r}). \tag{4.4}$$

The benefit of this basis expansion of the field is that the required information to completely describe the optical field is reduced to a set of complex expansion coefficients $\{c_n\}$: this is sufficient to characterize every possible field in amplitude and phase propagating in the optical system. The mode coefficients are uniquely determined by the inner product, or overlap integral, of the optical field U and the corresponding mode field distributions

$$c_n = \langle \psi_n | U \rangle = \iint_{\mathbb{R}^2} d^2 r \, \psi_n^*(\mathbf{r}) U(\mathbf{r}). \tag{4.5}$$

It is the basic idea of the presented technique to optically perform this inner product and to directly measure the amplitude and phase of the complex mode coefficients $c_n = \rho_n \exp(i\Delta\phi_n)$ (cf. Section 4.2.2). For the squared absolute values of c_n, the following relation holds:

$$\sum_{n=1}^{n_{max}} |c_n|^2 = \sum_{n=1}^{n_{max}} \rho_n^2 = 1, \tag{4.6}$$

which is due to the completeness of $\{\psi_n\}$ and a consequence of energy conservation. The influence of modal amplitudes ρ_n and intermodal phase differences $\Delta\phi_n$ on the resulting beam U, especially on its intensity distribution, is discussed in detail in Section 4.3.1.

In what follows, we introduce the selected field distributions of four mode sets and start with the well-known eigenfunctions of laser resonators with simple geometries. These wave fields are solutions of Equation 4.1 in the homogeneous media of the free space [$n(\mathbf{r}) = \text{const} \approx 1$]. Depending on the choice of the coordinate system, one obtains Hermite–Gaussian (HG) modes (in Cartesian coordinates) or Laguerre–Gaussian (LG) modes (in cylindrical coordinates), as discussed in Chapter 1. The solutions in other geometries are covered in Chapter 8 and will not be discussed here. The number of modes being actually supported strongly depends on the geometry and quality of the resonator. However, in general, as eigenfunctions of free space, an infinite number of modes belongs to the HG or LG mode sets and an infinite number of mode coefficients is required to completely express arbitrary free space fields. Hence, for these cases, $n_{max} \to \infty$ (Equation 4.4). Figure 4.1 exemplarily depicts the intensity and phase distribution of selected HG modes (a) and LG modes (b).

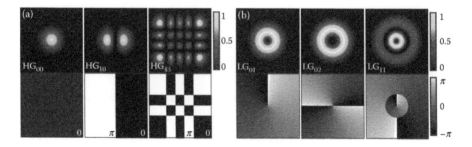

FIGURE 4.1 Examples of intensity and phase distribution of HG modes (a) and LG modes (b).

Scalar modes guided in optical fibers have to fulfill the following eigenvalue problem (cf. Equation 4.1) [2]:

$$\left[\Delta_\perp + k_0^2 n^2\left(\mathbf{r}\right)\right]\psi_n\left(\mathbf{r}\right) = \beta_n^2\psi_n\left(\mathbf{r}\right),\tag{4.7}$$

with the transverse Laplace operator $\Delta_\perp$ and the squared propagation constants β_n^2 as eigenvalues. The fiber's refractive index distribution $n\left(\mathbf{r}\right)$ in Equation 4.7 acts similar to the potential well of the stationary Schrödinger equation. Hence, the set of eigensolutions contains a *finite* number of modes $n_{\max}$ and every beam exiting the fiber is completely described by a countable set of $n_{\max}$ complex numbers.

For the cases of highly symmetric refractive index distributions such as, for example, step-index fibers, guided modes can be analytically calculated that result in the well-known set of LP modes [2]. Figure 4.2a shows intensity and phase distributions of selected LP modes. For more sophisticated waveguiding structures such as, for example, photonic crystal fibers or multicore fibers, numerical mode solver schemes, such as finite differences, and finite element methods, exist yielding approximately linearly polarized (scalar) mode sets. As an example for these types of microstructured optical fibers, selected modes are depicted in Figure 4.2b guided in a multicore fiber consisting of 19 hexagonally arranged cores.

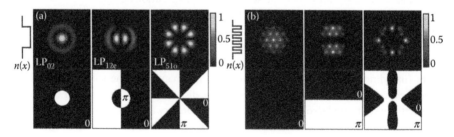

FIGURE 4.2 Scheme of the refractive index distribution as well as examples of intensity and phase distribution of three step-index fiber modes (a) and three modes guided in a multicore fiber (b).

4.2.2 Optical Inner Product Measurements

To illustrate the simplicity of the approach to "calculate" all inner products (Equation 4.5) optically, consider the scenario depicted in Figure 4.3, where a single mode (a), (b), and a multimode beam (c) is to be analyzed, respectively; in our example, the beam U comprises some unknown mode compositions (Equation 4.4) from the set LP_{01}, LP_{02}, and LP_{11e}. Now, it is well-known that if a match filter [3] or a correlation filter is set in the front focal plane of a lens, then in the far field (back focal plane), the signal on the optical axis (at the origin of the detector plane) is proportional to the power guided by the respective mode. To be specific, if the transmission function of the match filter was set to be

$$T(\mathbf{r}) = \psi_n^*(\mathbf{r}),\qquad(4.8)$$

then the signal returned would be proportional to $|\langle\psi_n|U\rangle|^2$, thus, proportional to ρ_n^2 (Equation 4.5). To return *all* the modal weightings simultaneously, the linearity property of optics can be exploited: simply multiplex each required match filter (one for each mode to be detected) with a spatial carrier frequency (grating) to spatially separate the signals at the Fourier plane [4,5].

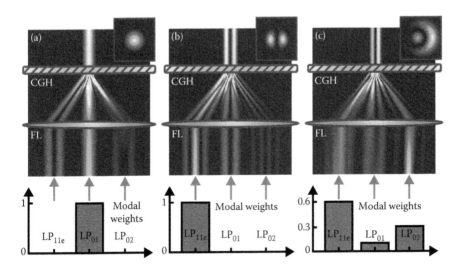

FIGURE 4.3 Simulation of the working principle of an optical inner product for detecting the modal weights of modes LP_{11e}, LP_{01}, and LP_{02} in the far-field diffraction pattern (from left to right). CGH, computer generated hologram; FL, Fourier lens with focal length f. (a) Pure fundamental mode illumination. The intensity on the optical axes of the diffracted far-field signals (correlation signals) denoted by arrows results in the stated modal power spectrum. (b) Pure LP_{11e} mode illumination. (c) The illuminating beam is a mixture of three modes. According to the beam's composition, different intensities are detected on the corresponding correlation answers that result in the plotted modal power spectrum.

Returning to the figure, we conceptually implement the match filter with a digital holo-gram or computer-generated hologram (CGH) and monitor the on-axis signal (pointed out by the arrows) in the Fourier plane of a lens (FL). We illustrate that in the case of illuminating the filter with the pure fundamental mode Figure 4.3a, only the on-axis intensity for the LP_{01} mode is nonzero, whereas the other two have no signals due to the zero overlap with the incoming mode and the respective match filters. In Figure 4.3b, another pure mode illumination is simulated, now with mode LP_{11e}. An on-axis inten-sity is measurable only on the optical axis detecting the corresponding modal weight. In the third case, Figure 4.3c, the converse is shown, where all the modes have a non-zero weighting, and thus, all the match filters have a nonzero overlap with the incoming beam. These intensity measurements return the desired coefficient, ρ_n^2, for each mode.

For the measurement of the intermodal phase difference $\Delta\phi_n$ of mode ψ_n to a certain reference mode ψ_0, two further transmission functions have to be implemented into the CGH representing an interferometric superposition [4,6]

$$T_n^{\cos}(\mathbf{r}) = \left[\psi_0^*(\mathbf{r}) + \psi_n^*(\mathbf{r})\right]/\sqrt{2}, \tag{4.9}$$

$$T_n^{\sin}(\mathbf{r}) = \left[\psi_0^*(\mathbf{r}) + i\psi_n^*(\mathbf{r})\right]/\sqrt{2}. \tag{4.10}$$

From the resulting intensity signals $I_n^{\cos}$, $I_n^{\sin}$, and the corresponding modal amplitude signals $I_0\rho_0^2$, $I_n\rho_n^2$, the intermodal phase difference can be determined unambiguously

$$\Delta\phi_n = -\tan^{-1}\left[\frac{2I_n^{\sin} - I_0\left(\rho_n^2 - \rho_0^2\right)}{2I_n^{\cos} - I_0\left(\rho_n^2 - \rho_0^2\right)}\right] \in [-\pi, \pi]. \tag{4.11}$$

Using the described holographic measurement technique, direct experimental access is provided to modal amplitudes $|c_n|$ and to intermodal phase states $\arg(c_n)$. Hence, the composition of Equation 4.4 can be executed and the complete field information is at hand [4].

4.2.3 Realization of the Correlation Filter

Today, optical correlation filters have been successfully realized as phase-only or amplitude-only digital holograms [4,7]. Pure amplitude or phase filters fabricated via laser lithography can be designed to analyze a large number of mode coefficients simul-taneously due to the use of high spatial carrier frequencies [4,6]. However, such stationary elements are specifically designed to decompose beams into a fixed mode set. To overcome this limitation, complex amplitude-modulation-encoded digital holograms on a liquid crystal on silicon-based phase-only spatial light modulator (SLM) can be employed. The advantages of replacing the stationary holograms by flexible SLMs can be found in the rapid change of the transmission function and real-time switching of the digital holograms [8–10]. Hence, with a single SLM, an arbitrary and unknown laser resonator or fiber can be investigated since the user is able to iteratively adapt the mode set for the decomposition; if, for example, the exact fiber properties and thus the spatial

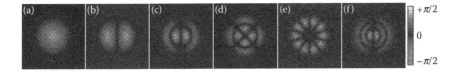

FIGURE 4.4 **(See color insert.)** (a)–(f) Phase modulations $\Psi(\mathbf{r})$ for measuring the mode powers of LP_{01}, LP_{11e}, LP_{12e}, LP_{22e}, LP_{61e}, and LP_{13e} with the SLM in the first order of diffraction. (Adapted from Flamm, D. et al. 2012. *Opt. Lett.* 37: 2478–2480; Flamm, D. et al. 2013. *J. Lightwave Technol.* 3: 387–389.)

information about the modes are unknown, or if the decomposition is intended to be done in distorted mode fields as in the case of fiber bending [11]. Finally, the ubiquitous nature of SLMs nowadays in combination with a simple experimental setup makes the presented procedure outstandingly suitable for analyzing the modes of arbitrary laser beams. By way of example a phase-coding technique is explained in the following, allowing a simple and direct calculation of the required transmission functions.

To encode the transmission function $T(\mathbf{r}) = \psi_l^*(\mathbf{r})$ (Equation 4.8) onto the SLM, the technique proposed by Arrizón et al. can be used [7]. Here, the complex-valued function $T(\mathbf{r}) = A(\mathbf{r})\exp[i\Phi(\mathbf{r})]$, with normalized amplitude $A \in [0,1]$ and phase $\Phi \in [-\pi,\pi]$ is encoded into a phase hologram $H(\mathbf{r}) = \exp[i\Psi(\mathbf{r})]$ with a given unit amplitude transmittance and a certain phase modulation $\Psi(A,\Phi)$. In the literature, different phase modulations $\Psi(\mathbf{r})$ are discussed [7], providing the same information as the original transmission function $T(\mathbf{r})$ in a certain diffraction order. For measurements recorded in the first order of diffraction, one can use $\Psi(A,\Phi) = f(A)\sin(\Phi)$, where $f(A)$ results from $J_1[f(A)] \cong 0.58A$, with the first-order Bessel function $J_1(x)$ [7]. To achieve the spatial separation of the diffracted signal, a sinusoidal grating as a phase carrier is required. The resulting phase modulations for measuring the modal powers of LP modes of low order are exemplarily depicted in Figure 4.4a–f.

4.3 Measurement of Laser Beam Properties

The previous section outlined the basics of the correlation filter technique to perform a modal decomposition of a laser beam. The following section illustrates how to apply this technique to measure the fundamental beam properties, such as mode content, intensity, beam propagation ratio, wave front, polarization, and orbital angular momentum (OAM) density.

4.3.1 Modal Content and Intensity

The field of a normalized, scalar, and transverse mode of order n can be written as

$$\psi_n(\mathbf{r}) = \sqrt{I_n(\mathbf{r})}\exp[i\chi_n(\mathbf{r})], \tag{4.12}$$

with the mode's normalized intensity $I_n(\mathbf{r})$ and phase distribution $\chi_n(\mathbf{r})$. As stated by Equation 4.4, the beam is coherently composed of the corresponding set of modes $\{\psi_n(\mathbf{r})\}$. Using the definitions from Section 4.2.1, the resulting beam intensity $I(\mathbf{r})$ is calculated from

$$I(\mathbf{r}) = U^*(\mathbf{r}) U(\mathbf{r}) = \sum_{n,m}^{n_{max}} \rho_n \rho_m \sqrt{I_n(\mathbf{r}) I_m(\mathbf{r})} \exp\{i[\Delta\phi_{mn} + \Delta\chi_{mn}(\mathbf{r})]\}, \qquad (4.13)$$

where $\Delta\phi_{mn} = \phi_m - \phi_n$ is the intermodal phase difference and $\Delta\chi_{mn}(\mathbf{r}) = \chi_m(\mathbf{r}) - \chi_n(\mathbf{r})$ results from the differences of the modal phase distributions. As known from statistical optics, we can take advantage of splitting the sum into an incoherent $I_{incoh}(\mathbf{r})$ and an interference term $I_{inter}(\mathbf{r})$

$$I(\mathbf{r}) = I_{incoh}(\mathbf{r}) + \gamma I_{inter}(\mathbf{r}), \qquad (4.14)$$

with γ as the absolute value of the complex degree of temporal coherence [1], see Chapter 10 for more details on coherence. Both intensity distributions read as follows:

$$I_{incoh}(\mathbf{r}) = \sum_{n}^{n_{max}} \rho_n^2 I_n(\mathbf{r}), \qquad (4.15)$$

$$I_{inter}(\mathbf{r}) = \sum_{\substack{n,m \\ n \neq m}}^{n_{max}} \rho_n \rho_m \sqrt{I_n(\mathbf{r}) I_m(\mathbf{r})} \cos[\Delta\phi_{nm} + \Delta\chi_{nm}(\mathbf{r})]. \qquad (4.16)$$

The interference term $I_{inter}(\mathbf{r})$ is contributing to the entire intensity by means of the phase-dependent cosine modulation that is weighted by γ. It is known from statistical optics that γ is proportional to the Fourier transform of light's spectral density [12] and can take the following values:

$$\begin{array}{ll} \gamma = 1 & \text{Complete coherence,} \\ 0 < \gamma < 1 & \text{Partial coherence,} \\ \gamma = 0 & \text{Complete incoherence.} \end{array} \qquad (4.17)$$

For the case of a complete incoherent mode composition as it is, for example, the case for classical resonator modes, the interference term vanishes and the beam equals the weighted sum of the mode's single intensities (Equation 4.15) and is independent from modal phase states. A fundamental property of such compositions is that the beam's barycenter or first-order moment [13] equals its optical axis. The resulting intensity is distributed symmetrically around this axis, see Figure 4.5a. The inversion is the case for a coherent mode composition. Depending on the spatial distribution of the phase differences $\Delta\chi_{mn}(\mathbf{r})$, the cosine modulation (Equation 4.16) shifts the first-order moment away from the optical axis.

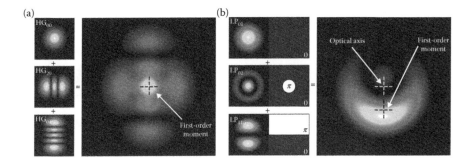

FIGURE 4.5 Mode superposition and its consequence on the beam's first-order moments (beam center) for (a) an incoherent superposition of Hermite-Gaussian modes, and (b) a coherent superposition of LP fiber modes.

4.3.2 Beam Propagation Ratio M^2

Among the various methods for characterizing the "quality" of a laser beam, the beam propagation ratio or M^2 parameter as introduced by Siegman [14] has become the most common and accepted measurement standard. Using this single parameter, almost all laser systems may be specified and categorized. The standardized M^2 measurement procedure is based on the determination of the ten second-order moments of the Wigner distribution function described in the ISO-standard 11146-2 [13].

Fortunately, owing to their high symmetry, most laser beams of practical interest require less than these 10 independent parameters to be measured for a complete characterization: most lasers emit beams that are simply astigmatic or are even stigmatic because of their resonator design. In this case, the M^2 determination is based on a caustic measurement that includes the determination of the beam width as a function of the propagation distance. To do this, the beam width has to be determined at least in 10 positions along the propagation axis, completed by a hyperbolic fit. According to ISO 11146-1 [27], the beam width determination has to be carried out using the three spatial second-order moments of the intensity distribution. In case of generally astigmatic beams, an additional cylindrical lens must be used to infer the remaining beam moments [13].

Despite its experimental simplicity, the caustic measurement is still quite time consuming and requires a careful treatment of background intensity, measuring area, and noise, which makes high demands on the temporal stability of a continuous wave laser or the pulse-to-pulse stability of a pulsed source. Therefore, caustic measurements are unsuitable to characterize fast dynamics of a laser system.

To overcome these drawbacks, many alternative methods have been suggested that try to avoid the traditional scan in the propagation direction [15–17]. Furthermore, the monitoring of the beam quality based on a decomposition of the laser beam into its constituent eigenmodes has proved to be an effective technique [18]. Many laser resonators

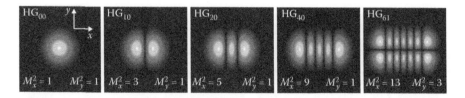

FIGURE 4.6 Relation of selected pure HG modes and M^2 parameters in the stated coordinate system.

possess rectangular or circular symmetry and can be described completely using a super-position of HG or LG modes. Regarding the beam quality, they represent a natural choice of description since the fundamental Gaussian mode has the lowest possible value of $M^2 = 1$. Any deviation from the ideal diffraction-limited Gaussian beam profile can be attributed to the contribution of higher-order modes, leading to $M^2 > 1$. Here, any excited higher-order mode contributes with a certain value to the beam propagation parameter. This value can be readily calculated for every mode in a way that is conformal to the International Standard Organization (ISO) standard [13]. Examples of HG modes with assigned M^2 parameters are depicted in Figure 4.6. In general, different modes of a resonator possess slightly dissonant frequencies so that the interference term (Equation 4.16) vanishes and an HG beam composition according to Equation 4.15 results in a known M^2 parameter due to the following simple relations:

$$M_x^2 = \sum_{mn} \rho_{mn}^2 (2m + 1), \quad M_y^2 = \sum_{mn} \rho_{mn}^2 (2n + 1) \qquad (4.18)$$

in the coordinate system as defined in Figure 4.6.

To illustrate the simplicity of this approach, consider the beam shown in Figure 4.7. This multimode beam was analyzed using the correlation filter method and decomposed into a set of HG modes. The measured modal power spectrum is depicted in Figure 4.7 and confirms the dominating higher-order mode content for HG_{20} and HG_{30}. Applying Equation 4.18 to this mode content results in $M_x^2 = 6.3$, $M_y^2 = 1.05$. This procedure yields beam propagation ratios that conform to the ISO standard, but are monitored in real-time, without performing time-consuming and error-prone caustic measurements.

As stated by Equations 4.14 through 4.17, for the general description of a multimode beam composition (the coherent case), the intermodal phase differences $\Delta\phi_{mn}$ play a crucial role. An example for a coherent mode superposition is a beam from a multimode optical fiber, seeded with a narrow line width laser. In such a case, it is not sufficient to only know how strongly a certain mode is excited in the fiber. The resulting properties of the beam will strongly depend on the relative phase differences between all excited modes. This is illustrated in Figure 4.8. Here, the excitation of a higher-order mode (10% of total power) with varying phase delays to the fundamental mode leads to a variation of M^2 of more than 30% [5].

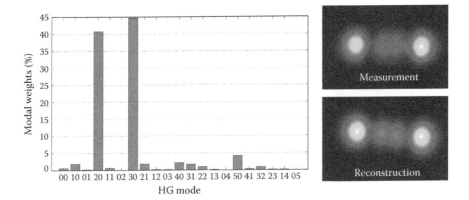

FIGURE 4.7 Measured mode spectrum of a higher-order mode beam (left) with the corresponding measured and reconstructed near-field intensity distributions (right). The resulting beam quality was determined to be $M_x^2 = 6.3$, $M_y^2 = 1.05$.

FIGURE 4.8 Numerical M^2 determination of a beam superposition of step-index fiber modes LP_{01} and LP_{02}: intensity and phase distributions with stated weightings and different intermodal phase differences $\Delta\phi$.

The procedure of determining the beam quality of such beams is based on the holographic procedure described in Section 4.2.2, where the mode coefficients can be directly measured in amplitude and phase. From this knowledge, the optical field under investigation is reconstructed and its free space propagation is simulated using the angular spectrum method [3]. Subsequently, the ten second-order moments of the Wigner distribution function are determined as being necessary for an ISO conformal M^2 characterization in the case of general astigmatic beams [13]. According to Ref. [5], this seminumerical procedure is known as *virtual* caustic measurement (VCM). Similar to the incoherent case (Equation 4.18), the outstanding advantage of a VCM can be found in the enormous reduction of temporal and experimental effort. For this reason, the virtual M^2 characterization technique is of particular interest in fiber optics where coherent mode superpositions are at hand and where fast-fluctuating modal processes need to be investigated. This includes simple fiber-coupling processes that can be optimized by means of the beam propagation ratio M^2 using VCMs [5]. Figure 4.9 exemplarily depicts the real-time M_{eff}^2 determination during the alignment of a single mode fiber with respect to

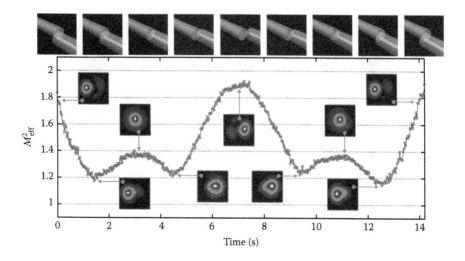

FIGURE 4.9   Real-time M_{eff}^2 determination during a fiber-to-fiber coupling process. The fiber movement is schematically shown on top of the graph. After the horizontal coupling process was completed at ≈ 7 s, the scanning movement is reversed and returned to the starting position. Insets in the graph depict reconstructed beam intensities at the corresponding positions.

a multimode fiber. The current mode content and phase differences strongly depend on the coupling position and affect the beam quality.

4.3.3 Wave Front

Wave front reconstruction of optical fields has become an important task in many domains of optics: astronomy, for example, is nowadays inconceivable without wave front measurements. In combination with deformable mirrors for correction of distorted wave fronts, wave front reconstruction has become a fundamental part of adaptive optics systems that enable high-quality terrestrial observations. The correction and control of wave fronts is not less essential in the fields of microscopy, such as 2-photon microscopy and confocal microscopy, and ophthalmology, such as optical coherence tomography and scanning laser ophthalmoscopy. Another field of application for wave front control is laser material processing, in particular, processes that require a high beam quality, such as laser cutting and drilling, often performed by fiber lasers. Concerning the wave front measurement itself, there exists a variety of different types of sensors, including laser ray tracing [19], pyramid sensors [20], interferometric approaches [21], and the widely used Shack-Hartmann sensor (SHS) [22]. Recent advances have seen the use of CGHs to encode certain aberrations to determine the Zernike coefficients [23], use of ring-shaped phase masks for decomposition in azimuthal modes [24], nonlinear approaches for high-intensity light pulses [25], and the application of correlation filters (CGH) to reconstruct phase and wave front by modal decomposition [26]. In this chapter, we will focus on last-mentioned approach.

Any optical field can be described as a superposition of orthonormal modes (Section 4.2). Modal decomposition using correlation filters is an elegant approach to directly measure the power and phase spectrum of the modes and hence, to reconstruct the optical field (Equation 4.4). The Poynting vector $\mathbf{P}$ is computable from the knowledge of the optical field by

$$\mathbf{P}(\mathbf{r}) = \frac{1}{2}\Re\left[\frac{i}{\omega\epsilon_0}\epsilon^{-1}(\mathbf{r})[\nabla \times \mathbf{U}(\mathbf{r})] \times \mathbf{U}^*(\mathbf{r})\right], \tag{4.19}$$

for vector fields $\mathbf{U}$, and by

$$\mathbf{P}(\mathbf{r}) = \frac{\epsilon_0\omega}{4}\left[i(U\nabla U^* - U^*\nabla U) + 2k|U|^2\mathbf{e}_z\right], \tag{4.20}$$

for scalar fields U, where ω is the angular frequency, ϵ_0 is the vacuum permittivity, and ϵ is the permittivity distribution. According to the ISO-standard 15367-1, the wave front is defined as the continuous surface that is normal to the time average direction of energy propagation, that is, normal to the time average Poynting vector $\mathbf{P}$ [27]:

$$w(\mathbf{r}, z) \perp \mathbf{P}(\mathbf{r}, z), \tag{4.21}$$

where z denotes the position of the measurement plane. Since there might be no continuous surface that fulfills this condition, the wave front can be more generally defined as the continuous surface that minimizes the power density-weighted deviations of the direction of its normal vectors to the direction of energy flow in the measurement plane [27]:

$$\iint |\mathbf{P}|\left|\frac{\mathbf{P}_t}{|\mathbf{P}|} - \nabla_t w\right|^2 dxdy \to \min, \tag{4.22}$$

where $\mathbf{P}_t = [P_x, P_y, 0]'$. In the simple case of scalar, that is, linearly polarized beams, the wave front is equal to the phase distribution $\Phi(\mathbf{r})$ of the beam, except for a proportionality factor [27]:

$$w(\mathbf{r}) = \frac{\lambda}{2\pi}\Phi(\mathbf{r}). \tag{4.23}$$

It is important to note that this expression is only valid as long as there are no phase jumps or phase singularities because the wave front is always considered to be a continuous surface.

To measure the phase and wave front of a beam, a generalized setup as in Figure 4.10 can be used. The aberrated wave front is relay imaged (4f imaged) to the correlation filter, which performs the modal decomposition on the beam. The diffraction pattern of the first order is observed with a changed coupled device (CCD) camera (CCD$_2$) in the far field of the correlation filter using a single lens in a 2f configuration. In the case of vector fields, a set of a quarter-wave plate and a polarizer is used to determine the polarization state in addition to the modal power spectrum [28]. Using a beam splitter (nonpolarizing), the aberrated wave front is simultaneously imaged onto the correlation filter and a CCD camera to capture the beam intensity directly (CCD$_1$, cf. Figure 4.10). As stated in Section 4.2, the hologram can be of a different type, for example, a phase-only SLM as used by Flamm et al. [8] or a solid amplitude-only filter fabricated via laser lithography as used by Kaiser et al. [4].

To ensure that the measured wave front is not influenced by the optical setup (lenses, beam splitters, etc.), some kind of calibration is necessary. To avoid the impact of especially defocusing and to ensure proper imaging of the plane of the wave front to the CGH, the focal length of each used lens can be measured with an SHS by entering the lenses with

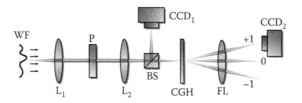

FIGURE 4.10 Schematic measurement setup: WF, aberrated wave front to be relay imaged onto the correlation filter (CGH); P, quarter-wave plate and polarizer; L$_{1,2}$, lenses; FL, Fourier lens; BS, beam splitter; CCD$_{1,2}$, CCD cameras.

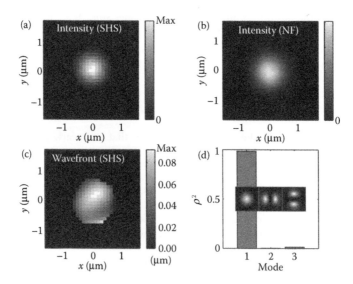

FIGURE 4.11 Fundamental mode illumination of CGH and SHS for calibration. (a) Intensity measured with the SHS. (b) Intensity measured with the CCD camera. (c) Wave front measured with the SHS (scale in μm). (d) Modal power spectrum (the insets depict the respective mode intensities).

a collimating beam and using the Zernike coefficient of defocus to determine the focal length and principal plane. Placing the lenses at the proper distances, the experimental setup as a whole is calibrated using a reference beam with a flat wave front. Accordingly, the setup is adjusted to yield a wave front as flat as possible in the detection plane. The outcome of this procedure is shown in Figure 4.11. Here, an optical fiber in single-mode operation is used to produce a Gaussian beam with a flat wave front [26], which was proved using an SHS, as seen in Figure 4.11a and c. Obviously, the recorded wave front is very flat with a maximum deviation from the plane below 0.09 μm, which is less than $\lambda/10$ in this experiment (used wavelength $\lambda = 1064$ nm). Additionally, the intensities measured with the SHS (Figure 4.11a) and with the CCD camera (Figure 4.11b) are in good agreement, which is, alongside with the flat wave front, another proof that the optical setup itself adds no severe aberrations. The modal spectrum shows (Figure 4.11d) a pureness of the fundamental mode of 99% of total power.

As a first example, consider the beam emanating from an optical step-index fiber (core diameter 7.7 μm, numerical aperture NA 0.12), which is seeded by an Nd:YAG laser ($\lambda =1064$ nm) yielding a three-mode system. Hence, the aberrated wave front is generated at the fiber end face by multimode interference. Higher-order modes can be excited by transverse misalignment or by using a phase mask at the fiber front [29].

Figure 4.12 depicts the results for an LP beam composed of three fiber modes. According to the measured modal power spectrum (Figure 4.12c) the beam consists of 44%

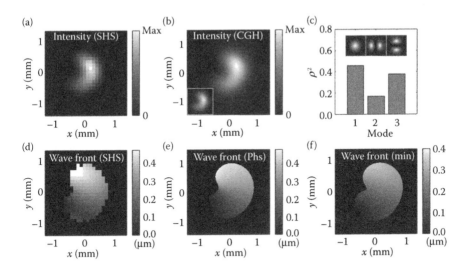

FIGURE 4.12 Wave front reconstruction for a higher-order-mode scalar beam. (a) Intensity measured with the wave front sensor (SHS). (b) Reconstructed intensity (the inset depicts the directly measured intensity with CCD₁). (c) Modal power spectrum (the insets depict mode intensities). (d) Wave front measured with the SHS (scale in μm). (e) Wave front determined from the phase reconstruction according to Equation 4.23 (scale in μm). (f) Wave front from the minimization according to Equation 4.22 (scale in μm).

fundamental mode, 18% mode 2, and 38% mode 3 (modal intensities are depicted in insets of Figure 4.12). Figure 4.12a and b shows the intensities measured with the SHS, reconstructed from the modal decomposition, and directly measured with a CCD camera.

The wave front is inferred from the reconstructed optical field, first, by deduction from phase (Equation 4.23), and second, by employing the minimization (Equation 4.22). The corresponding results are shown in Figure 4.12e and Figure 4.12f. For reference purposes the wave front is additionally measured with an SHS (Figure 4.12d). Note that the wave front reconstructed with the CGH (Figure 4.12e and f) is artificially truncated to values where the corresponding intensity exceeds 5% of its maximum value, to enable better comparison with the SHS measurement.

All wave front results are in good agreement concerning the shape and scale. Obviously, in this case where there are no phase jumps or singularities, Equation 4.23 is valid for wave front reconstruction. Since aberrations added by the optical setup would have given rise to deviations in the wave fronts, a significant influence can be excluded. This is because a wave front reconstructed using the CGH cannot incorporate such aberrations, when the decomposition is done in only fiber modes since such modes depict a complete set only in terms of characterizing fields that are formed by multimode interference. As a consequence, only aberrations emerging from mode superpositions can be detected. In general, however, aberrated wave fronts can be reconstructed from decomposition in LG modes or HG modes, which depict a complete set in free space [30].

4.3.4 Polarization

The polarization of a laser beam is a fundamental quantity and as discussed in Chapter 9, it may be exploited to create novel optical fields for many applications. Several analysis techniques exist for the characterization of such laser beams, mostly based on the measurement of the Stokes parameters of the beam using quarter-wave plates and polarizers [31], polarization grating [32], or by employing a quarter-wave plate and a combination of mirrors [33]. Additionally, the combination of the classical Stokes measurement using a quarter-wave plate and a polarizer with a modal decomposition was shown to yield the polarization modally resolved in the case of multimode beams [28]. In the following, we will focus on the last-mentioned approach and demonstrate its principle by characterizing a vector beam emanating from an optical fiber. According to the classical measurement of the Stokes parameters, six intensity measurements are required to determine the polarization state of the beam:

$$\mathbf{S} = \begin{bmatrix} S_0 \\ S_1 \\ S_2 \\ S_3 \end{bmatrix} = \begin{bmatrix} |U_x|^2 + |U_y|^2 \\ |U_x|^2 - |U_y|^2 \\ 2|U_x||U_y|\cos\Phi_r \\ 2|U_x||U_y|\sin\Phi_r \end{bmatrix} = \begin{bmatrix} I(0°) + I(90°) \\ I(0°) - I(90°) \\ I(45°) - I(135°) \\ I_{\lambda/4}(45°) - I_{\lambda/4}(135°) \end{bmatrix}, \tag{4.24}$$

where $I(\alpha)$ is the measured (respectively reconstructed) intensity behind the polarizer at angular orientations $\alpha = 0°, 45°, 90°, 135°$ and $I_{\lambda/4}(\alpha)$ denotes two additional measurements with the polarizer placed at $\alpha = 45°, 135°$ and a preceding quarter-wave plate. By these means, the last equality relates to the theoretical and experimental definition of the Stokes parameters, which in addition, enables the calculation of the relative phase Φ_r:

$$\Phi_r = \tan^{-1}\left(\frac{S_3}{S_2}\right),\qquad(4.25)$$

which is the phase difference between x- and y-component, U_x and U_y, of the optical field. Using a CCD camera, these six intensity measurements yield the Stokes parameters and hence, the polarization state is spatially resolved. Moreover, in the case of multimode beams, this procedure can be extended by a combination with the modal decomposition introduced in Section 4.2. In this manner, the Stokes vector $\mathbf{S}_l$ can be measured for each mode individually [28]:

$$\mathbf{S}_l = \begin{bmatrix} S_0 \\ S_1 \\ S_2 \\ S_3 \end{bmatrix}_l = \begin{bmatrix} \varrho_{l,x}^2 + \varrho_{l,y}^2 \\ \varrho_{l,x}^2 - \varrho_{l,y}^2 \\ 2\varrho_{l,x}\varrho_{l,y}\cos\delta_l \\ 2\varrho_{l,x}\varrho_{l,y}\sin\delta_l \end{bmatrix},\qquad(4.26)$$

where $\varrho_{l,x}$ and $\varrho_{l,y}$ are the measured modal powers in the x- and y-direction and δ_l is the *intramodal* phase, which represents the phase difference between the two polarization components of one mode. A typical setup to combine the classical Stokes measurement with a modal decomposition experiment is shown in Figure 4.13. Here, a multimode optical step-index fiber (core diameter 7.7 µm, NA 0.12) is seeded with an Nd:YAG laser to excite a beam with complex polarization properties. At 1064 nm, the fiber guides three modes, the fundamental mode LP$_{01}$, and two higher-order modes LP$_{11e}$ and LP$_{11o}$, which only differ in angular orientation (cf. Figure 4.13). The beam plane at the fiber end face is relay imaged onto a camera with the preceding quarter-wave plate and a polarizer to perform the classical polarization measurement according to Equation 4.24. In addition, the fiber end face is relay imaged onto a correlation filter that executes the modal

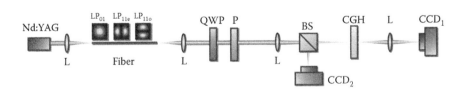

FIGURE 4.13 Scheme of the experimental setup. Nd:YAG, Nd:YAG laser ($\lambda = 1064$ nm); L, lens; QWP, quarter-wave plate; P, polarizer; BS, beam splitter; CCD1,2, CCD cameras; CGH, correlation filter. The insets depict the three modes guided by the fiber.

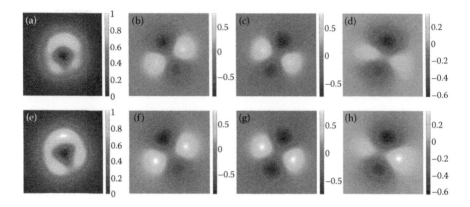

FIGURE 4.14 Spatially resolved Stokes parameters S_0, S_1, S_2, and S_3. (a)–(d) S_0, S_1, S_2, and S_3 are determined from the recorded beam intensity (CCD_2 in Figure 4.13). (e)–(h) S_0, S_1, S_2, and S_3 are determined from the reconstructed beam intensity.

decomposition for the field component selected by the quarter-wave plate and a polarizer. Figure 4.14 depicts the results for the spatially resolved Stokes parameters. The six intensities that appear on the right-hand side of Equation 4.24 are measured directly using a CCD camera (CCD_2) and are reconstructed from the modal decomposition experiment (Equation 4.13 in Section 4.3.1). Accordingly, all Stokes parameters are determined as spatially resolved. The good agreement between the Stokes parameters determined from direct intensity measurement and from reconstruction is a measure of fidelity of the modal decomposition experiment. Regarding the shape of the Stokes parameters, it can be seen that $S_1 \ldots S_3$ exhibit a four-leaved structure, which can be assigned to the petals of the modes due to the absence of the fundamental mode in this superposition as will be shown later. In the horizontal direction, S_1 and S_2 are basically positive; hence, we expect

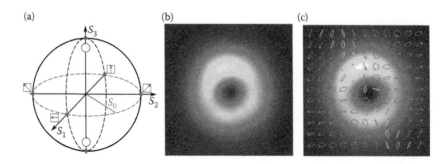

FIGURE 4.15 **(See color insert.)** Spatially resolved polarization. (a) Definition of Stokes parameters. (b) Measured beam intensity. (c) Reconstructed intensity with local polarization ellipses. The red and yellow color denote left- and right-handed sense of orientation. The bluish dots mark the position of the electric field vector at a fixed time.

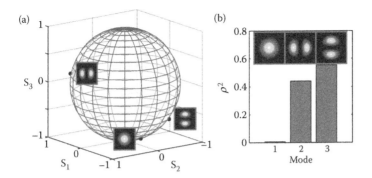

FIGURE 4.16 Modal polarization. (a) Poincaré representation of the Stokes vectors of each fiber mode. (b) The corresponding modal power spectrum (in %). The insets depict mode intensities.

a dominantly horizontal orientation of the polarization ellipses with a slight tilt toward 45°, whereas in the vertical direction, they are negative hence we mostly expect vertical alignment of the polarization ellipses with a slight tilt toward 135° (cf. Figure 4.15a). The sense of rotation will mostly be left-handed since S_3 is dominantly negative. Figure 4.15 illustrates the corresponding beam intensity, $|U_x|^2 + |U_y|^2$, measured (Figure 4.15b), and reconstructed with associated polarization ellipses at distinct points on top (Figure 4.15c). In conformance with Figure 4.14, the polarization ellipses are mainly left handed and orientated horizontally in the horizontal direction and vertically in the vertical direction. From the modal decomposition results, the Stokes vector can be obtained for each mode of the superposition. Accordingly, Figure 4.16a depicts the modal polarization on the Poincaré sphere with the corresponding modal power spectrum in Figure 4.16b. It can be seen that both higher-order modes sit on opposite points on the sphere. Hence, while the LP_{11e} mode is nearly linearly polarized in the horizontal direction, the LP_{11o} mode is elliptically polarized with left-handed and vertical orientation. This is in agreement with the observed four-leaved structure of Figure 4.14 and the illustration of local polarization ellipses in Figure 4.15b, and indicates that, in this case, the two higher-order modes determine the spatial polarization pattern of the beam.

4.3.5 Orbital Angular Momentum

Laser beams carrying orbital angular momentum (OAM) have attained a lot of interest in recent times and are discussed in detail in Chapter 11. Despite a lot of work that has been spent to generate these beams and to investigate their nature to search for new physical effects and applications, only a few approaches concentrate on the detection of OAM. In this section, we will continue with the modal decomposition technique and illustrate its use for the measurement of the OAM of light. As we have shown earlier, the Poynting vector may be deduced from the field of the modal decomposition procedure

(a)   (b)

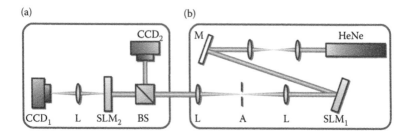

FIGURE 4.17   Experimental setup for the (a) characterization and (b) generation of LG beams of different order. HeNe, helium-neon laser; L, lens; M, mirror; $SLM_{1,2}$, spatial light modulators; A, aperture; $CCD_{1,2}$, CCD cameras; BS, beam splitter.

(Equations 4.19 and 4.20). It is now a straightforward step to determine the OAM density:

$$\mathbf{j} = \mathbf{r} \times \frac{\mathbf{P}}{c^2}. \tag{4.27}$$

In the following, the, z-component j_z, which is of most relevance for application, is only considered and termed "OAM density." To illustrate the principle, we apply it to various classes of laser beams, namely, LG beams and modes in fibers.

LG beams depict a well suited example for studies since the OAM can be easily calculated (Equations 4.20 and 4.27). Moreover, their generation is straightforward using phase masks or appropriate holograms to induce the necessary helical phase structure, see Chapters 6 and 11. A typical setup for the generation and subsequent beam characterization of laser beams is shown in Figure 4.17. An LP helium-neon laser ($\lambda = 633$ nm) is expanded by a telescope and illuminates an SLM to approximate a plane wave. The field generated with the SLM is filtered in its Fourier plane to enhance the quality of the beam and is relay imaged onto a second SLM (SLM_2 in Figure 4.17), which is used as a correlation filter [8]. Together with a single lens and a CCD camera (CCD_1), SLM_2 performs the modal decomposition, which yields the OAM density. The modal decomposition is done by displaying the transmission functions on SLM_2 subsequently and recording the correlation signal in the Fourier plane at CCD_1, which yields the modal powers ϱ_l^2 and phases $\Delta\varphi_l$, and hence, the optical field (Section 4.2), the Poynting vector (Equations 4.19, 4.20), and the OAM density (Equation 4.27). To record the beam intensity directly, a second CCD camera (CCD_2) and a beam splitter are used. As a first example a pure LG mode $LG_{p=0,l=1}$ was generated with the radial index p and the azimuthal index l, whereas last mentioned index indicates the order of the helical phase and the topological charge, respectively [34,35]. Figure 4.18 depicts the results of the characterization of this beam. Figure 4.18a shows the measured near-field intensity, which reveals the typical donut shape with a hole in its center indicating the phase singularity. The results of the modal decomposition are shown in Figure 4.18b and d. As expected, the modal power spectrum reveals a correlation signal only for the $LG_{0,1}$ mode, whereas all other mode powers in the range $p = 0 \ldots 2$ and $l = 0 \ldots 2$ are zero. Additionally, all phases

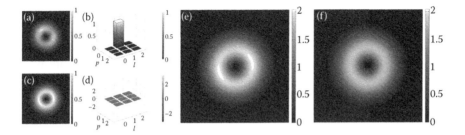

FIGURE 4.18 Characterization of an $LG_{0,1}$ mode beam. (a) Measured near-field intensity. (b) Modal power spectrum. (c) Reconstructed intensity. (d) Modal phase spectrum. (e) Measured OAM density. (f) Calculated OAM density. The OAM densities are in 10^{-22} Ns/m^2.

are measured to be zero, which are, however, in the case of a single mode, meaningless. From the measurement of modal powers and phases, the beam intensity is reconstructed according to Equation 4.13 in Section 4.3.1, as shown in Figure 4.18c. The comparison of the reconstructed (Figure 4.18c) and measured beam intensity (Figure 4.18a) reveals very good agreement. Following Equations 4.20 and 4.27, the OAM density is inferred (Figure 4.18e) and compared to the theoretical density (Figure 4.18f), revealing a good agreement. The similarity of the intensity and OAM density can be viewed as a characteristic of LG beams and is a logical consequence of Equation 4.20 and 4.27. Remarkably, the OAM density is exclusively positive. However, this fact can be easily understood, considering that the z-component of the OAM density is proportional to the φ-component of the Poynting vector and hence, to the azimuthal index l (here $l = +1$), provided the beam carries a helical phase [36]. Moreover, it becomes clear that the presented technique is capable of measuring the OAM density quantitatively. Here, the maximum is found to be 2.5×10^{-22} Ns/m^2.

The results of Figure 4.18 showed that a beam having a helical phase structure has a nonzero OAM density distribution, which mainly follows the intensity distribution.

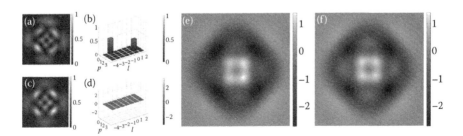

FIGURE 4.19 Characterization of an $LG_{2,1} + LG_{0,-3}$ beam. (a) Measured near-field intensity. (b) Modal power spectrum. (c) Reconstructed intensity. (d) Modal phase spectrum. (e) Measured OAM density. (f) Calculated OAM density. The OAM densities have a scale of 10^{-22} Ns/m^2.

However, LG beams with an azimuthal index $l = 0$ will have a vanishing φ-component of the Poynting vector P_φ [36] and consequently, a vanishing OAM density.

Similarly, a mode superposition of mixed l values will provide an OAM density of both positive and negative values. As an example, Figure 4.19 depicts the OAM density measurement on an in-phase superposition of two LG beams $LG_{2,1} + LG_{0,-3}$. Clearly, the modal power spectrum (Figure 4.19b) reflects the existence of the two involved modes. From the phase spectrum, it can be seen that both modes are in phase. Figure 4.19e and f depicts the measured and calculated OAM density, which correlates very well. Since the azimuthal indices of the involved modes are of positive and negative sign ($l = 1$ and $l = -3$), the resulting OAM density varies around zero. Regarding the scale of the OAM density, the pure $LG_{0,1}$ mode and the superposition $LG_{2,1} + LG_{0,-3}$ are of the same order of magnitude of 10^{-22} Ns/m^2. The comparison of the measured OAM densities for different LG beams with the theoretical predictions proved the presented technique to yield accurate OAM density measurements regarding the shape and absolute scale. Hence, the method can be applied to unknown laser beams, for which no reference can be provided. A suitable example is a beam emanating from an optical multimode fiber. Instead of programming modes and mode superpositions on a seperate SLM for beam generation, the optical field here is naturally created by multimode interference inside the fiber. By way of an example, a multimode fiber (core diameter 25 μm, NA = 0.064), guiding six transverse modes, is used. Correspondingly, the experimental setup is changed, as shown in Figure 4.20. The fiber was seeded with the fundamental Gaussian beam of an Nd:YAG laser ($\lambda = 1064$ nm) and the output was relay imaged (4f imaged) to the correlation filter, which was in this experiment a static binary amplitude-only CGH, as used by Kaiser et al. [4]. To perform the polarization measurement, as described in Section 4.3.4, a quarter-wave plate and a polarizer preceded the CGH. As in the previous experiment, a 2f setup behind the CGH (previously SLM) enabled the correlation measurement with a CCD camera (CCD_1). Again, a beam splitter in front of the hologram and a second CCD camera (CCD_2) provided the recording of the beam intensity.

Figure 4.21 illustrates an example of a fiber vector beam consisting of 27% LP_{01}, 0% LP_{02}, 41% LP_{11e}, 30% LP_{11o}, 2% LP_{21e}, and 0% LP_{21o} (LP modes, cf. [2]). The comparison

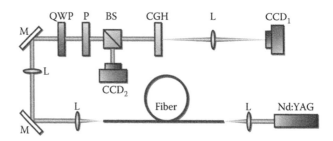

FIGURE 4.20 Experimental setup for measuring the OAM density of a fiber beam. Nd:YAG, Nd:YAG laser; L, lens; M, mirror; QWP, quarter-wave plate; P, polarizer; BS, beam splitter; CGH, computer-generated hologram; $CCD_{1,2}$, CCD cameras.

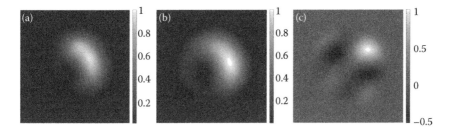

FIGURE 4.21 Characterization of a fiber beam consisting of a superposition of six modes (27% LP_{01}, 0% LP_{02}, 41% LP_{11e}, 30% LP_{11o}, 2% LP_{21e}, and 0% LP_{21o}). (a) Measured near-field intensity. (b) Reconstructed intensity. (c) Measured OAM density. The OAM density is in 10^{-16} Ns/m^2.

of measured (Figure 4.21a) and reconstructed (Figure 4.21b) beam intensity reveals a very good agreement. Note that both intensities represent the sum of intensities recorded in the *x*- and *y*-direction of the polarizer. Figure 4.21c illustrates the inferred OAM density of this beam. Apparently, the OAM density is much larger than in the previous cases, with a maximum of 1×10^{-16} Ns/m^2. This is a logical consequence of the much smaller dimension of the beam, which is roughly extended within 30 μm of diameter.

4.4 Conclusion

In this chapter, we have outlined a modern technique for the characterization of laser beams. The technique, modal decomposition, uses a match filter and a lens to decode the modal information of the field from which all the desired optical parameters may be inferred. We have illustrated how the intensity, wave front, beam propagation ratio, polarization, and OAM may be determined and have shown the versatility of the tool by applying it to vector beams, scalar beams, OAM carrying beams, fiber modes, and free space modes. More details on the properties of these laser beams may be found in the other chapters.

References

1. Saleh, B.E.A. and Teich, M.C. 1991. *Fundamentals of Photonics*. New York: Wiley.
2. Snyder, A.W. and Love, J.D. 1996. *Optical Waveguide Theory*. London: Chapman & Hall.
3. Goodman, J.W. 2005. *Introduction to Fourier Optics*. Englewood, CO: Roberts & Company Publishers.
4. Kaiser, T., Flamm, D., Schröter, S., and Duparré, M. 2009. Complete modal decomposition for optical fibers using CGH-based correlation filters, *Opt. Express* 17: 9347–56.

5. Flamm, D., Schulze, C., Brüning, R., Schmidt, O.A., Kaiser, T., Schröter, S., and Duparré, M. 2012. Fast M^2 measurement for fiber beams based on modal analysis, *Appl. Opt.* 51: 987–93.
6. Soifer, V. 2002. *Methods for Computer Design of Diffractive Optical Elements*. New York: John Wiley & Sons.
7. Arrizón, V., Ruiz, U., Carrada, R., and González. L.A. 2007. Pixelated phase computer holograms for the accurate encoding of scalar complex fields, *J. Opt. Soc. Am. A* 24: 3500–7.
8. Flamm, D., Naidoo, D., Schulze, C., Forbes, A., and Duparré. M. 2012. Mode analysis with a spatial light modulator as a correlation filter, *Opt. Lett.* 37: 2478–80.
9. Flamm, D., Schulze, C., Naidoo, D., Schröter, S., Forbes, A., and Duparré, M. 2013. All-digital holographic tool for mode excitation and analysis in optical fibers, *J. Lightwave Technol.* 3: 387–89.
10. Schulze, C., Ngcobo, S., Duparré, M., and Forbes, A. 2012. Modal decomposition without a priori scale information, *Opt. Express* 20: 27866–73.
11. Schulze, C., Lorenz, A., Flamm, D., Hartung, A., Schröter, S., Bartelt, H., and Duparré, M. 2013. Mode resolved bend loss in few-mode optical fibers, *Opt. Express* 12: 19714–25.
12. Born, M. and Wolf, E. 1991. *Principles of Optics*. Exeter, Great Britain: Pergamon Press.
13. ISO-11146-2. 2005. Test methods for laser beam widths, divergence angles and beam propagation ratios. Part 2: General astigmatic beams.
14. Siegman, A.E. 1990. New developments in laser resonators, *Proc. SPIE* 1224: 2–14.
15. Nemes, G. and Siegman, A. E. 1994. Measurement of all ten second-order moments of an astigmatic beam by the use of rotating simple astigmatic (anamorphic) optics, *J. Opt. Soc. Am. A* 11: 2257–64.
16. Lambert, R.W., Cortés-Martínez, R., Waddie, A.J., Shephard, J.D., Taghizadeh, M.R., Greenaway, A.H., and Hand, D.P. 2004. Compact optical system for pulse-to-pulse laser beam quality measurement and applications in laser machining. *Appl. Opt.* 43: 5037–46.
17. Schulze, C., Flamm, D., Duparré, M., and Forbes, A. 2012. Beam-quality measurements using a spatial light modulator, *Opt. Lett.* 37: 4687–89.
18. Schmidt, O.A., Schulze, C., Flamm, D., Brüning, R., Kaiser, T., Schröter, S., and Duparré, M. 2011. Real-time determination of laser beam quality by modal decomposition, *Opt. Express* 19: 6741–6748.
19. Navarro, R. and Moreno-Barriuso, E. 1999. Laser ray-tracing method for optical testing. *Opt. Lett.* 24: 951–53.
20. Chamot, S.R., Dainty, C., and Esposito, S. 2006. Adaptive optics for ophthalmic applications using a pyramid wave front sensor, *Opt. Express* 14: 518–26.
21. Velghe, S., Primot, J., Guérineau, N., Cohen, M., and Wattellier, B. 2005. Wave-front reconstruction from multidirectional phasederivatives generated by multilateral shearing interferometers, *Opt. Lett.* 30: 245–47.
22. Lane, R.G. and Tallon, M. 1992. Wave-front reconstruction using a Shack–Hartmann sensor, *Appl. Opt.* 31: 6902–8.

23. Changhai, L., Fengjie, X., Shengyang, H., and Zongfu, J. 2012. Performance analysis of multiplexed phase computer-generated hologram for modal wave front sensing. *Appl. Opt.* 50: 1631–39.

24. Litvin, I.A., Dudley, A., Roux, F.S., and Forbes, A. 2012. Azimuthal decomposition with digital holograms, *Opt. Express* 20: 10996–11004.

25. Borrego-Varillas, R., Romero, C., de Aldana, J.R.V., Bueno, J.M., and Roso, L. 2011. Wave front retrieval of amplified femtosecond beams by second-harmonic generation, *Opt. Express* 19: 22851–62.

26. Schulze, C., Naidoo, D., Flamm, D., Schmidt, O.A., Forbes, A., and Duparré, M. 2012. Wave front reconstruction by modal decomposition, *Opt. Express* 20: 19714–25.

27. ISO-15367-1. 2003. Lasers and laser-related equipment—test methods for determination of the shape of a laser beam wave front—Part 1: Terminology and fundamental aspects.

28. Flamm, D., Schmidt, O.A., Schulze, C., Borchardt, J., Kaiser, T., Schröter, S., and Duparré, M. 2010. Measuring the spatial polarization distribution of multimode beams emerging from passive step-index large-mode-area fibers, *Opt. Lett.* 35: 3429–31.

29. Grosjean, T., Sabac, A., and Courjon, D. 2005. A versatile and stable device allowing the efficient generation of beams with radial, azimuthal or hybrid polarizations, *Opt. Commun.* 252: 12–21.

30. Schulze, C., Dudley, A., Flamm, D., Duparré, M., and Forbes, A. 2013. Reconstruction of laser beam wavefronts based on mode analysis, *Appl. Opt.* 52: 5312–5317.

31. Kihara, T. 2011. Measurement method of stokes parameters using a quarter-wave plate with phase difference errors. *Appl. Opt.* 50: 2582–87.

32. Gori, F. 1999. Measuring Stokes parameters by means of a polarization grating, *Opt. Lett.* 24: 584–86.

33. Fridman, M., Nixon, M., Grinvald, E., Davidson, N., and Friesem, A.A. 2010. Real-time measurement of unique space-variant polarizations, *Opt. Express* 18: 10805–12.

34. Yao, A.M. and Padgett, M.J. 2011. Orbital angular momentum: Origins, behavior and applications. *Adv. Opt. Photon.* 3: 161–204.

35. Soskin, M.S., Gorshkov, V.N., Vasnetsov, M.V., Malos, J.T., and Heckenberg, N.R. 1997. Topological charge and angular momentum of light beams carrying optical vortices, *Phys. Rev. A* 56: 4064–75.

36. Litvin, I.A., Dudley, A., and Forbes, A. 2011. Poynting vector and orbital angular momentum density of superpositions of Bessel beams, *Opt. Express* 19: 16760–16771.

5

Characterization of Time Domain Pulses

Thomas Feurer
University of Bern

This chapter aims to present a structured overview of the numerous methods and techniques employed to characterize ultrashort laser pulses. The variety of techniques is quite substantial and the common goal of all these techniques is to determine the electric field of an ultrashort laser pulse as precisely as possible. This quest is not as easy as it sounds and it took quite a number of years before practical, robust, and accurate methods emerged.

Henceforth, I assume that the coherent laser pulses propagate preferentially in one direction and that their electric field vector has only one transverse component, that is, the pulses are linearly polarized. Additionally, I assume that the relevant real-valued electric field component can be decomposed in a product separating the spatial from the temporal part, that is, $E(\mathbf{x}, t) = f(\mathbf{x})E(t)$. While the spatial intensity distribution $\propto |f(\mathbf{x})|^2$ can often be obtained by means as simple as a charge-coupled device (CCD) camera, the temporal part is more difficult to measure. Before describing suitable techniques, I introduce the analytic form of the electric field

$$E(t) = E^+(t) + E^-(t). \tag{5.1}$$

Moreover, the analytic field with amplitude $|E^+(t)|$ and phase $\varphi(t)$ is split in a slowly varying envelope times an oscillating carrier frequency

$$E^+(t) = |E^+(t)| \, e^{i\varphi(t)} = \frac{1}{2} \, \mathcal{E}^+(t) \, e^{i\omega_0 t}. \tag{5.2}$$

For ultrashort pulses, both the carrier phase $\omega_0 t$ and the slowly varying envelop $\mathcal{E}^+(t)$ vary so rapidly with time that no device is able to measure their time dependence directly. While the carrier frequency ω_0 can be measured by a spectrometer, the slowly varying envelope requires indirect methods. The first derivative of the phase $\varphi(t)$ with respect to time is the instantaneous frequency $\omega(t)$

$$\omega(t) = \frac{d\varphi(t)}{dt} = \omega_0 + \frac{d\arg(\mathcal{E}^+(t))}{dt}. \tag{5.3}$$

The spectral fields

$$E^+(\omega) = |E^+(\omega)| \, e^{i\phi(\omega)} = \frac{1}{2} \, \mathcal{E}^+(\omega - \omega_0) \tag{5.4}$$

are obtained through Fourier transformation. The spectral phase is often expanded in a Taylor series:

$$\phi(\omega) = \sum_{m=0}^{\infty} \frac{1}{m!} \left. \frac{d\phi}{d\omega} \right|_{\omega_0} (\omega - \omega_0)^m. \tag{5.5}$$

The lowest-order phase derivative ($m = 0$) is called absolute phase, $m = 1$ linear phase, $m = 2$ quadratic phase (linear chirp), $m = 3$ cubic phase (quadratic chirp), and so on. To illustrate the techniques, I present analytic results for a Gauss-shaped input pulse with a quadratic phase modulation, that is

$$E_{\text{Gauss}}^+(\omega) = \exp\left[-\frac{(\omega - \omega_0)^2}{\Delta\omega^2} (1 + iC) \right]$$
$$\mathcal{E}_{\text{Gauss}}^+(\omega) = \frac{1}{2} \exp\left[-\frac{\omega^2}{\Delta\omega^2} (1 + iC) \right] \tag{5.6}$$

with a spectral bandwidth $\Delta\omega$, a carrier frequency ω_0, and a quadratic phase (linear chirp) C. Note that the spectral bandwidth is unaffected by the phase modulation. The corresponding temporal electric fields are

$$E_{\text{Gauss}}^+(t) \propto \exp\left[-\frac{\Delta\omega^2 t^2}{4(1 + C^2)} (1 - iC) + i\omega_0 t \right]$$
$$\mathcal{E}_{\text{Gauss}}^+(t) \propto \exp\left[-\frac{\Delta\omega^2 t^2}{4(1 + C^2)} (1 - iC) \right]. \tag{5.7}$$

The analytic results are useful to verify numerical codes simulating signals for arbitrary pulses. Moreover, I present simulated results for an asymmetric double pulse, a pulse with a quadratic, a cubic, and a quartic spectral phase modulation, and a pulse that has undergone self-phase modulation. Such pulses are frequently encountered in the daily lab work and the simulations help to identify them.

In the following, I classify the different methods to characterize laser pulses in the general framework of filter theory. This approach is very similar to that published by Rolf Gase [1], but is used in many other contexts as well. Each element within a measurement setup is represented by a transfer function and there are rules how to calculate the electric field after the element from that before the element. A sequence of such elements plus a detector then comprises a single measurement setup. I also consider interferometric-type measurement techniques, where the incoming pulse is split in two or more replicas each traveling through a different sequence of elements before being recombined and analyzed by the detector. Next to classifying, my second goal is to indicate what type of information (intensity, electric field amplitude, phase etc.) is to be expected from a specific measurement technique and to show generic experimental setups.

5.1 Basics

Before starting, I briefly introduce the concept of joint time–frequency distribution functions. They are useful, first, to visualize rather intuitively the dynamic evolution of the electric field and, second, because some of the measurement techniques produce such joint time–frequency distributions. So far, we have described the electric field as a function either of time or of frequency alone. Joint time–frequency distributions aim to display both aspects simultaneously. However, the time–energy (frequency) uncertainty principle of waves prohibits that both aspects can be shown with infinite resolution. Still, we will see that even under this constraint, the graphic information gain is extremely valuable.

5.1.1 Two-Dimensional Distribution Functions

Figure 5.1 shows one of the oldest known joint time–frequency distributions. As time goes on (*x*-axis), the frequency (*y*-axis) increases and so does the amplitude. What we see here is the acoustic analog of a linearly chirped pulse train.

The joint time–frequency distribution functions introduced here are only a subset of those known—the Wigner distribution, because it is one of the most widely used, and the spectrogram and the Page distribution, because they are related to some of the measurement techniques. An excellent review on the subject was published by Leon Cohen [2]. A further important question to pose in the context of pulse characterization is the following. Suppose a measurement has produced a specific time–frequency distribution, say a spectrogram; is it then possible to extract all information on the electric field from it?

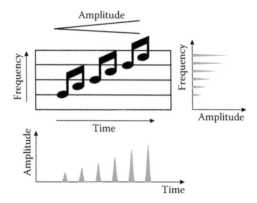

FIGURE 5.1   An ancient time–frequency distribution still in use. The time is running from the left to the right, the frequency is increasing in discrete steps from the bottom to the top, and the amplitude is indicated above the tones. Clearly, the time–frequency distribution reveals much more information on the dynamic evolution than either the temporal intensity (bottom plot) or the spectrum (right plot).

5.1.1.1 General Approach

Ideally, a generic joint time–frequency distribution function, $C(t, \omega)$, should relate to the intensity per time and per frequency interval and fulfill the following constraints:

$$\int d\omega \, C(t, \omega) \propto |E^+(t)|^2 \tag{5.8a}$$

$$\int dt \, C(t, \omega) \propto |E^+(\omega)|^2 \tag{5.8b}$$

$$\iint dt \, d\omega \, C(t, \omega) \propto F. \tag{5.8c}$$

Integrating $C(t, \omega)$ with respect to frequency should yield a function proportional to the temporal intensity and integrating with respect to time should result in a function proportional to the spectrum. Finally, integration over both variables should yield the total fluence F. A very general form that fulfills all these requirements is

$$C(t, \omega) = \frac{1}{4\pi^2} \iiint du \, d\tau \, d\theta \, e^{-i\theta t - i\tau\omega + i\theta u} \, \Phi(\theta, \tau)$$
$$\times E^-\left(u - \frac{\tau}{2}\right) E^+\left(u + \frac{\tau}{2}\right), \tag{5.9}$$

where $\Phi(\theta, \tau)$ is referred to as the kernel. It is the kernel that defines a specific distribution and it is the kernel that determines the properties of the joint time–frequency

TABLE 5.1 Three Joint Time–Frequency Distributions with Their Corresponding Kernel and Analytic Forms

Distribution	Kernel $\Phi(\theta, \tau)$	$C(t, \omega)$
Wigner	1	$\frac{1}{2\pi} \int d\tau \, E^-(t - \tau/2) \, E^+(t + \tau/2) \, e^{-i\omega\tau}$
Page	$e^{i\theta|\tau|/2}$	$\frac{1}{2\pi} \frac{\partial}{\partial t} \left| \int_{-\infty}^{t} d\tau \, E^+(\tau) \, e^{-i\omega\tau} \right|^2$
Spectrogram	$\int du \, h^*(u - \tau/2) \, e^{-i\theta u} \, h(u + \tau/2)$	$\frac{1}{2\pi} \left| \int d\tau \, h(\tau - t) \, E^+(\tau) \, e^{-i\omega\tau} \right|^2$

distribution. In addition, with the help of the kernel, it is possible to transform each distribution to others. The kernel may be a function of time, of frequency, or even of the signal itself. Table 5.1 lists the kernels of the three joint time–frequency distributions discussed here.

The general approach to the joint time–frequency distributions allows to study their properties in a standardized way. As stated above, an essential question is whether it is possible to extract the electric field once a distribution has been measured. Unfortunately, the answer is only partially yes. In general, the electric field is obtained through the following procedure:

$$E^+(t) = \frac{1}{2\pi E^-(0)} \int d\theta \, \frac{M(\theta, t)}{\Phi(\theta, t)} \, e^{-i\theta t/2}, \tag{5.10}$$

with the characteristic function

$$M(\theta, \tau) \doteq \iint dt \, d\omega \, C(t, \omega) \, e^{i\theta t + i\tau\omega}, \tag{5.11}$$

which is just the two-dimensional inverse Fourier transformation of the time–frequency distribution $C(t, \omega)$. The first thing to notice is that it is only possible to extract the electric field up to a constant phase factor. Second, in order to recover the electric field, the characteristic function has to be divided by the kernel prior to an integration. It has been shown that the electric field may be recovered uniquely if the kernel has only isolated zeros. That is to say, the kernel should not be zero in regions where the characteristic function is nonzero. More importantly, however, the kernel must be known, which is trivial for the Wigner and the Page distribution but problematic for the spectrogram. The kernel of the spectrogram requires a gate function $h(t)$, which is frequently identical to the electric field $E^+(t)$ itself. Under those circumstances it is obvious that the reconstruction of the electric field through Equation 5.10 is impossible. But even if direct electric field reconstruction is possible, it is usually extremely susceptible to statistical and systematic noise in the experimental data. A more robust method is to extract the relevant information from the conditional moments. The nth-order conditional moments are defined as

$$c_t^{(n)}(t) \;\doteq\; \int d\omega \; \omega^n \, C(t,\omega) \tag{5.12a}$$

$$c_\omega^{(n)}(\omega) \;\doteq\; \int dt \; t^n \, C(t,\omega). \tag{5.12b}$$

The zeroth-order conditional moments coincide with the constraints (5.8a) imposed earlier on the joint time–frequency distributions. In other words, the temporal intensity and, thus, the amplitude of the electric field can be extracted from $c_t^{(0)}(t)$. It will be shown later that the phase (up to a constant phase) may often be extracted from the first-order conditional moment $c_t^{(1)}(t)$.

5.1.1.2 Wigner Distribution

The Wigner distribution has the simplest kernel of all distributions in Table 5.1; it is equal to one. As a consequence, it is possible to invert the Wigner distribution and to recover the electric field. Unfortunately, there is no direct experimental technique that can measure the Wigner distribution for a short laser pulse. For the linearly chirped Gauss pulse (5.6), we find

$$W_{\text{Gauss}}(t,\omega) \propto \exp\left[-\frac{\Delta\omega^2}{2}\,t^2 + 2Ct(\omega - \omega_0) - \frac{2(1 + C^2)}{\Delta\omega^2}(\omega - \omega_0)^2\right]. \tag{5.13}$$

Figure 5.2 shows the Wigner distribution for a Gauss-shaped pulse with no phase modulation, a double pulse, a pulse with a quadratic, a cubic, and a quartic phase modulation, and a pulse that has experienced self-phase modulation.

Already, such simple examples nicely demonstrate the strength of joint time–frequency distribution functions. Simply by inspection one can infer whether the pulse has a phase modulation, whether it is linear, quadratic, and so on. The unmodulated Gauss pulse is one of very few pulse shapes for which the Wigner distribution is always positive. Usually, the Wigner distribution has regions with negative amplitude, which is also a strong indication that the Wigner distribution cannot be measured directly with a square law detector.

5.1.1.3 Spectrogram

The spectrogram is somewhat reminiscent of a windowed Fourier transformation. The gate function $h(t)$ slices out small parts of the electric field and each slice is then Fourier transformed and its spectrum is measured. Although the spectrogram is very intuitive, it has a number of disadvantages. Assume it is necessary to increase the temporal resolution, that is to decrease the width of the gate function. The price to pay is a dramatic loss in spectral resolution because the slices become shorter and less of the electric field contributes to the spectrum; in fact, most of the electric field does not contribute at all. This loss in spectral resolution has nothing to do with the electric field itself; it is

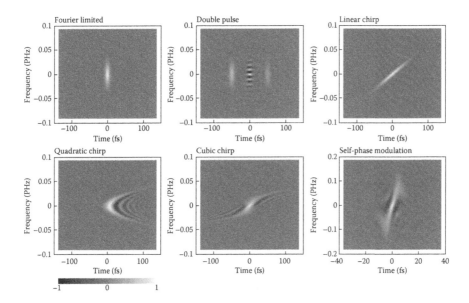

FIGURE 5.2 The Wigner distribution for a Gauss-shaped pulse with no phase modulation, a double pulse, a pulse with a quadratic, a cubic, and a quartic phase modulation, and a pulse that has experienced self-phase modulation.

merely a consequence of the way the spectrogram is calculated. In this sense, every signal has an optimal gate function. This insight immediately leads to the next problem. Suppose the electric field is a double pulse and each pulse has a different optimal gate function. In this case, it is almost impossible to find an optimal gate function. Irrespective of all drawbacks, the spectrogram is a very important time–frequency distribution for the experimentalist as there exist techniques that measure the spectrogram of ultrashort laser pulses. Unfortunately, the gate function is typically the pulse itself and we already know that this renders the direct inversion process impossible. When we lack all knowledge of the electric field, we cannot calculate the kernel and, hence, there is no direct way to extract the electric field from the spectrogram. To overcome the problem of inversion, Rick Trebino and coworkers [4] were the first to point out that iterative algorithms may be used to reconstruct the electric field from a measured spectrogram. Figure 5.3 shows the spectrograms for the same pulses as in Figure 5.2.

Reconstruction is trivial if the gate function is well characterized, as it is the case in many cross-correlation measurement. Knowing the gate function allows us to calculate the kernel and the inversion can be performed. For a Gauss-shaped gate with a temporal width $1/b$

$$h(t) = \exp(-2b^2 t^2) \tag{5.14}$$

and a Gauss pulse the spectrogram $S(t, \omega)$ becomes

$$S_{\text{Gauss}}(t, \omega) \propto \exp[-\alpha^2 t^2 + \beta t(\omega - \omega_0) + \gamma^2 (\omega - \omega_0)^2] \tag{5.15}$$

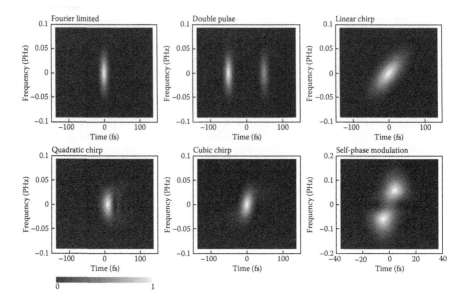

FIGURE 5.3 The spectrogram for a Gauss-shaped pulse with no phase modulation, a double pulse, a pulse with a quadratic, a cubic, and a quartic phase modulation, and a pulse that has experienced self-phase modulation.

with

$$\alpha^2 \doteq \frac{4a^2 b^2 (b^2 + 2a^2 + a^2 C^2)}{(a^2 + 2b^2)^2 + a^4 C^2} \qquad \beta \doteq \frac{4a^2 b^2 C}{(a^2 + 2b^2)^2 + a^4 C^2}$$

$$\gamma^2 \doteq \frac{(a^2 + 2b^2)/2}{(a^2 + 2b^2)^2 + a^4 C^2} \qquad a^2 \doteq \frac{\Delta\omega}{4(1 + C^2)}.$$

5.1.1.4 Page Distribution

Similar to the Wigner distribution, the Page distribution has a relatively simple kernel; at least, it does not depend on the electric field itself. Hence, it is possible to invert the Page distribution and to extract the electric field. Rewriting the definition of the Page distribution given in Table 5.1 leads to

$$P(t, \omega) = \frac{1}{2\pi} \frac{\partial}{\partial t} \left| \int d\tau \, \Theta(t - \tau) \, E^+(\tau) \, e^{-i\omega\tau} \right|^2. \tag{5.16}$$

The integration now extends from negative infinity to positive infinity and $\Theta(t)$ is the Heaviside step function. With this, it is easy to derive an experimental method to measure the Page distribution. Suppose an ultrafast transmission shutter is scanned across the

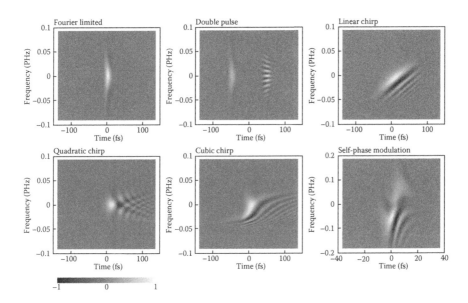

FIGURE 5.4 The Page distribution for a Gauss-shaped pulse with no phase modulation, a double pulse, a pulse with a quadratic, a cubic, and a quartic phase modulation, and a pulse which has experienced self-phase modulation.

unknown pulse. If the transmitted electric field is recorded by a spectrometer, then the time–frequency distribution

$$S(t, \omega) \propto \left| \int d\tau \, \Theta(t - \tau) \, E^+(\tau) \, e^{-i\omega\tau} \right|^2 \tag{5.17}$$

is measured. From Equation 5.17, the Page distribution is readily derived by calculating the derivative of $S(t, \omega)$ with respect to time. Figure 5.4 shows examples of the Page distribution.

Once the Page distribution is known, the electric field of the pulse may be extracted using the procedure described above. Alternatively, we may use the method based on conditional moments. For the Page distribution $P(t, \omega)$, we find

$$c_t^{(0)}(t) = |E^+(t)|^2 \tag{5.18a}$$

$$c_t^{(1)}(t) = |E^+(t)|^2 \, \frac{\partial}{\partial t} \varphi(t) \tag{5.18b}$$

assuming that the electric field is $E^+(t) = |E^+(t)| \exp[i\varphi(t)]$. Thus, the zeroth conditional moment $c_t^{(0)}$ allows to extract the amplitude of the electric field $|E^+(t)|$, and with $|E^+(t)|$ and the first conditional moment $c_t^{(1)}$, the derivative of the phase with respect to time

$\partial \varphi(t)/\partial t$ can be extracted. A simple integration then leads to the phase $\varphi(t)$ up to an unknown integration constant, that is, the absolute phase.

5.1.2 Filter Theory

Next, we formulate the mathematical framework for our goal to decompose each measurement setup into a sequence of filters. Each filter is represented by a response function $M_n(t_1, t_2)$ and the electric field after the nth filter is related to the electric field before the filter through

$$E_n^+(t_1) = \int dt_2 \, M_n(t_1, t_2) \, E_{n-1}^+(t_2). \tag{5.19}$$

With the Fourier transform pair

$$M_n(\omega_1, \omega_2) = \iint dt_1 dt_2 \, M_n(t_1, t_2) \, e^{-i\omega_1 t_1 - i\omega_2 t_2} \tag{5.20a}$$

$$M_n(t_1, t_2) = \frac{1}{(2\pi)^2} \iint d\omega_1 d\omega_2 \, M_n(\omega_1, \omega_2) \, e^{i\omega_1 t_1 + i\omega_2 t_2}, \tag{5.20b}$$

one can show that

$$E_n^+(\omega_1) = \int d\omega_2 \, M_n(\omega_1, -\omega_2) \, E_{n-1}^+(\omega_2). \tag{5.21}$$

Table 5.2 summarizes the transfer functions for the most frequently encountered filters.

Typically, a *time delay* is realized through a mechanical translation stage with a retro-reflector mounted on top. Such a time-delay shifts the slowly varying envelope as well as the carrier phase by the same time delay τ. Alternatively, a *time delay* can be introduced by a programmable pulse shaper in which case it is possible to shift the slowly varying envelope while leaving the carrier phase untouched. A *spectral bandpass* can be an interference filter or even a monochromator. Ideally, it has an infinite spectral resolution in which case the resulting temporal electric field is simply a harmonic wave oscillating at the bandpass frequency ω_s with a spectral amplitude $E_{n-1}^+(\omega_s)$. Also, we allow for a finite width of the spectral filter with a spectral response given by $m(\omega - \omega_s)$. For a Gauss-shaped bandpass $m(\omega - \omega_s) = \exp[-(\omega - \omega_s)^2/\delta\omega^2]$, we find $m(t) = \delta\omega/(2\sqrt{\pi}) \exp[-\delta\omega^2 t^2/4 + i\omega_s t]$. A *spectral shear* is the frequency domain analog of a time delay τ; however, it is not as easy to realize. A spectral shear requires some non-linear optical process as will be shown later. *Time gates* are nonlinear operations as the electric field is multiplied by a single gate function $G(t - \tau)$ or a product of two gate functions $G_1(t - \tau_1)G_2(t - \tau_2)$. For simplicity, the time delays are not explicitly introduced through the corresponding transfer function, rather we assume that the gate functions

TABLE 5.2   Filters and Transfer Functions of Several Building Blocks

Filter	Transfer Functions and Fields
Time delay (envelop and carrier)	$M_n(t_1, t_2) = \delta(t_1 - t_2 - \tau)$ $M_n(\omega_1, -\omega_2) = 2\pi\, e^{-i\omega_1\tau}\, \delta(\omega_1 - \omega_2)$ $E_n^+(t_1) = E_{n-1}^+(t_1 - \tau)$ $\mathcal{E}_n^+(t_1) = e^{-i\omega_0\tau}\, \mathcal{E}_{n-1}^+(t_1 - \tau)$ $E_n^+(\omega_1) = e^{-i\omega_1\tau}\, E_{n-1}^+(\omega_1)$ $\mathcal{E}_n^+(\omega_1 - \omega_0) = e^{-i\omega_1\tau}\, \mathcal{E}_{n-1}^+(\omega_1 - \omega_0)$
Time delay (envelop only)	$M_n(t_1, t_2) = \delta(t_1 - t_2 - \tau)\, e^{i\omega_0\tau}$ $M_n(\omega_1, -\omega_2) = 2\pi\, e^{-i(\omega_1 - \omega_0)\tau}\, \delta(\omega_1 - \omega_2)$ $E_n^+(t_1) = e^{i\omega_0\tau}\, E_{n-1}^+(t_1 - \tau)$ $\mathcal{E}_n^+(t_1) = \mathcal{E}_{n-1}^+(t_1 - \tau)$ $E_n^+(\omega_1) = e^{-i(\omega_1 - \omega_0)\tau}\, E_{n-1}^+(\omega_1)$ $\mathcal{E}_n^+(\omega_1 - \omega_0) = e^{-i(\omega_1 - \omega_0)\tau}\, \mathcal{E}_{n-1}^+(\omega_1 - \omega_0)$
Spectral bandpass (infinite resolution)	$M_n(t_1, t_2) = \dfrac{1}{2\pi}\, e^{i\omega_s(t_1 - t_2)}$ $M_n(\omega_1, -\omega_2) = 2\pi\, \delta(\omega_1 - \omega_s)\, \delta(\omega_1 - \omega_2)$ $E_n^+(t_1) = \dfrac{1}{2\pi}\, e^{i\omega_s t_1}\, E_{n-1}^+(\omega_s)$ $\mathcal{E}_n^+(t_1) = \dfrac{1}{2\pi}\, e^{i(\omega_s - \omega_0)t_1}\, \mathcal{E}_{n-1}^+(\omega_s - \omega_0)$ $E_n^+(\omega_1) = E_{n-1}^+(\omega_1)\, \delta(\omega_1 - \omega_s)$ $\mathcal{E}_n^+(\omega_1 - \omega_0) = \mathcal{E}_{n-1}^+(\omega_1 - \omega_0)\, \delta(\omega_1 - \omega_s)$
Spectral bandpass (finite resolution)	$M_n(t_1, t_2) = m(t_1 - t_2)$ $M_n(\omega_1, -\omega_2) = 2\pi\, m(\omega_1 - \omega_s)\, \delta(\omega_1 - \omega_2)$ $E_n^+(t_1) = \int dt_2\, m(t_1 - t_2)\, E_{n-1}^+(t_2)$ $\mathcal{E}_n^+(t_1) = \int dt_2\, m(t_1 - t_2)\, e^{-i\omega_0(t_1 - t_2)}\, \mathcal{E}_{n-1}^+(t_2)$ $E_n^+(\omega_1) = m(\omega_1 - \omega_s)\, E_{n-1}^+(\omega_1)$ $\mathcal{E}_n^+(\omega_1 - \omega_0) = m(\omega_1 - \omega_s)\, \mathcal{E}_{n-1}^+(\omega_1 - \omega_0)$
Spectral shear	$M_n(t_1, t_2) = e^{i\alpha t_1}\, \delta(t_1 - t_2)$ $M_n(\omega_1, -\omega_2) = 2\pi\, \delta(\omega_1 - \omega_2 - \alpha)$ $E_n^+(t_1) = e^{i\alpha t_1}\, E_{n-1}^+(t_1)$ $\mathcal{E}_n^+(t_1) = e^{i\alpha t_1}\, \mathcal{E}_{n-1}^+(t_1)$ $E_n^+(\omega_1) = E_{n-1}^+(\omega_1 - \alpha)$ $\mathcal{E}_n^+(\omega_1 - \omega_0) = \mathcal{E}_{n-1}^+(\omega_1 - \omega_0 - \alpha)$
Single time gates	$M_n(t_1, t_2) = G(t_1 - \tau)\, \delta(t_1 - t_2)$ $M_n(\omega_1, -\omega_2) = G(\omega_1 - \omega_2)\, e^{-i(\omega_1 - \omega_2)\tau}$ $E_n^+(t_1) = G(t_1 - \tau)\, E_{n-1}^+(t_1)$ $\mathcal{E}_n^+(t_1) = G(t_1 - \tau)\, e^{i(\omega_{0,n-1} - \omega_{0,n})t_1}\, \mathcal{E}_{n-1}^+(t_1)$ $E_n^+(\omega_1) = \dfrac{1}{2\pi} \int d\omega_2\, G(\omega_1 - \omega_2)\, e^{-i(\omega_1 - \omega_2)\tau}\, E_{n-1}^+(\omega_2)$ $\mathcal{E}_n^+(\omega_1 - \omega_{0,n}) = \dfrac{1}{2\pi} \int d\omega_2\, G(\omega_1 - \omega_2)\, e^{-i(\omega_1 - \omega_2)\tau}\, \mathcal{E}_{n-1}^+(\omega_2 - \omega_{0,n-1})$

TABLE 5.2 **(continued)** Filters and Transfer Functions of Several Building Blocks

Double time gates	$M_n(t_1, t_2) = G_1(t_1 - \tau_1)\, G_2(t_1 - \tau_2)\, \delta(t_1 - t_2)$

$$M_n(\omega_1, -\omega_2) = \frac{1}{2\pi} \int d\omega_3\, G_1(\omega_3)\, G_2(\omega_1 - \omega_2 - \omega_3)\, e^{-i\omega_3\tau_1 - i(\omega_1 - \omega_2 - \omega_3)\tau_2}$$

$$E_n^+(t_1) = G_1(t_1 - \tau_1)\, G_2(t_1 - \tau_2)\, E_{n-1}^+(t_1)$$

$$\mathcal{E}_n^+(t_1) = G_1(t_1 - \tau_1)\, G_2(t_1 - \tau_2)\, e^{i(\omega_{0,n-1} - \omega_{0,n})t_1}\, \mathcal{E}_{n-1}^+(t_1)$$

$$E_n^+(\omega_1) = \frac{1}{(2\pi)^2} \iint d\omega_2\, d\omega_3\, G_1(\omega_3)\, G_2(\omega_1 - \omega_2 - \omega_3) \\ \times e^{-i\omega_3\tau_1 - i(\omega_1 - \omega_2 - \omega_3)\tau_2}\, E_{n-1}^+(\omega_2)$$

$$\mathcal{E}_n^+(\omega_1 - \omega_{0,n}) = \frac{1}{(2\pi)^2} \iint d\omega_2\, d\omega_3\, G_1(\omega_3)\, G_2(\omega_1 - \omega_2 - \omega_3) \\ \times e^{-i\omega_3\tau_1 - i(\omega_1 - \omega_2 - \omega_3)\tau_2}\, \mathcal{E}_{n-1}^+(\omega_2 - \omega_{0,n-1})$$

may be time delayed by τ, τ_1, or τ_2, respectively. While *single time gates* are typically realized through second-order nonlinear processes in suitable nonlinear crystals, *double time gates* require a third-order nonlinear process. Henceforth, we assume that the nonlinear crystals are always thin enough to guarantee close to perfect phase matching over the entire spectral bandwidths. Often, the gate function is another pulse in which case we may use the slowly varying envelope approximation, that is $G(t) = 1/2\, g(t)\, \exp[i\omega_G t]$ and $G(\omega) = 1/2\, g(\omega - \omega_G)$.

The time gates are summarized in Table 5.3. In difference frequency generation (DFG) or sum frequency generation (SFG), the gate function is another laser pulse $E_G^{\mp}(t)$ mixed with the incident field $E_{n-1}^+(t)$ such that the resulting field has a carrier frequency $\omega_{0,n} = \omega_{0,n-1} \mp \omega_G$. Second-harmonic generation (SHG) is a special case of SFG with the gate pulse being a replica of the incident pulse. A *time shutter* is an amplitude modulation whose transmission switches from one to zero at a given delay time. Such a time shutter can be approximated, for example, through a rapid ionization process; before ionization, the sample is transparent, and after ionization, the plasma density is above critical density and the transmission goes to zero. Examples for double time gates are third-harmonic generation (THG), polarization gating (PG), self-diffraction (SD), or transient grating (TG).

5.1.3 Square Law Detector

The most commonly used detector is the square law detector. Its response to the incident electric field of an isolated pulse is proportional to the integral over the instantaneous intensity $I(t) = \eta\, |E^+(t)|^2$, where η includes the sensitivity of the detector material. For a small area detector, the signal is approximately proportional to the fluence

$$F = \eta \int dt\, |E^+(t)|^2 = \frac{\eta}{2\pi} \int d\omega\, |E^+(\omega)|^2, \tag{5.22}$$

TABLE 5.3   Single and Double Time Gates

Time Gates	Gate Functions
Single Time Gates	
Difference frequency generation (DFG)	$G(t) = E_G^-(t) = \frac{1}{2}\,\mathcal{E}_G^-(t)\,e^{-i\omega_G t}$
$\omega_{0,n} = \omega_{0,n-1} - \omega_G$	$G(\omega) = E_G^-(\omega) = \frac{1}{2}\,\mathcal{E}_G^-(\omega + \omega_G)$
Sum frequency generation (SFG)	$G(t) = E_G^+(t) = \frac{1}{2}\,\mathcal{E}_G^+(t)\,e^{i\omega_G t}$
$\omega_{0,n} = \omega_{0,n-1} + \omega_G$	$G(\omega) = E_G^+(\omega) = \frac{1}{2}\,\mathcal{E}_G^+(\omega - \omega_G)$
Second harmonic generation (SHG)	$G(t) = E_{n-1}^+(t) = \frac{1}{2}\,\mathcal{E}_{n-1}^+(t)e^{i\omega_{0,n-1}t}$
$\omega_{0,n} = 2\omega_{0,n-1}$	$G(\omega) = E_{n-1}^+(\omega) = \frac{1}{2}\,\mathcal{E}_{n-1}^+(\omega - \omega_{0,n-1})$
Time shutter	$G(t) = 1 - \Theta(t) = \Theta(-t)$
$\omega_{0,n} = \omega_{0,n-1}$	$G(\omega) = \pi\,\delta(\omega) - \dfrac{1}{i\omega}$
Time shutter (finite rise time δt)	$G(t) = \int dt_3\;\Theta(t - t_3)\;\exp\left[-\dfrac{t_3^2}{\delta t^2}\right]$
$\omega_{0,n} = \omega_{0,n-1}$	$G(\omega) = \left(\pi\,\delta(\omega) - \dfrac{1}{i\omega}\right)\exp\left[-\dfrac{\delta t^2}{4}\omega^2\right]$
Double Time Gates	
Third harmonic generation (THG)	$G_1(t) = E_{n-1}^+(t) = \frac{1}{2}\,\mathcal{E}_{n-1}^+(t)e^{i\omega_{0,n-1}t}$
$\tau_1 = \tau_2 = \tau$	$G_2(t) = G_1(t)$
$\omega_{0,n} = 3\omega_{0,n-1}$	$G_1(\omega) = E_{n-1}^+(\omega) = \frac{1}{2}\,\mathcal{E}_{n-1}^+(\omega - \omega_{0,n-1})$
	$G_2(\omega) = G_1(\omega)$
Polarization gating (PG)	$G_1(t) = E_{n-1}^+(t) = \frac{1}{2}\,\mathcal{E}_{n-1}^+(t)\,e^{i\omega_{0,n-1}t}$
$\tau_1 = \tau_2 = \tau$	$G_2(t) = E_{n-1}^-(t) = \frac{1}{2}\,\mathcal{E}_{n-1}^-(t)\,e^{-i\omega_{0,n-1}t}$
$\omega_{0,n} = \omega_{0,n-1}$	$G_1(\omega) = E_{n-1}^+(\omega) = \frac{1}{2}\,\mathcal{E}_{n-1}^+(\omega - \omega_{0,n-1})$
	$G_2(\omega) = E_{n-1}^-(\omega) = \frac{1}{2}\,\mathcal{E}_{n-1}^-(\omega + \omega_{0,n-1})$
Self diffraction (SD)	$G_1(t) = E_{n-1}^+(t) = \frac{1}{2}\,\mathcal{E}_{n-1}^+(t)\,e^{i\omega_{0,n-1}t}$
$\tau_1 = 0$ and $\tau_2 = \tau$	$G_2(t) = E_{n-1}^-(t) = \frac{1}{2}\,\mathcal{E}_{n-1}^-(t)\,e^{-i\omega_{0,n-1}t}$
$\omega_{0,n} = \omega_{0,n-1}$	$G_1(\omega) = E_{n-1}^+(\omega) = \frac{1}{2}\,\mathcal{E}_{n-1}^+(\omega - \omega_{0,n-1})$
	$G_2(\omega) = E_{n-1}^-(\omega) = \frac{1}{2}\,\mathcal{E}_{n-1}^-(\omega + \omega_{0,n-1})$
Transient grating (TG)	$G_1(t) = E_{n-1}^-(t) = \frac{1}{2}\,\mathcal{E}_{n-1}^-(t)\,e^{-i\omega_{0,n-1}t}$
$\tau_1 = 0$ and $\tau_2 = \tau$	$G_2(t) = E_{n-1}^+(t) = \frac{1}{2}\,\mathcal{E}_{n-1}^+(t)\,e^{i\omega_{0,n-1}t}$
$\omega_{0,n} = \omega_{0,n-1}$	$G_1(\omega) = E_{n-1}^-(\omega) = \frac{1}{2}\,\mathcal{E}_{n-1}^-(\omega + \omega_{0,n-1})$
	$G_2(\omega) = E_{n-1}^+(\omega) = \frac{1}{2}\,\mathcal{E}_{n-1}^+(\omega - \omega_{0,n-1})$

Note: The time delays and the energy conservation relations are indicated.

where we have made use of Parseval's theorem. In case the detector area is much larger than the beam area, the detector measures the pulse energy

$$Q = \eta \iint dA\, dt\, |f(\mathbf{x})\, E^+(t)|^2. \tag{5.23}$$

There exist some photodiodes that show a quadratic response on the incident intensity because two photons are required to cross the band gap and to generate a single electron–hole pair. In this case, the signal is

$$S \propto \int dt\, |E^+(t)|^4. \tag{5.24}$$

Such diodes can sometimes replace a second-harmonic generation crystal and a square law detector.

5.2 Single-Path Detection Schemes

In single-path detection schemes, as shown in Figure 5.5, the incoming pulse passes through a sequence of different filters before being measured by a square law detector.

We start with no filter (energy and power measurement), then proceed with one filter (spectrometer, intensity correlations), with two filters (sonogram and spectrogram, specifically FROG), and with three filters (STRUT) followed by a detector. Note that a filter that applies a pure phase to the pulse, for example, a time delay, has no effect whatsoever on the measurement results if it is the last element just before the square law detector.

5.2.1 Energy- and Power-Meter

Assume the pulse is incident on a square law detector, without passing through any filter at all. Further assume that the response time of the detector is so fast that it is possible to distinguish between individual pulses. Then the signal obtained is simply the pulse fluence or, if the detector area is much larger than the beam area, the pulse energy as described above. Should the pulse repetition rate of the laser be larger than the detector's response time, the square law detector will measure the average power that is given by the pulse energy multiplied by the laser repetition rate.

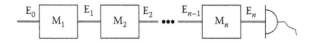

FIGURE 5.5 Arrangement of filters and a detector in a generic single-path detection scheme.

Spectrometer

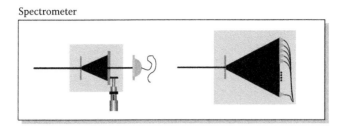

FIGURE 5.6 Schematic of a spectrometer realized through a spectral bandpass filter and a square law detector or a dispersive element and a square law detector array.

5.2.2 Spectrometer

A spectrometer is described by a single filter, that is, a spectral bandpass with finite or infinite resolution, followed by a square law detector. Assuming a Gauss-shaped bandpass characteristic centered around the bandpass frequency ω_s the detected signal is

$$S(\omega_s) \propto \int d\omega_1 \left| \exp\left[-\left(\frac{\omega_1 - \omega_s}{\delta\omega}\right)^2 \right] E_0^+(\omega_1) \right|^2, \tag{5.25}$$

and for the Gauss pulse we find

$$S_{\text{Gauss}}(\omega_s) \propto \exp\left[-2\frac{(\omega_s - \omega_0)^2}{\delta\omega^2 + \Delta\omega^2} \right]. \tag{5.26}$$

The width of the measured spectrum is broadened by the finite resolution and is given by $\sqrt{\delta\omega^2 + \Delta\omega^2}$. In ultrafast optics, most lasers have bandwidths in excess of 10 nm while spectrometers offer a resolution of 0.5 nm or better. Therefore, in many cases, it is well justified to assume infinite resolution for any monochromator or spectrometer used. Figure 5.6 shows the schematic of a spectrometer realized through a spectral bandpass filter and a square law detector. Today, the bandpass filter and the square law detector are mostly replaced by a dispersive element and a square law detector array.

5.2.3 *nth*-Order Intensity Correlation

Somewhat more interesting in terms of temporal pulse characterization are single time gates followed by a detector in which case the signal is

$$S(\tau) \propto \int dt_1 \, |G(t_1 - \tau) E_0^+(t_1)|^2. \tag{5.27}$$

This type of arrangement is called a correlation measurement, specifically a cross-correlation measurement if $G(t)$ is different from $E_0^+(t)$ and an autocorrelation measurement if $G(t) = E_0^+(t)$. These techniques were among the first used in ultrafast optics to characterize laser pulses [3]. In the following, we will inspect three commonly used examples. First, a time-gate based on second-harmonic generation. The measured signal is

$$S(\tau) \propto \int dt_1\ |E_0^+(t_1 - \tau)\ E_0^+(t_1)|^2 \propto \int dt_1\ I_0(t_1 - \tau)\ I_0(t_1). \qquad (5.28)$$

Such a measurement yields the so-called second-order background-free intensity autocorrelation. The detector measures a signal when the two replicas fully or partially overlap in time and no signal when they are well separated. For a linearly chirped Gauss pulse, we obtain

$$S_{\text{Gauss}}(\tau) \propto \exp\left[-\frac{\Delta\omega^2}{4(1 + C^2)}\tau^2\right]. \qquad (5.29)$$

The duration of the measured signal depends on the linear chirp C (but not its sign) and, therefore, some information on the Gauss pulse can be extracted. Figure 5.7 shows the simulated second-order background-free intensity autocorrelation for different input pulses.

A major disadvantage of this method is that the pulse duration extracted depends on the pulse shape assumed. This is because the ratio of the full width at half maximum

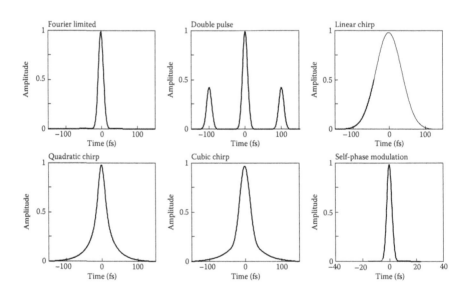

FIGURE 5.7 The second-order background-free intensity autocorrelation for a Gauss-shaped pulse with no phase modulation, a double pulse, a pulse with a quadratic, a cubic, and a quartic phase modulation, and a pulse that has experienced self-phase modulation.

TABLE 5.4 Temporal Intensity, FWHM of the Temporal Intensity, Autocorrelation Signal, FWHM of the Autocorrelation Signal, and the Ratio of the Two FWHM for Several Analytic Pulse Shapes

Pulse Shape	Intensity	FWHM$_I$	Autocorrelation	FWHM$_S$	FWHM$_S$/FWHM$_I$
Square	1	$2a$	$2a - |\tau|$	$2a$	1
Parabolic	$1 - \dfrac{t^2}{a^2}$	$\sqrt{2}a$	$\dfrac{32a^5 - |\tau|^5 + 20|\tau|^3 a^2 - 40\tau^2 a^3}{30a^4}$	$1.6226a$	1.1474
Cosine	$\cos\dfrac{\pi t}{2a}$	$4a/3$	$\dfrac{2\pi a \cos\xi - \pi|\tau|\cos\xi + 2a\sin\xi}{2\pi}$ $\xi \doteq \dfrac{\pi|\tau|}{2a}$	$1.5736a$	1.1802
Gauss	$\exp\left(-2\dfrac{t^2}{a^2}\right)$	$\sqrt{2\ln 2a}$	$\dfrac{\sqrt{\pi}a}{2}\exp\left(-\dfrac{\tau^2}{a^2}\right)$	$\sqrt{4\ln 2a}$	$\sqrt{2}$
Lorentz	$\dfrac{a^4}{(\tau^2 + a^2)^2}$	$2\sqrt{\sqrt{2}-1}a$	$\pi a^5 \dfrac{\tau^2 + 20a^2}{(\tau^2 + 4a^2)^3}$	$2.1290a$	1.6540
Hyp Sec	$\text{sech}^2\left(\dfrac{\tau}{a}\right)$	$1.7627a$	$\dfrac{4a\left[\dfrac{\tau}{a}\coth\left(\dfrac{\tau}{a}\right)-1\right]}{\sinh^2\left(\dfrac{\tau}{a}\right)}$	$2.7196a$	1.5429

Note: For the square, the parabolic, and the cosine pulse shape the intensity and the autocorrelation signal have a temporal support of $|t| \le a$ and $|\tau| \le 2a$, respectively.

(FWHM) of the temporal intensity and the FWHM of the autocorrelation function differs for different analytic pulse shapes. Several examples are listed in Table 5.4.

Note that the second-order intensity autocorrelation is symmetric with respect to time, that is, $S(\tau) = S(-\tau)$, with the effect that asymmetries in pulse shape or the sign of the phase cannot be determined without ambiguity. This problem is sometimes circumvented by converting the autocorrelation setup into a cross-correlation setup through introducing a piece of dispersive medium in one of the two beam paths. But even then it is not possible to extract the entire electric field information. However, such an intensity cross-correlation measurement together with a measurement of the fundamental spectrum can be used to determine the electric field through an optimization procedure [5]. The algorithm assumes that the fundamental spectrum $I_0(\omega_s)$ and the second-order intensity correlation $S_{exp}(\tau)$ have been measured. The unknown phase $\arg[E_n^+(t)]$ is optimized through the following steps:

$$(1) \quad E_{n+1}^+(\omega_s) = \sqrt{I_0(\omega_s)}\, e^{\arg[E_n^+(\omega_s)]} \tag{5.30}$$

$$(2) \quad G_{n+1}^+(\omega_s) = E_{n+1}^+(\omega_s)\, e^{-i\Phi} \tag{5.31}$$

$$(3) \quad E_{n+1}^+(t) = \int d\omega_s\, E_{n+1}^+(\omega_s)\, e^{i\omega_s t} \tag{5.32}$$

$$(4) \quad G_{n+1}^+(t) = \int d\omega_s \, G_{n+1}^+(\omega_s) \, e^{i\omega_s t} \tag{5.33}$$

$$(5) \quad S_{n+1}(\tau) = \int dt |E_{n+1}^+(t)|^2 \, |G_{n+1}^+(t-\tau)|^2, \tag{5.34}$$

where Φ is the phase modulation of the pulse after passing through the dispersive medium, until the deviation from the experimental trace is minimized, that is

$$RMS \doteq \sqrt{\frac{1}{N} \sum \left(S_{exp}(\tau) - S_{n+1}(\tau)\right)^2} = min. \tag{5.35}$$

With a double time gate based on third-harmonic generation and $\tau_1 = \tau_2 = \tau$, we obtain the third-order background-free intensity autocorrelation

$$S(\tau) \propto \int dt_1 \, |E_0^+(t_1-\tau)^2 \, E_0^+(t_1)|^2 \propto \int dt_1 \, I_0^2(t_1-\tau) \, I_0(t_1). \tag{5.36}$$

For a linearly chirped Gauss pulse, the third-order autocorrelation is

$$S_{Gauss}(\tau) \propto \exp\left[-\frac{\Delta\omega^2}{3(1+C^2)}\tau^2\right]. \tag{5.37}$$

Figure 5.8 shows the simulated third-order background-free intensity autocorrelation for different input pulses.

In principle, third-harmonic generation allows for two independent time delays in which case we obtain the so-called triple correlation [6]

$$S(\tau_1, \tau_2) \propto \int dt_1 \, I_0(t_1-\tau_1) \, I_0(t_1-\tau_2) \, I_0(t_1). \tag{5.38}$$

The triple correlation of a linearly chirped Gauss pulse is

$$S_{Gauss}(\tau_1, \tau_2) \propto \exp\left[-\frac{\Delta\omega^2}{3(1+C^2)} (\tau_1^2 - \tau_1\tau_2 + \tau_2^2)\right]. \tag{5.39}$$

Figure 5.9 shows the simulated triple correlation for different input pulses.

By measuring the triple correlation of an ultrashort laser pulse, it is possible to directly determine its temporal intensity; no *a priori* assumptions on the pulse shape are required. The recipe is as follows. From the triple correlation, the so-called bi-spectrum is calculated:

$$B(\omega_1, \omega_2) \propto \iint d\tau_1 d\tau_2 \, S(\tau_1, \tau_2) \, e^{-i\omega_1\tau_1 - i\omega_2\tau_2} \tag{5.40}$$

$$= A_0(\omega_1 + \omega_2) \, A_0(-\omega_1) \, A_0(-\omega_2), \tag{5.41}$$

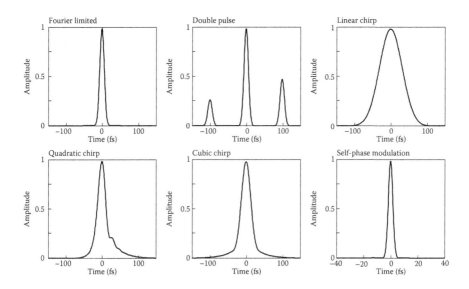

FIGURE 5.8 The third-order background-free intensity autocorrelation for a Gauss-shaped pulse with no phase modulation, a double pulse, a pulse with a quadratic, a cubic, and a quartic phase modulation, and a pulse that has experienced self-phase modulation.

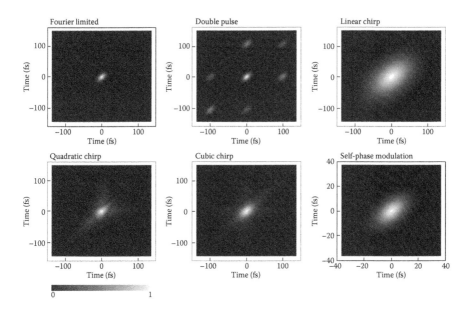

FIGURE 5.9 The triple correlation for a Gauss-shaped pulse with no phase modulation, a double pulse, a pulse with a quadratic, a cubic, and a quartic phase modulation, and a pulse that has experienced self-phase modulation.

with $A_0(\omega)$ being the Fourier transform of $I_0(t)$. Then, Equation 5.41 is used to determine $A_0(\omega)$ and an additional inverse Fourier transformation yields the temporal intensity $I_0(t)$. We start by separating Equation 5.41 in an amplitude and a phase part

$$|B(\omega_1, \omega_2)| = |A_0(\omega_1 + \omega_2)|\,|A_0(-\omega_1)|\,|A_0(-\omega_2)| \tag{5.42a}$$

$$\beta(\omega_1, \omega_2) = \phi(\omega_1 + \omega_2) + \phi(-\omega_1) + \phi(-\omega_2). \tag{5.42b}$$

Note that the intensity $I_0(t)$ is real and, therefore, we can make use of $A_0(-\omega) = A_0^*(\omega)$, that is, $|A_0(-\omega)| = |A_0(\omega)|$ and $\phi(-\omega) = -\phi(\omega)$. If we set $\omega_1 = \omega_2 \equiv 0$, we can determine the dc components to

$$|A_0(0)| = \sqrt[3]{B(0,0)} \tag{5.43a}$$

$$\phi(0) = 0. \tag{5.43b}$$

With Equations 5.43 and setting $\omega_2 \equiv 0$, the amplitude, and from arbitrary frequency pairs, the phase

$$|A_0(\omega_1)| = \sqrt{\frac{B(\omega_1, 0)}{A_0(0)}} \tag{5.44a}$$

$$\phi(\omega_1 + \omega_2) = \beta(\omega_1, \omega_2) + \phi(\omega_1) + \phi(\omega_2) \tag{5.44b}$$

for all nonzero frequencies can be extracted. The phase reconstruction requires some more explanation. Suppose the sampling is in discrete steps, that is, $\omega_1 = p\,\Delta\omega$ and $\omega_2 = q\,\Delta\omega$ with $p, q \in \mathcal{N}$. With this we see that

$$\phi_2 = \beta_{1,1} + 2\phi_1$$

$$\phi_3 = \beta_{2,1} + \phi_2 + \phi_1 = \beta_{2,1} + \beta_{1,1} + 3\phi_1$$

$$\vdots$$

$$\phi_n = \sum_{k=1}^{n-1} \beta_{k,1} + n\phi_1$$

holds. Apparently, the phase may be determined except for the linear phase $n\phi_1$. Because ϕ_1 is not known, we may arbitrarily set it to zero. The phase reconstruction can be done via several paths through the $\beta_{p,q}$ plane and the different paths may be used for averaging purposes. Once the amplitude $|A_0(\omega)|$ and the phase $\phi(\omega)$ are known, the temporal intensity $I_0(t)$ is obtained from a Fourier transformation. Combining this result with a measurement of the fundamental spectrum and a Gerchberg–Saxton algorithm [7] allows for an iterative reconstruction of the electric field of the laser pulse [8]. The algorithm assumes that the fundamental spectrum $I_0(\omega_s)$ has been measured and the intensity $I_0(t)$

has been reconstructed following the procedure outlined above and starts with a random phase $\arg[E_1^+(t)]$.

$$(1)\ E_{n+1}^+(t) = \sqrt{I_0(t)}\ e^{\arg[E_n^+(t)]} \tag{5.45}$$

$$(2)\ E_{n+1}^+(\omega_s) = \int dt\ E_{n+1}^+(t)\ e^{-i\omega_s t} \tag{5.46}$$

$$(3)\ E_{n+2}^+(\omega_s) = \sqrt{I_0(\omega_s)}\ e^{\arg[E_{n+1}(\omega_s)]} \tag{5.47}$$

$$(4)\ E_{n+2}^+(t) = \int d\omega_s\ E_{n+2}^+(\omega_s)\ e^{i\omega_s t} \tag{5.48}$$

If the algorithm converges after N iterations the unknown phase $\arg[E_N^+(t)]$ or $\arg[E_N^+(\omega_s)]$ can be readily extracted and the electric field of the laser pulse is known.

Figure 5.10 shows simplified schematics of a second-order intensity, a third-order intensity, and triple correlation measurement setup. In a second-order intensity autocorrelator, the incoming beam is split into two replicas and one of them is time delayed by a moveable retroreflector. Both replicas are then focused through the same lens in a suitable nonlinear crystal and the generated second-harmonic signal is measured by a square law detector. In a third-order intensity autocorrelator, the third-harmonic generation is usually cascaded in a second harmonic plus a sum frequency mixing process. The second-harmonic radiation produced in one arm is overlapped with the time-delayed fundamental pulse from the second arm to produce the sum frequency, that is the third harmonic, in a second nonlinear crystal. In a triple correlation measurement, the second-harmonic generation is done by two time-delayed pulses, the resulting second harmonic is combined with another time-delayed fundamental pulse to produce the third harmonic.

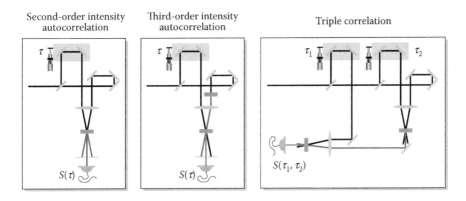

Second-order intensity autocorrelation

Third-order intensity autocorrelation

Triple correlation

FIGURE 5.10 Schematic of a second-order intensity, third-order intensity, and triple correlation measurement setup.

5.2.4 Sonogram

Next, we analyze a spectral bandpass filter $m(t)$ followed by a time gate $G(t)$. The signal is calculated as

$$S(\tau, \omega_s) \propto \int dt_1 \left| G(t_1 - \tau) \int dt_2 \, m(t_1 - t_2) \, E_0^+(t_2) \right|^2.$$

(5.49)

As an example, we assume that the time gate is SHG and that the duration of the original pulse is much shorter than the replica substantially broadened by the spectral bandpass, that is, we may safely approximate $G(t) \approx \delta(t)$. With this, Equation 5.49 simplifies to

$$S(\tau, \omega_s) \propto \left| \int dt_2 \, m(\tau - t_2) \, E_0^+(t_2) \right|^2.$$

(5.50)

Equation 5.50 represents a sonogram. For a Gauss-shaped spectral bandpass and a Gauss pulse, the sonogram is

$$S_{\text{Gauss}}(\tau, \omega_s) \propto \exp\left[-\frac{2}{(\Delta\omega^2 + \delta\omega^2)^2 + \delta\omega^4 C^2} \left(\begin{array}{c} \left[\Delta\omega^2 + \delta\omega^2(1 + C^2)\right] (\omega_s - \omega_0)^2 \\ -\Delta\omega^2 \delta\omega^2 C(\omega_s - \omega_0)\tau + \frac{1}{4}(\Delta\omega^4 \delta\omega^2 + \Delta\omega^2 \delta\omega^4)\,\tau^2 \end{array} \right) \right].$$

(5.51)

Note that a spectral filter with infinite resolution would yield a signal independent of τ. Consequently, a tunable broadband interference filter is better suited than a narrowband monochromator. Figure 5.11 shows the simulated sonogram for different input pulses.

Figure 5.12 shows a simplified schematic of a sonogram measurement setup. While one replica is frequency filtered the other is time delayed before both are focused in a nonlinear crystal.

5.2.5 Spectrogram: FROG

If the order of the two filters from the previous section is reversed, the signal is

$$S(\tau, \omega_s) \propto \int dt_1 \left| \int dt_2 \, m(t_1 - t_2) \, G(t_2 - \tau) E_0^+(t_2) \right|^2.$$

(5.52)

Here, the spectral filter is assumed to be a monochromator with infinite resolution and Equation 5.52 simplifies to

$$S(\tau, \omega_s) \propto \left| \int dt_1 \, G(t_1 - \tau) \, E_0^+(t_1) \, e^{-i\omega_s t_1} \right|^2.$$

(5.53)

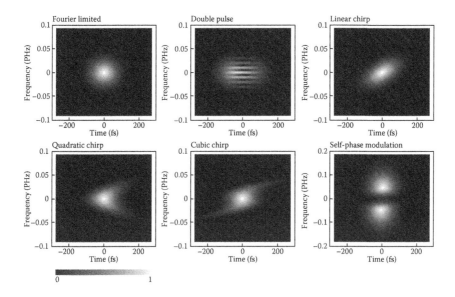

FIGURE 5.11 The sonogram for a Gauss-shaped pulse with no phase modulation, a double pulse, a pulse with a quadratic, a cubic, and a quartic phase modulation, and a pulse that has experienced self-phase modulation.

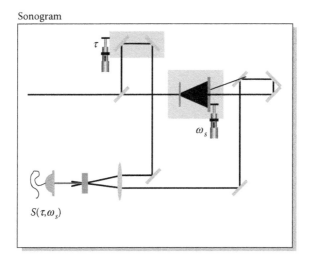

FIGURE 5.12 Schematic of a setup measuring a sonogram.

Upon comparison, we realize that Equation 5.53 is identical to a spectrogram, assuming that $G(t)$ is equivalent to the gate function $h(t)$. In the ultrafast community, such a measurement is well known under the acronym FROG which stands for frequency-resolved optical gating. From what was said above on joint time–frequency distributions, it is clear that no direct electric field reconstruction is possible. However, it has been shown that iterative algorithms may be used to retrieve the original electric field [4]. One of the simplest algorithms works as follows. We define

$$\xi(\tau, t_1) \doteq G(t_1 - \tau) \, E_0^+(t_1) \tag{5.54}$$

and its one-time Fourier transform

$$\xi(\tau, \omega_s) = \int dt_1 \, \xi(\tau, t_1) \, e^{-i\omega_s t_1}. \tag{5.55}$$

From Equations 5.53 and 5.55 follows that the spectrogram is given by $S(\tau, \omega_s) \propto |\xi(\tau, \omega_s)|^2$. With the help of Equation 5.54, the electric field can be obtained through

$$E_0^+(t) \propto \int d\tau \, \xi(\tau, t), \tag{5.56}$$

with $\int d\tau \, G(t - \tau)$ yielding a constant. In each loop, the iterative algorithm determines the $(n + 1)$th electric field from the nth electric field by running through the following steps:

(1) $E_n^+(t)$ $\tag{5.57}$

(2) $\xi_n(\tau, t) = G_n(t - \tau) \, E_n^+(t)$ $\tag{5.58}$

(3) $\xi_n(\tau, \omega_s) = \int dt \, \xi_n(\tau, t) \, e^{-i\omega_s t}$ $\tag{5.59}$

(4) $\xi_{n+1}(\tau, \omega_s) = \sqrt{S(\tau, \omega_s)} \, \exp\left[i \arg(\xi_n(\tau, \omega_s))\right]$ $\tag{5.60}$

(5) $\xi_{n+1}(\tau, t) = \int d\omega_s \, \xi_{n+1}(\tau, \omega_s) \, e^{i\omega_s t}$ $\tag{5.61}$

(6) $E_{n+1}(t) = \int d\tau \, \xi_{n+1}(\tau, t)$ $\tag{5.62}$

In each loop, the square root of the measured FROG trace $\sqrt{S(\tau, \omega_s)}$ replaces the amplitude of $\xi_n(\tau, \omega_s)$ while keeping its phase (step 4). The last two steps may be interchanged, thereby saving a lot of computation time as the number of Fourier transforms is reduced to one. In most cases, the electric field converges to the sought-after electric field. Notoriously problematic cases are double pulses or multipulse waveforms where at least two subpulses have similar amplitudes. For those pulses, more sophisticated iterative algorithms must be used.

If the time gate is second-harmonic generation, the signal is the SHG FROG:

$$S(\tau, \omega_s) \propto \left| \int dt \, E_0^+(t - \tau) \, E_0^+(t) \, e^{-i\omega_s t} \right|^2, \tag{5.63}$$

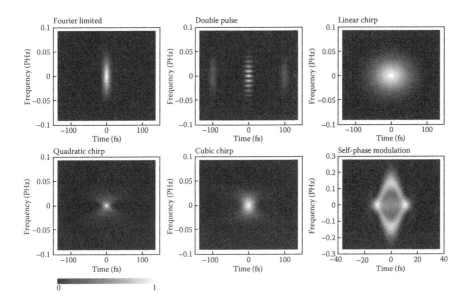

FIGURE 5.13 The SHG FROG traces for a Gauss-shaped pulse with no phase modulation, a double pulse, a pulse with a quadratic, a cubic, and a quartic phase modulation, and a pulse which has experienced self-phase modulation.

which for the linearly chirped Gauss pulse is

$$S_{Gauss}(\tau, \omega_s) \propto \exp\left[-\frac{\Delta\omega^2}{4(1+C^2)}\tau^2 - \frac{(\omega_s - 2\omega_0)^2}{\Delta\omega^2}\right]. \tag{5.64}$$

The SHG FROG trace is centered around the second-harmonic frequency $2\omega_0$ and symmetric with respect to $\tau = 0$. The linear chirp C affects the width of the trace along the time axis but does not change the spectral width. Figure 5.13 shows the simulated SHG FROG for different input pulses.

Other time gates may be used instead leading to expressions for the PG FROG

$$S_{Gauss}(\tau, \omega_s) \propto \exp\left[\begin{array}{c} -\dfrac{(3+C^2)\Delta\omega^2}{(9+C^2)(1+C^2)}\tau^2 + \dfrac{4C}{9+C^2}\tau(\omega_s - \omega_0) \\ -\dfrac{6(1+C^2)}{(9+C^2)\Delta\omega^2}(\omega_s - \omega_0)^2 \end{array}\right], \tag{5.65}$$

the SD or TG FROG

$$S_{Gauss}(\tau, \omega_s) \propto \exp\left[-\frac{3\Delta\omega^2}{(9+C^2)}\tau^2 + \frac{8C}{9+C^2}\tau(\omega_s - \omega_0) - \frac{6(1+C^2)}{(9+C^2)\Delta\omega^2}(\omega_s - \omega_0)^2\right],$$

$$\tag{5.66}$$

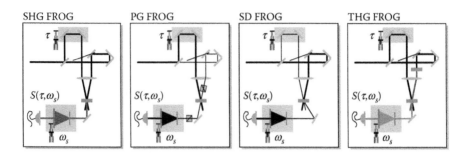

FIGURE 5.14 Schematic of different FROG measurement schemes.

or the THG FROG

$$S_{\text{Gauss}}(\tau, \omega_s) \propto \exp\left[-\frac{\Delta\omega^2}{3(1 + C^2)}\tau^2 - \frac{2}{3\Delta\omega^2}(\omega_s - 3\omega_0)^2\right]. \qquad (5.67)$$

Figure 5.14 shows simplified schematics of an SHG, a PG, an SD, and a THG FROG measurement setup. The SHG FROG is a variation of the second-order background-free autocorrelation setup with a spectrometer introduced just before the detector. The PG FROG uses a transient birefringence caused by the intense, time-delayed pulse as a time gate. The polarization of the weaker pulse is rotated by 45° with respect to the strong pulse and is passed through two crossed polarizers with a thin glass plate in between. If the strong pulse does not overlap with the weak pulse, the transmission through the second polarizer is zero. If the two pulses overlap, the transient birefringence causes a proportional change in the state of polarization and the transmission through the second polarizer is nonzero. The SD FROG makes use of a transient "prism" which is generated by the strong pulse in a thin glass plate. If the two pulses overlap, this "prim" causes a refraction of the weaker pulse, which is then analyzed by a monochromator followed by a square law detector. The THG FROG typically makes use of a cascaded third-order non-linearity. One replica is frequency doubled before it is overlapped with the time-delayed third pulse in a second nonlinear crystal in which the sum frequency of the fundamental and the second harmonic, that is, the third harmonic, is generated and subsequently analyzed.

If the time gate is an ultrafast time shutter, the signal becomes

$$S(\tau, \omega_s) \propto \left|\int dt_1 \, \Theta(\tau - t_1) \, E_0^+(t_1) \, e^{-i\omega_s t_1}\right|^2, \qquad (5.68)$$

which is identical to Equation 5.17 and, thus, the Page distribution can be derived from such a measurement. Moreover, it is possible to directly invert the Page distribution and to extract the electric field [9].

5.2.6 Spectrally and Temporally Resolved Frequency Conversion: STRUT

The only configuration with a series of three filters analyzed here is a spectral band pass (typically an interference filter) followed by a time gate and a second spectral bandpass (typically a monochromator) [11]. For an SHG time gate and assuming infinite resolution for the monochromator, the signal is

$$S(\tau, \omega_{s3}) \propto \left| \int d\omega_1 \, E_0^+(\omega_{s3} - \omega_1) \, E_0^+(\omega_1) \, m_1(\omega_1 - \omega_{s1}) \, e^{-i(\omega_{s3}-\omega_1)\tau} \right|^2, \quad (5.69)$$

where $m_1(\omega - \omega_{s1})$ describes the bandpass characteristic of the first spectral bandpass. Separating the electric field in amplitude and phase, that is, $E_0^+(\omega) = |E_0^+(\omega)| \exp[i\phi(\omega)]$, we find approximately

$$S(\tau, \omega_{s3}) \approx \left| E_0^+(\omega_{s3} - \omega_{s1}) \, E_0^+(\omega_{s1}) \right|^2 \left| m_1\left(\tau + \left. \frac{d\phi}{d\omega} \right|_{\omega_{s3}} \right) \right|^2. \quad (5.70)$$

For a fixed ω_{s1}, we obtain a two-dimensional distribution $S(\tau, \omega_{s3})$ and the maxima of the distribution appear at delays where the spectral slice temporally overlaps with the corresponding portion of the reference spectrum. The spectral phase $\phi(\omega)$ of the original

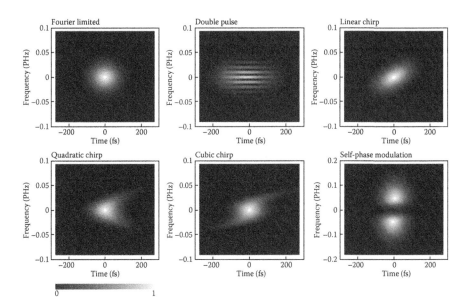

FIGURE 5.15 The STRUT traces for a Gauss-shaped pulse with no phase modulation, a double pulse, a pulse with a quadratic, a cubic, and a quartic phase modulation, and a pulse that has experienced self-phase modulation.

STRUT

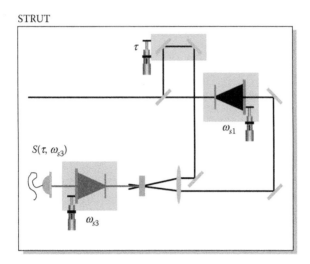

FIGURE 5.16   Schematic of a STRUT measurement scheme.

pulse can be readily extracted because the delay $\tau_{max}(\omega)$ where the maximum signal appears is determined by $d\phi/d\omega$. Contrary to an SHG FROG measurement, there is no time ambiguity, that is, the so-called STRUT trace is generally asymmetric with respect to time and the phase is obtained through integration. Figure 5.15 shows the simulated STRUT traces for different input pulses.

Figure 5.16 shows a simplified schematic of a STRUT measurement setup. The incident pulse is split in two replica, one being time delayed and the other spectrally filtered before being focused to the nonlinear crystal. The second harmonic generated in the crystal is spectrally analyzed by a spectrometer.

5.3 Two-Path Detection Schemes

In principle, two-path detection schemes are nothing but variations of an interferometer. The incoming electric field is split in two replicas by a beam splitter. Each replica passes through one arm of the interferometer before both are recombined at a second beam splitter and measured by the detector. Each path A and B may contain a number of different filters. The recombination is such that the two replica propagate parallel and their individual electric fields add coherently. After the second beam splitter, the beam may pass through further filters (Figure 5.17).

5.3.1 Linear Interferometry

The simplest two-path detection scheme is linear interferometry with a time delay in one of the two arms. The detector measures the sum of the original and the time-delayed

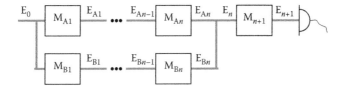

FIGURE 5.17 Arrangement of filters in a two-path detection scheme.

electric field, that is

$$S(\tau) \propto \int dt_1 \, |E_0^+(t_1) + E_0^+(t_1 - \tau)|^2 \tag{5.71}$$

$$\propto \int dt_1 \, |E_0^+(t_1)|^2 + \int dt_1 \, |E_0^+(t_1 - \tau)|^2 + 2\int dt_1 \, \Re\left[E_0^+(t_1)E_0^-(t_1 - \tau)\right]. \tag{5.72}$$

The first two terms are equal and contribute a constant background irrespective of time delay τ. The third term $S_3(\tau)$ is more interesting and can be rewritten in the form

$$S_3(\tau) \propto \frac{1}{\pi} \int d\omega_1 \, |E_0^+(\omega_1)|^2 \, \cos(\omega_1 \tau). \tag{5.73}$$

That is, the interference term $S_3(\tau)$ is related to the Fourier (cosine) transform of the spectral intensity $|E_0^+(\omega)|^2$. Of course, this is the basis of Fourier transform spectroscopy, where $S_3(\tau)$ is measured and the spectrum is obtained by inverse Fourier transformation of the measured signal. If we invoke the slowly varying envelope approximation, the interference term becomes

$$S_3(\tau) \propto \cos(\omega_0 \tau) \int dt_1 \, \Re\left[\mathcal{E}_0^+(t_1)\mathcal{E}_0^+(t_1 - \tau)\right]. \tag{5.74}$$

Now, it is obvious why $S_3(\tau)$ is labeled interference term, it oscillates as a function of time delay with the center frequency of the pulse. For the Gauss pulse with an initial linear chirp, we obtain

$$S_{\text{Gauss}}(\tau) \propto 1 + \exp\left[-\frac{\Delta\omega^2\tau^2}{8}\right] \cos(\omega_0 \tau). \tag{5.75}$$

No influence of the chirp is observed, that is to say the only information one can gain from such an interference measurement is the spectral amplitude.

5.3.2 *n*th-Order Interferometric Correlation

Next, we discuss an interferometer, that is, two arms with a time delay in one, followed by a time gate and a detector. Note that the intrinsic time delay of the time gate is set to zero since it is not needed.

$$S(\tau) \propto \int dt_1 \left| G(t_1) \left[E_0^+(t_1) + E_0^+(t_1 - \tau) \right] \right|^2. \tag{5.76}$$

If the gate pulse is identical to the incident pulse, this technique is called second-order interferometric autocorrelation; otherwise, interferometric cross-correlation. The interferometric autocorrelation signal is

$$S(\tau) \propto \int dt_1 \left| \left[E_0^+(t_1) + E_0^+(t_1 - \tau) \right]^2 \right|^2. \tag{5.77}$$

After some rearrangements, Equation 5.77 may be rewritten as

$$\begin{aligned}
S(\tau) \propto & \int dt_1 \left| E_0^+(t_1) \right|^4 + \int dt_1 \left| E_0^+(t_1 - \tau) \right|^4 \\
& + 4 \int dt_1 \left| E_0^+(t_1) \right|^2 \left| E_0^+(t_1 - \tau) \right|^2 \\
& + 4 \int dt_1 \left[\left| E_0^+(t_1) \right|^2 \left| E_0^+(t_1 - \tau) \right|^2 \right] \Re \left\{ E_0^+(t_1) E_0^-(t_1 - \tau) \right\} \\
& + 2 \int dt_1 \, \Re \left\{ E_0^{+2}(t_1) E_0^{-2}(t_1 - \tau) \right\}.
\end{aligned} \tag{5.78}$$

The contrast ratio, that is, $S(0)/S(\pm\infty)$, is 8:1. More generally, an *n*th-order interferometric measurement yields a contrast of 2^{2n-1}:1. For the linearly chirped Gauss pulse, we find

$$\begin{aligned}
S_{\text{Gauss}}(\tau) = & 1 + 2 \exp \left[-\frac{\Delta\omega^2}{4(1 + C^2)} \tau^2 \right] \\
& + 4 \exp \left[-\frac{(3 + C^2)\Delta\omega^2}{16(1 + C^2)} \tau^2 \right] \cos \left[\frac{\Delta\omega^2 C \tau^2}{8(1 + C^2)} \right] \cos(\omega_0 \tau) \\
& + \exp \left[-\frac{\Delta\omega^2}{4} \tau^2 \right] \cos(2\omega_0 \tau).
\end{aligned} \tag{5.79}$$

The interferometric autocorrelation has a dc contribution and two oscillating parts; one oscillates at the center frequency of the fundamental and the other at the center frequency of the second harmonic. Figure 5.18 shows the simulated second-order interferometric autocorrelation for different input pulses.

It is not possible to extract the entire electric field information from an interferometric measurement alone. However, an interferometric autocorrelation measurement together

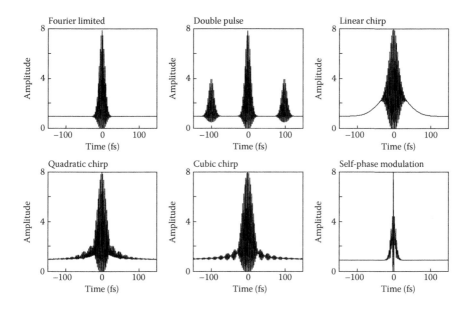

FIGURE 5.18 The second-order interferometric autocorrelation for a Gauss-shaped pulse with no phase modulation, a double pulse, a pulse with a quadratic, a cubic, and a quartic phase modulation, and a pulse that has experienced self-phase modulation.

with a measurement of the fundamental spectrum can be used to determine the electric field through an iterative procedure [11]. When the time gate is THG, the third-order interferometric autocorrelation is obtained:

$$S(\tau) \propto \int dt_1 \left| \left[E_0^+ (t_1) + E_0^+ (t_1 - \tau) \right]^3 \right|^2. \tag{5.80}$$

For the Gauss pulse, we obtain

$$S_{\text{Gauss}}(\tau) = 1 + 9 \, \exp\left[-\frac{\Delta\omega^2}{3(1+C^2)} \, \tau^2 \right] + 9 \, \exp\left[-\frac{(9+C^2)\Delta\omega^2}{24(1+C^2)} \, \tau^2 \right] \cos(\omega_0 \tau)$$

$$+ 6 \, \exp\left[-\frac{(5+C^2)\Delta\omega^2}{24(1+C^2)} \, \tau^2 \right] \cos\left[\frac{C\Delta\omega^2 \tau^2}{6(1+C^2)} \right] \cos(\omega_0 \tau)$$

$$+ 6 \, \exp\left[-\frac{(2+C^2)\Delta\omega^2}{6(1+C^2)} \, \tau^2 \right] \cos\left[\frac{C\Delta\omega^2 \tau^2}{6(1+C^2)} \right] \cos(2\omega_0 \tau)$$

$$+ \exp\left[-\frac{3\Delta\omega^2}{8} \, \tau^2 \right] \cos(3\omega_0 \tau). \tag{5.81}$$

In addition to the oscillations at the fundamental and the second-harmonic frequencies, we also find a contribution oscillating at the third-harmonic frequency. Figure 5.19 shows the simulated third-order interferometric autocorrelation for different input pulses.

Figure 5.20 shows the simplified schematics of a second- and a third-order interferometric correlation measurement. The two time-delayed replicas are combined collinear and are focused in a nonlinear crystal where they generate the second- and the third-harmonic radiation, respectively.

5.3.3 Spectral Interferometry

In spectral interferometry, the two paths of an interferometer are recombined and the resulting field is analyzed by a monochromator. In such an arrangement, the measured signal is a function of the time delay and the spectral bandpass frequency, that is,

$$S(\tau, \omega_s) \propto \int d\omega_1 \left| m(\omega_1 - \omega_s) \left[E_0^+(\omega_1) + E_0^+(\omega_1) e^{-i\omega\tau} \right] \right|^2. \tag{5.82}$$

Assuming infinite spectral resolution, that is, $m(\omega_1 - \omega_s) = \delta(\omega_1 - \omega_s)$, yields

$$S(\tau, \omega_s) \propto \left| E_0^+(\omega_s) \right|^2 \cos^2\left(\frac{\omega_s \tau}{2} \right). \tag{5.83}$$

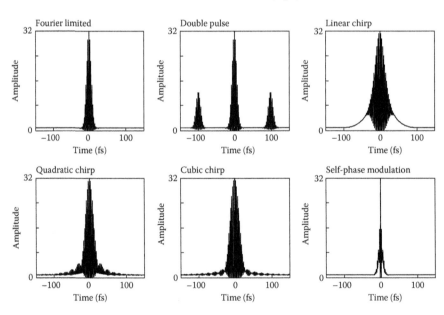

FIGURE 5.19 The third-order interferometric autocorrelation for a Gauss-shaped pulse with no phase modulation, a double pulse, a pulse with a quadratic, a cubic, and a quartic phase modulation, and a pulse that has experienced self-phase modulation.

Second-order interferometric
autocorrelation

Third-order interferometric
autocorrelation

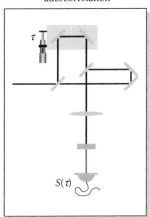
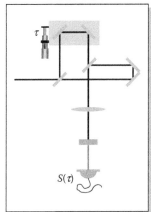

FIGURE 5.20 Schematic of a second-order and a third-order interferometric correlation measurement setup.

The spectrum is that of the original pulse modulated with a squared cosine function. The modulation frequency is proportional to the time delay τ and, thus, such a measurement is an easy linear method to determine the time delay between two pulses. It may seem surprising that time delays much greater than the pulse duration can be measured. The reason is that the pulse after the spectral filter, that is, the monochromator, is much longer as its spectral bandwidth is substantially reduced. The maximum time delay, which can be measured, is consequently determined by the spectral resolution of the monochromator and not the pulse duration.

5.3.4 Spectral Shearing Interferometry: SPIDER

In spectral shearing interferometry, one arm of the interferometer produces a frequency-shifted replica (frequency shift α) and the second arm produces a time-delayed replica (fixed time delay τ). The resulting field is analyzed by a monochromator and a detector.

$$S(\omega_s) \propto \int \mathrm{d}\omega_1 \, \left| m(\omega_1 - \omega_s) \left[E_0^+ (\omega_1 - \alpha) + E_0^+(\omega_1)\mathrm{e}^{-\mathrm{i}\omega_1 \tau} \right] \right|^2 . \qquad (5.84)$$

Assuming infinite spectral resolution yields

$$S(\omega_s) \propto \left| E_0^+ (\omega_s) \right|^2 + \left| E_0^+ (\omega_s - \alpha) \right|^2 + 2 \left| E_0^+ (\omega_s) \right|$$
$$\left| E_0^+ (\omega_s - \alpha) \right| \, \cos[\omega_s \tau - \phi(\omega_s) + \phi(\omega_s - \alpha)]. \qquad (5.85)$$

SPIDER

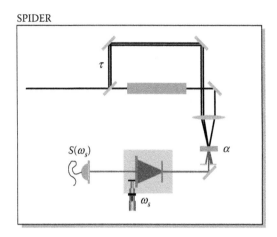

FIGURE 5.21   Schematic of a SPIDER measurement setup.

If the spectral shear is small, that is $|E_0^+(\omega_s)| \approx |E_0^+(\omega_s - \alpha)|$ and $\phi(\omega_s - \alpha) = \phi(\omega_s) - d\phi/d\omega\,\alpha + \cdots$, we find

$$S(\omega_s) \propto \left|E_0^+(\omega_s)\right|^2 \left[1 + \cos\left(\omega_s\tau + \left.\frac{d\phi}{d\omega}\right|_{\omega_s} \alpha\right)\right]. \tag{5.86}$$

The differential phase can be extracted by Fourier transforming $S(\omega_s)$ and by filtering out the contribution around the time delay τ before Fourier transforming back to the frequency domain. The spectral phase $\phi(\omega_s)$ then follows from integrating $d\phi/d\omega$. Spectral shearing is at the heart of a short pulse characterization technique called SPIDER [12]. The trick in SPIDER is the technique through which the spectral shear is realized. This is done by mixing two time-delayed replicas of the incident pulse with a strongly chirped copy of the pulse. Due to the large linear chirp, each replica is effectively mixed with a different quasi-monochromatic frequency component of the chirped pulse and as a result the center frequencies of the two upconverted replicas are different; just as it is required by spectral shearing interferometry. Figure 5.21 shows a simplified schematics of a SPIDER setup. The first beam splitter produces a pair of reflected pulses separated by a time delay, which is given by the thickness of the beam splitter. The transmitted pulse passes through a dispersive element producing a stretched and strongly phase-modulated replica of the incident pulse. The pulse pair and the stretched replica are focused in a type II nonlinear crystal and the generated second-harmonic spectrum is recorded by a spectrometer.

5.4 Summary

After reading this chapter, you should

- Know how to calculate the measured signal for virtually any experimental setup consisting of filters based on the transfer functions given in Table 5.2

TABLE 5.5 Comparison of the Different Techniques

Method	Order	Spectral Amplitude	Spectral Phase	Temporal Amplitude	Temporal Phase	Exact	Comments
Spectrometer	1	Yes	No	No	No	Yes	
Second-order AC	2	No	No	Estimate	No	No	Time ambiguity
Third-order AC	3	No	No	Estimate	No	No	
Triple correlation	3	No	No	Yes	No	Yes	Time consuming
Sonogram	2						
SHG FROG	2	Yes	Yes	Yes	Yes	Iterative	Time ambiguity
PG FROG	3	Yes	Yes	Yes	Yes	Iterative	
SD/TG FROG	3	Yes	Yes	Yes	Yes	Iterative	
THG FROG	3	Yes	Yes	Yes	Yes	Iterative	
Time shutter	n	Yes	Yes	Yes	Yes	Yes	Difficult to realize
STRUT	2	Yes	Yes	Yes	Yes	Yes	Time consuming
Interferometer	1	Yes	No	No	No	Yes	
Second-order IAC	2	No	No	Estimate	No	No	Time ambiguity
Third-order IAC	3	No	No	Estimate	No	No	
SPIDER	2	Yes	Yes	Yes	Yes	Yes	Settings must be carefully adjusted

- Be able to understand and build your version of the most commonly used ultrafast characterization tools
- Understand what information on the electric field of an ultrashort laser pulse can be extracted from a specific measurement technique

It is not my intent to praise one technique over the others, all have their niche of application. Some are very simple to implement but yield limited information. If this limited information is all you need, do not make the setup more complicated than necessary. In Table 5.5, I summarize the different techniques with some basic remarks. A general comment is that the higher the nonlinear process, the more intense pulses are required to generate a detectable signal.

Something I have not mentioned here is that many of the techniques that require a variable-frequency filter or a variable time delay can be parallelized by some modifications of the experimental setup. The mathematics and information extraction remain unaffected. Frequency filters just before a square law detector can be replaced by a spectrometer equipped with a diode array or a CCD camera, as mentioned above. Variable time delays can be parallelized by a technique called time-to-space mapping in which the time delay is mapped onto a spatial axis. The easiest way to realize such a scheme is by overlapping two line foci with an angle in between.

References

1. Gase, R. 1991. Time-dependent spectrum of linear optical systems, *J. Opt. Soc. Am. A* 8: 850.
2. Cohen, L. 1989. Time-frequency distributions—a review, *Proc. IEEE* 77: 941.

3. Trebino, R., 2000. *Frequency-Resolved Optical Gating: The Measurement of Ultrashort Laser Pulses.* Boston: Kluwer Academic Publishers.

4. Weber, H.P. 1967. Method for pulsewidth measurement of ultrashort light pulses generated by phase-locked lasers using nonlinear optics, *J. Appl. Phys.* 38: 2231.

5. Nicholson, J.W., Jasapara, J., Rudolph, W., Omenetto, F.G., and Taylor, A.J. 1999. Full-field characterization of femtosecond pulses by spectrum and cross-correlation measurements, *Opt. Lett.* 24: 1774–76.

6. Feurer, T., Niedermeier, S., and Sauerbrey, R. 1998. Measuring the temporal intensity of ultrashort laser pulses by triple correlation, *Appl. Phys. B* 66: 163–68.

7. Gerchberg, R.W. and Saxton, W.O. 1971. Phase determination for image and diffraction plane pictures in the electron microscope, *Optik* 34: 275.

8. Liu, T.-M., Huang, Y.-C., Chern, G.-W., Lin, K.-H., Lee, C.-J., Hung, Y.-C., and Suna, C.-K. 2002. Triple-optical autocorrelation for direct pulse-shape measurement, *Appl. Phys. Lett.* 81: 1402–4.

9. Michelmann, K., Wagner, U., Feurer, T., Teubner, U., Foerster, E., and Sauerbrey, R. 2001. Measurement of the Page function of an ultrashort laser pulse, *Opt. Commun.* 198: 163–170.

10. Chilla, J.L.A. and Martinez, O.E. 1991. Direct determination of the amplitude and phase of femtosecond light pulses, *Opt. Lett.* 16: 39.

11. Diels, J.-C.M., Fontaine, J.J., McMichael, I.C., and Simoni, F. 1985. Control and measurement of ultrashort pulse shapes (in amplitude and phase) with femtosecond accuracy, *Appl. Opt.* 24: 1270.

12. Iaconis, C. and Walmsley, I.A. 1998. Spectral phase interferometry for direct electric-field reconstruction of ultrashort optical pulses, *Opt. Lett.* 23: 792–4.

6

Generation of Laser Beams by Digital Holograms

Jeffrey A. Davis
San Diego State University

Ignacio Moreno
Universidad Miguel

6.1 Introduction

In this chapter, we review the use of spatial light modulators SLMs to display reconfigurable digital holograms capable of modifying and shaping laser beams. In order to achieve this goal, we concentrate on liquid-crystal displays (LCDs) as SLM devices. We will review the physical properties of these devices, and how to achieve specific optical modulation configurations. Our major goal is to achieve phase-only modulation and we examine their diffraction efficiency implications. We then introduce an encoding technique that allows amplitude information to be encoded onto phase-only filters.

175

This technique permits the design of reconfigurable digital holograms, useful for beam-shaping applications, in particular for the generation of laser modes. The same technique can be applied to two orthogonal polarization components to additionally allow a full control of the state of polarization. This approach should then be useful for the generation of polarized laser beams.

6.2 Liquid-Crystal Displays

LCDs are electro-optical devices that change the polarization of light [1,2] when controlled by an applied voltage. Although initially only expensive SLMs were available such as the Hughes liquid-crystal light valve, the field was energized when *Radio Shack* introduced [3] the first low-cost device in 1985. Since then, twisted-nematic liquid-crystal displays (TN-LCDs) have become common components of cell phones, commercial displays, and projection devices. Because they can be changed in real time, they are very useful as programmable diffractive optical elements, and they have been employed in optics and photonics laboratories for numerous applications such as beam shaping, optical trapping and manipulation, optical image processing, optical pattern recognition, and so on [3–7]. There are also new applications where the LCD can control the two-dimensional polarization states of the output beam. Other less common types of LCDs are zero-twist (or parallel-aligned-nematic) LCDs and ferroelectric LCDs. The zero-twist devices are less commercially available, but they are easier to operate for many of the applications mentioned above. Ferroelectric LCDs are faster devices than nematic LCDs, but restricted to binary modulation modes.

Currently, LCD devices specifically fabricated to develop SLM applications in these fields, with many different characteristics of modulation range, resolution, speed, and so on, are commercially available from a number of companies such as Meadowlark, Holoeye, Hamamatsu, Forth-Dimension, among others (Figure 6.1).

In the usual mode of operation, LCDs are inserted between two linear polarizers in order to achieve intensity modulation. For display applications, high contrast intensity

FIGURE 6.1   Picture of some commercial liquid-crystal SLM devices. (Courtesy of Holoeye.)

modulation can be easily obtained, and the phase modulation is not a concern. However, when used for displaying diffractive elements and digital holograms, phase modulation is usually more important than amplitude modulation [8–13], and a more general fully complex modulation of both amplitude and phase is most desirable.

Different LCD technologies are capable of producing different types of optical modulation onto a coherent light beam. Figure 6.2 shows a typical graphical representation, where the modulation domains are represented in the complex plane [14]. In principle, the realization of general diffractive optical elements requires a full complex modulation capability in the display (Figure 6.2a). However, this is not possible to achieve in a direct way, and a combination of two displays would usually be required.

Phase-only modulation (Figure 6.2b) is much more useful for several reasons: (1) it can be directly generated with nematic LCDs (with a linear polarization configuration in parallel aligned displays, or with an elliptical polarization configuration in TN displays), (2) it provides the maximum transmission through the diffractive element, and (3) other types of modulation can be realized from phase-only modulation by means of simple Fourier analysis.

In addition, full complex modulation can be encoded onto a device with phase-only modulation, as we show in the next section. Full phase-only modulation is obtained when the maximum modulation depth reaches 2π radians. Then the circle is completed in the complex diagram (Figure 6.2b).

Amplitude-only modulation (Figure 6.2c) can be approximately obtained with TN displays. Other devices, such as ferroelectric LCDs or magneto-optic devices are capable of producing binary phase (Figure 6.2d) or binary amplitude (Figure 6.2e) modulation.

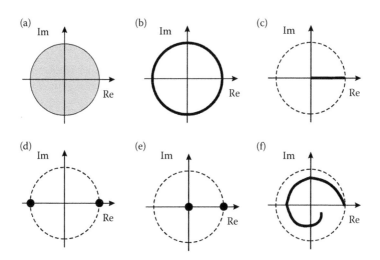

FIGURE 6.2 Representation in the complex plane of different coding domains useful for optical image processing. (a) Fully complex. (b) Phase-only. (c) Amplitude only. (d) Binary phase. (e) Binary amplitude. (f) Coupled.

In TN-LCDs, a restricted coupled modulation of both the amplitude and phase of the transmitted light is in general achieved, leading to a curved trajectory in the complex plane (Figure 6.2f). Both binary-amplitude and binary phase modulation can also be obtained with TN-LCDs using two properly oriented linear polarizers.

Next, we will review the construction of the LCD and a Jones matrix model that allows examination of the various ways of operating these devices. We will focus on nematic LCDs since they are the most common type and they can provide phase-only modulation.

6.3 Optical Phase Modulation in LCDs

Nematic LCDs are constructed as a two-dimensional array of pixels, each of which is filled with nematic liquid-crystal materials. These molecules can be represented by a cigar-shaped index ellipsoid having an extraordinary index of refraction n_e and an ordinary index of refraction n_o [15].

The orientations of the extraordinary indices of refraction for the liquid-crystal molecules (called the *director* axes) are fixed at both sides by alignment grooves etched onto the glass plates that constrain the liquid-crystal molecules. These alignment grooves can be made parallel on each side of the device, producing a parallel-aligned or zero-twist device (Figure 6.3a), or can be rotated by a twist angle, that is usually about 90° producing a TN device (Figure 6.3b). In the first case the device acts as a linear wave-plate. However, in the second case, the TN liquid-crystal structure is created, where the orientation of the molecules rotates from the input director axis to the output director axis.

TNLC SLM are inexpensive SLMs. Their original use was for display purposes where intensity modulation is important. However, for applications such as programmable digital holography and diffractive optics, the transmitted intensity must remain constant while the optical birefringence is varied, that is a phase-only modulation response is more adequate. This is difficult to achieve with TN devices because their transmission consists of coupled amplitude and phase modulation, and require configurations with elliptically polarized light. On the contrary, phase-only operation can be easily achieved using devices with the parallel aligned or zero-twist displays.

In both types, a voltage is applied across each liquid-crystal pixel through transparent ITO electrodes, and a computer typically controls this voltage through a video card that

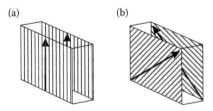

(a)   (b)

FIGURE 6.3 Representation of (a) a parallel-aligned and (b) a twisted-nematic liquid-crystal display.

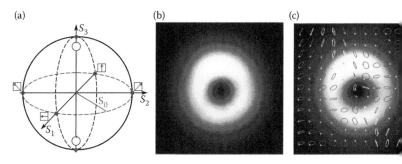

FIGURE 4.4 (a)–(f) Phase modulations $\Psi(\mathbf{r})$ for measuring the mode powers of LP_{01}, LP_{12e}, LP_{22e}, LP_{61e}, and LP_{13e} with the SLM in the first order of diffraction. (Adapted from D. et al. 2012. *Opt. Lett.* 37: 2478–2480; Flamm, D. et al. 2013. *J. Lightwave Technol.* 3: 387–.

(a) (b) (c)

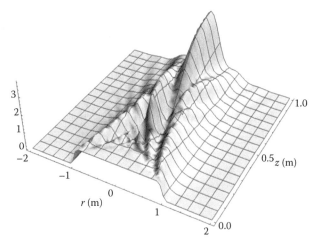

FIGURE 4.15 Spatially resolved polarization. (a) Definition of Stokes parameters. (b) Me beam intensity. (c) Reconstructed intensity with local polarization ellipses. The red and yello denote left- and right-handed sense of orientation. The bluish dots mark the position of the field vector at a fixed time.

FIGURE 7.3 A 3D plot showing the free-space propagation of a flat-top beam using the FC approximation.

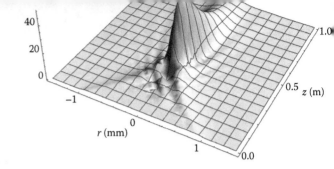

FIGURE 7.5 A 3D plot showing a flat-top beam after passing through a lens, as towards and through the focus.

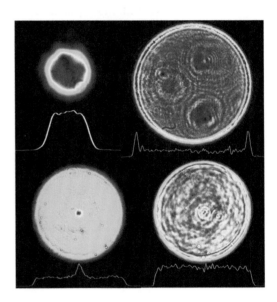

FIGURE 7.8 Examples of flat-top beams created in the laboratory (clockwise from onator method, two refractive elements, single diffractive element and phase-only ho to a spatial light modulator. The peak in the center of the diffractive optic examp undiffracted zeroth order.

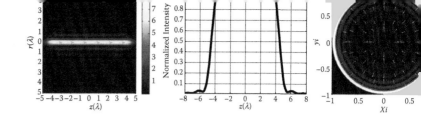

FIGURE 9.13 (a) Total intensity distribution through the focus region and (b) axial inten tribution for an optical needle field with extended DOF. The corresponding required incid in the pupil plane is illustrated in (c). (From Wang, J., Chen, W., and Zhan, Q. 2010. Engin high purity ultra-long optical needle field through reversing the electric dipole array radiat *Express* 18: 21965–21972. With permission of Optical Society of America.)

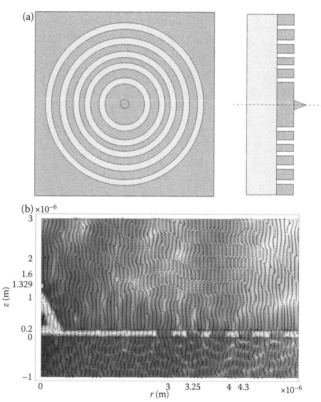

FIGURE 9.15 (a) Diagram of a combined structure of bulls-eye and a conical tip. (b) Log plot of the 2-D electric energy density and the electric field (show a by the arrow) distributi illuminated by radial polarization. A hot spot with extremely high field enhancement is c at the tip apex. (From Rui, G. et al., 2010. Plasmonic nearfield probe using the combin concentric rings and conical tip under radial polarization illumination. *J. Opt.* 12: 03500 permission.)

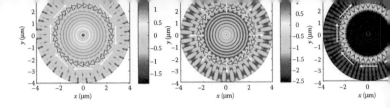

FIGURE 9.16 (a) Logarithmic electric energy density distribution at the air/silver i 32 isosceles triangles arranged in an antisymmetric mode under azimuthally polari tion. A highly confined focal spot is obtained at the center. (b) Logarithmic electric e distribution of the same structure under radially polarized illumination. The solid sp field enhancement is observed. (c) Logarithmic electric energy density distribution w les triangles arranged in a symmetric mode under azimuthally polarized illumination. dark center is observed. (From Chen, W., Nelson, R. L., and Zhan, Q. 2012. Geometri surface plasmon focusing with azimuthal polarization. *Opt. Lett.* 37: 581–583. With Optical Society of America.)

gives 256 gray levels. The number and dimensions of these pixels is constantly evolving toward higher numbers and smaller sizes.

6.3.1 Parallel-Aligned Devices

This type of SLM uses a homogeneous liquid-crystal layer in which the molecules are aligned parallel to each other. Therefore, the device acts as a wave-plate with phase shift given by

$$\beta = \frac{\pi}{\lambda}(n_e - n_o)d, \tag{6.1}$$

where d is the thickness of the liquid-crystal layer, λ is the wavelength, and n_e and n_o are the extraordinary and ordinary indices of refraction.

When an electric field is applied across the pixel, the liquid-crystal director tilts toward the electric field direction, and this causes a reorientation of the index ellipsoid of the material (Figure 6.4). Here, the LCD plane is considered the xy plane, and light propagates along the z direction. The effective extraordinary index of refraction depends on the tilt angle θ, as given by

$$\frac{1}{n_e^2(\theta)} = \frac{\cos^2(\theta)}{n_e^2} + \frac{\sin^2(\theta)}{n_o^2}. \tag{6.2}$$

Consequently, the extraordinary index of refraction $n_e(V)$ decreases under the applied electric field across each pixel while the ordinary index of refraction n_o remains constant. Therefore, the device acts as a programmable wave-plate, where the birefringence is voltage dependent $\beta(V)$. The tilt angle θ is related to the applied voltage. These voltage levels are controlled with the gray levels addressed from the computer, and can be usually adjusted using the brightness and contrast controls for the device.

The Jones matrix for this display, therefore, corresponds to that of a linear wave-plate. If we assume that the liquid-crystal director is aligned with the reference framework, then

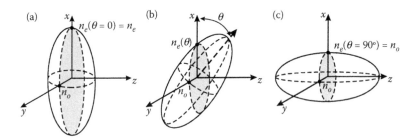

FIGURE 6.4 Representation of the index ellipsoid and its tilt when a voltage is applied at the liquid-crystal cell.

it is given by

$$\mathbf{M_{PAL}}(\beta) = \begin{pmatrix} \exp(-ik_e d) & 0 \\ 0 & \exp(-ik_o d) \end{pmatrix} = \exp(-ik_o d) \begin{pmatrix} \exp(-i2\beta) & 0 \\ 0 & 1 \end{pmatrix} \quad (6.3)$$

where $k_o = (2\pi/\lambda)n_o$ and $k_e = (2\pi/\lambda)n_e$ are the wave-numbers in the ordinary and extraordinary directions. Therefore, for each gray level g, there is a phase shift $\phi(g) = 2\beta(g)$ between the two orthogonal linear polarization components in the x and y directions. The device acts as a programmable linear wave-plate.

If the incident light beam is linearly polarized along the direction of the liquid-crystal director, then it is aligned with the principal axis of the wave-plate. In this situation, the Jones vector of the input light is $J_{in} = (1\ 0)^t$, the superscript t meaning the transposed vector. Thus, the output Jones vector is $J_{out} = \mathbf{M_{PAL}}J_{in}$, which, except for the constant phase factor $\exp(-ik_o d)$, equals to $J_{out} = \exp(-i2\beta)J_{in}$. This shows that a phase-only modulation is directly achieved, even without the second polarized analyzer. Some secondary effects like Fabry–Perot interferences might affect the modulation [16].

6.3.2 Twisted-Nematic Devices

TNLCDs are more common displays, but phase-only modulation with these devices requires a more complicated configuration. Early TN devices, which had relatively thick liquid-crystal layers, could provide phase-only modulation with linear polarization [17,18]. However, the thickness of these devices has been decreased to allow a faster response. Consequently, to obtain phase-only modulation, shorter wavelengths and elliptically polarized light configurations are required.

In TNLCDs, the alignment grooves are rotated from each other by a twist angle α that is usually about 90° and can be either clockwise or counterclockwise. The alignment direction at the input side of the LCD is oriented at some angle Ψ_d relative to the laboratory frame. In Figure 6.3b, Ψ_d is 45° and the twist is 90° in the clockwise direction. The birefringence β of the LCD is again defined as $\beta = \pi d(n_e tn_o)/\lambda$. The three relevant parameters that define the LCD (Ψ_d, α, and β are unfortunately not given by most LCD manufacturers, and a series of experiments can be conducted to measure them [19–23].

6.3.2.1 Jones Matrix Model

The Jones matrix for the LCD was initially introduced by Yariv and Yeh [24] and is based on the Jones matrix for a twisted anisotropic medium. The LCD is modeled as a stack of N birefringent layers, each one slightly twisted with respect to the previous one. Each layer is considered as a uniaxial birefringent plate. By forming the product of these N matrices and letting N approach infinity, the Jones matrix of the display can be written as the product of two matrices as

$$\mathbf{M_{LCD}}(\alpha, \beta) = \exp(-i\beta)\mathbf{R}(-\alpha)\mathbf{M}(\alpha, \beta) \quad (6.4)$$

Here, $\mathbf{R}(\theta)$ is the 2×2 rotation matrix

$$\mathbf{R}(\theta) = \begin{pmatrix} \cos\theta & \sin\theta \\ -\sin\theta & \cos\theta \end{pmatrix} \tag{6.5}$$

and $\mathbf{M}(\alpha, \beta)$ is the matrix

$$\mathbf{M}(\alpha, \beta) = \begin{pmatrix} X - iY & Z \\ Z & X + iY \end{pmatrix} \tag{6.6}$$

This matrix is defined relative to the orientation of the director axis of the molecules at the incident side of the LCD. The matrices depend on the twist angle α, and the birefringence β as defined earlier. The matrix elements in Equation 6.6 are defined as $X = \cos\gamma$, $Y = \beta \sin(\gamma)/\gamma$, $Z = \alpha \sin(\gamma)/\gamma$, and $\gamma^2 = \alpha^2 + \beta^2$. Lu and Saleh extended this simplified model to the LCD by assuming a voltage-dependent effective birefringence $\beta(V)$ [25]. The twist angle was considered as a constant. Note that if $\alpha = 0$, Equation 6.4 leads to the Jones matrix in Equation 6.1 corresponding to the parallel-aligned display.

As mentioned earlier, the parameters for this model can be accurately measured in the off state using two properly adjusted linear polarizers at the entrance and exit sides [19–23]. It is important to use several wavelengths in order to avoid the ambiguities caused by the sinusoidal nature of the parameters X, Y, and Z. When voltage is applied, however, this simple model begins to break down. Corrections of the model can be included accounting for the edge liquid-crystal layers [26–28], but the general form of the Jones matrix in Equation 6.4 is maintained.

LCDs are most often operated between two linearly polarizers oriented at angles φ_1 and φ_2 with respect to the input director axis. In this configuration, the output Jones vector is given by [27]

$$E_{OUT} = \mathbf{R}(-\varphi_2)\mathbf{P_0}\mathbf{R}(+\varphi_2)\mathbf{M_{LCD}}\mathbf{R}(-\varphi_1) \begin{pmatrix} 1 \\ 0 \end{pmatrix} = \sqrt{T}\exp(-i\delta) \begin{pmatrix} \cos(\varphi_2) \\ \sin(\varphi_2) \end{pmatrix} \tag{6.7}$$

Here, $\mathbf{P_0}$ corresponds to the Jones matrix of a linear polarizer oriented at $\varphi_2 = 0$, that is,

$$\mathbf{P_0} = \begin{pmatrix} 1 & 0 \\ 0 & 0 \end{pmatrix} \tag{6.8}$$

The intensity modulation T and the phase modulation δ can be obtained using usual Jones calculus leading to the following expressions:

$$T = (X\cos(\varphi_1 - \varphi_2 + \alpha)Z\sin(\varphi_1 - \varphi_2 + \alpha))^2 + (Y\cos(\varphi_1 + \varphi_2 - \alpha))^2 \tag{6.9}$$

$$\delta = \beta + \tan^{-1}\left(\frac{Y\cos(\varphi_1 + \varphi_2 - \alpha)}{X\cos(\varphi_1 - \varphi_2 + \alpha) + Z\sin(\varphi_1 - \varphi_2 + \alpha)}\right) \tag{6.10}$$

The birefringence $\beta(V)$ changes with voltage, producing in general a coupled amplitude and phase modulation as $T = T(V)$ and $\delta = \delta(V)$ as shown in Figure 6.2f. This analysis shows that amplitude-only modulation (with a small phase modulation) can be obtained when the first polarizer is oriented perpendicular to the director axis at the input side of the LCD ($\varphi_1 = 90°$) and the second linear polarizer is oriented perpendicular to the director axis at the output side of the LCD ($\varphi_2 = \alpha + 90°$) [25].

The same analysis revealed that phase-only modulation could be obtained for values of β much greater than α, a situation that happened in older devices with thicker liquid-crystal layers, such as the Epson model VP100-PS display [9]. This was achieved when the first polarizer is oriented parallel to the director axis at the input side ($\varphi_1 = 0$) and the second polarizer is oriented parallel to the director axis at the output side ($\varphi_2 = \alpha$) [25]. However, as we mentioned earlier, phase-only operation cannot be achieved using linear polarizers for the modern thinner displays under any circumstances.

In such thinner devices, phase-only modulation requires elliptical polarization configurations. Pezzaniti and Chipman introduced the polarization eigenvector concept for the TN devices [29]. We applied this concept to develop a technique based on creating and detecting the polarization rotated eigenvectors for the LCD (eigenvectors of the Jones matrix in Equation 6.6). This technique [30–32] leads to maximum transmission, and larger phase modulation. Later, it was demonstrated that other elliptical polarization configurations could be used to achieve phase-only modulation [33–35]. A proper linear combination of the two polarization eigenvectors can lead to a phase modulation with larger phase modulation depth, with the cost of reducing the average intensity transmission [36,37]. Next, we review these methods.

6.3.2.2 Eigenvectors Configuration to Achieve Phase-Only Modulation

The rotated eigenvectors $\mathbf{E}_{m\pm}$ and eigenvalues $\mu_\pm$ for the matrix $\mathbf{M}(\alpha, \beta)$ in Equation 6.4 are defined by

$$\mathbf{M}(\alpha, \beta)\mathbf{E}_{\mu\pm} = \mu_\pm\mathbf{E}_{\mu\pm} = \exp(\pm i\gamma)\mathbf{E}_{\mu\pm} \tag{6.11}$$

The solutions to this problem are elliptically polarized states given by [30]

$$\mathbf{E}_{\mu\pm} = \frac{1}{\sqrt{2(\gamma \pm \beta)\gamma}} \begin{pmatrix} \alpha \\ i(\beta \pm \gamma) \end{pmatrix} \tag{6.12}$$

As these elliptically polarized electric field eigenvectors propagate through the TN-LCD, they are rotated by the rotation matrix $\mathbf{R}(-\alpha)$ and receive phase delays of $\phi_\pm$ given by

$$\phi_\pm = \beta - (\pm\gamma). \tag{6.13}$$

These eigenvectors represent left- and right-hand elliptically polarized light whose major axes are aligned with the director axes for the TN-LCD. The transmitted eigenvector has the same ellipticity as the incident eigenvector. However, the major and minor axes of the elliptically polarized eigenvectors are rotated by the twist angle a of the TN-LCD. Because the birefringence changes with applied voltage, the ellipticity of the eigenvector also varies. Consequently, we defined an average eigenvector that approximates the exact eigenvector over the total applied voltage range.

The simplest (although not unique) optical system for generating and detecting the eigenvectors of the TN-LCD is shown in Figure 6.5. The input light comes from the left. The slow axes for the QWPs are aligned parallel to the director axes at the input and output sides. For the negative eigenvector in Figure 6.5a, the input linear polarizer (LP1) is rotated clockwise by an angle of $+\xi$ relative to the input director axis and the output linear polarizer (LP2) is rotated counterclockwise by an angle of $-\xi$ relative to the output director axis. Figure 6.5b shows the setup for generating and detecting the positive eigenvector, where the input and output polarizers are now rotated by $90°$ with respect to the scheme in Figure 6.5a.

Phase-only modulation achieved with these eigenvector configurations was first experimentally demonstrated in 1998 [31]. Since then, it has been demonstrated in several TN devices in several laboratories. As an example, we reproduce here the results [37] obtained using a CRL model XGA-3 TN-LCD having 1074×768 pixels, with illumination from the 488 nm Argon laser wavelength. In this device, maximum voltage is applied with zero gray level, and the voltage decreases as the gray level increases. The birefringence β was measured as a function of gray level g following methods that only require intensity mesurements [27,28]. This device showed a reduced maximum birefringence β_{max} of only $170°$ at the operating wavelength of 488 nm. The birefringence decreases for longer wavelengths.

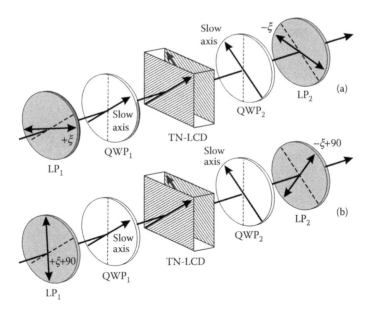

FIGURE 6.5 Schematic defining the generation of the (a) negative and (b) positive eigenvectors for the TN-LCD having a negative twist.

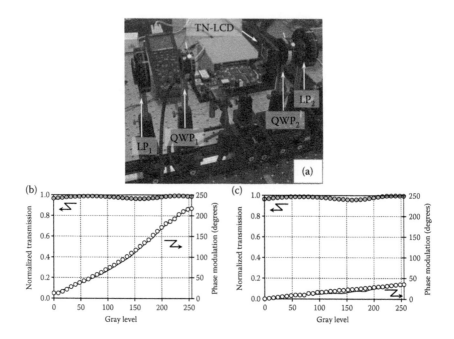

FIGURE 6.6 (a) Picture of the CRL-Opto TN-LCD in the eigenvector configuration. Intensity and phase modulation for (b) the negative eigenvector and (c) the positive eigenvector. (From Moreno, I., Martínez, J. L., Davis, J. A. 2007. Two-dimensional polarization rotator using a twisted-nematic liquid-crystal display. *Appl. Opt.* 46: 881–7. With permission of Optical Society of America.)

Figure 6.6a shows a picture of the device inserted in between the linear polarizer–QWP pair. Figures 6.6b and 6.6c show the output intensity and phase shift given by Equations 6.9 through 6.10 for both rotated eigenvector configurations from Figures 6.5a and 6.5b. The most uniform intensity transmission is obtained when the relative angles of the polarizers with respect to the wave-plates (see Figure 6.5) are $\xi = 51°$. Figures 6.6a and 6.6b correspond to the negative and positive eigenvectors, respectively. Both cases show that, when these average elliptically polarized eigenvectors are generated and detected, the maximum transmission is very close to 100% for the whole range of the addressed gray level. The phase data are also plotted in Figures 6.6a and 6.6b and show that the phase modulation for the negative average eigenvector (ϕ_-) is much greater than for the positive one (ϕ_+). Because the average eigenvector is generated and detected, transmission is not perfectly 100% for all gray level range, and a slight oscillation is produced.

Basically, operating in this eigenvector configuration can be viewed as a transformation of the system in order to operate the TN-LCD as a programmable wave-plate. In this case, the configuration is equivalent to the parallel-aligned display, where a programmable phase shift $\phi = \phi_- - \phi_+$ is introduced between the two orthogonal eigenvector states.

6.3.2.3 Other Generalized Configurations to Achieve Extended Phase-Only Modulation

Shortly after the rotated eigenvector configuration was demonstrated, it was recognized that this is not a unique solution to obtain a phase-only modulation response. First, the external wave-plates were allowed to differ from QWP, and could adopt orientations different from those parallel to the liquid-crystal director [33]. Other strategies adopted the Poincaré sphere representation [34,35], and were able to produce phase only modulation response with a single QWP, although the phase modulation depth was limited, not reaching 2π radians.

All these setups operate with elliptically polarized light, where an elliptically polarized light $\mathbf{E}_{in}$ is incident on the display, and where the light emerging from the display is projected onto the polarization state $\mathbf{E}_{out}$ transmitted by the second wave-plate–polarizer system.

The corresponding transmitted electric field E_T can be written as

$$E_T = \exp(-i\beta)\mathbf{E}_{out}^{\dagger} \cdot \mathbf{M} \cdot \mathbf{E}_{in}. \tag{6.14}$$

where the superscript † denotes the transposed complex-conjugated matrix.

This situation can be analyzed as a combination of the two rotated eigenvectors in Equation 6.12. The elliptical polarization $\mathbf{E}_{in}$ incident on the display can be decomposed as [36,37]

$$\mathbf{E}_{in} = g_+\mathbf{E}_{\mu_+} + g_-\mathbf{E}_{\mu_-} \tag{6.15}$$

where $g_\pm$ are complex numbers given by the inner product $g_\pm = \mathbf{E}_{m\pm}^{\dagger} \times \mathbf{E}_{in}$. Now, the propagation of this polarization through the display can be viewed as the transmission of each rotated eigenvector component, each one gaining a phase $\phi_\pm$ given by Equation 6.13, and being rotated an angle α. Finally these two components are projected onto the polarization state $\mathbf{E}_{out}$ by the inner product $t_\pm = \mathbf{E}_{out}^{\dagger} \times \mathbf{E}_{m\pm}$. Therefore, the electric field E_T can be written as

$$E_T = t_+g_+ \exp(-i\phi_+) + t_-g_- \exp(-i\phi_-) \tag{6.16}$$

This decomposition was employed [37] to analyze different modulation configurations, where we sought for a uniform intensity modulation, but we allowed the average transmission to decrease.

Figure 6.7 shows the intensity and phase modulation obtained with the same device as the one in Figure 6.6, for two illustrative relevant situations, showing how a reduction of the average intensity leads to higher phase modulation depth. These cases are two configurations designed to produce a flat intensity transmission with an average value of 50% (Figure 6.7a) and 5% (Figure 6.7b). Experimental data are depicted with symbols, while the expected theoretical results are depicted using continuous lines. This shows that the maximum phase depth can be dramatically increased by decreasing the average transmitted intensity. This can be very important to be able to reach 2π phase modulation, required to display diffractive elements with optimal diffraction efficiency (as we

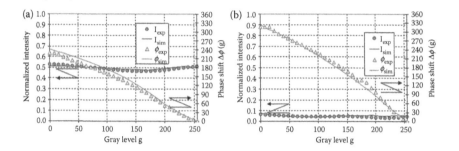

FIGURE 6.7 Intensity and phase modulation for (a) configuration with 50% average intensity transmission, and (b) configuration with 5% intensity transmission. Experimental data are represented by points and continuous lines depict theoretical predictions. Wavelength was 514 nm in these experiments. (From Martínez, J. L. et al. 2010. Extended phase modulation depth in twisted nematic liquid-crystal displays. *Appl. Opt.* 49: 5929–37. With permission of Optical Society of America.)

will analyze in the next section). This is especially relevant when working with modern displays, which usually have a thin liquid-crystal layer, thus providing a reduced value of β_{max}.

The major difficulty with the TN-LCD arises from the edge effects where the molecules cannot move. This introduces an effective wave-plate at both edges whose birefringence varies with the applied voltage [27].

All these results show how phase-only modulation can be obtained with LCDs. Next, we will analyze diffraction efficiency considerations in these devices.

6.4 Diffraction Efficiency Considerations

One key parameter in any diffractive element is its diffraction efficiency, that is the ratio of the intensity of the effective displayed diffractive optical element to the intensity of the input beam. The first sources of efficiency reduction are absorption and reflection losses (in transmissive displays). Another source of efficiency reduction is the pixelated structure of the displays, which produces a characteristic 2D diffraction grid pattern. Finally, the diffraction efficiency of the displayed elements is also dependent on the type of optical modulation provided by the SLM as well as on the type of encoding employed to display the diffractive element.

6.4.1 Effects of the Pixel Structure: Width-to-Period Ratio

Liquid-crystal SLMs are constructed of an array of individually addressable transmissive or reflective pixels. Figure 6.8a shows a picture of the pixels for the CRL Opto SLM

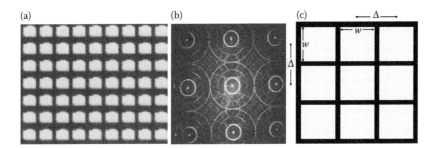

FIGURE 6.8 (a) Pixel structure for CRL-opto liquid-crystal display. (b) Diffraction pattern when a circular grating is addressed to the SLM. (c) Two-dimensional simplified model of the pixel grid structure.

mentioned earlier. This pixel structure of the LCD acts as a two-dimensional diffraction grating that produces a diffraction pattern of spots whose separation is proportional to the incident wavelength and inversely proportional to the pixel spacing [38]. Since the pixel size of SLMs is typically in the tens of microns, the efficiency analysis can be developed. within the scalar theory approximation of diffraction phenomena. Within this approximation, this diffraction pattern can be regarded as the multiplication of equally intense spots by the diffraction pattern produced by each pixel. This pixel function reduces the intensity of the higher-order spots.

Distribution of light into these diffraction orders plays an important role in diffractive optics. When a diffractive element is encoded onto the LCD, its response appears replicated on the location of each of these orders. Figure 6.8b shows the experimental pattern created by a circular binary grating, where the replicas located on each diffraction order created by the pixilated grid are clearly visible. The efficiency of these replicas is equivalent to the intensity of the diffraction orders in the absence of applied signal. In general, the central zero diffraction order is used, since it contains most of the energy.

Before continuing, we define the terminology regarding diffraction efficiency. These pixelated SLMs consist of regions (that we assume to be squares of width w) that transmit or reflect light and are surrounded by opaque areas that normally contain control electronics and wires. The spacing between pixels is defined as Δ (Figure 6.8c). There are two consequences of this structure. First, the fraction of the incident intensity that is transmitted is given by the fill factor $F = (w/\Delta)^2$. Second, the central principal diffracted order has the highest intensity and is roughly proportional to the square of the fill factor or $F^2 = (w/\Delta)^4$.

6.4.2 Fourier Series Analysis of Diffractive Elements

In general, one wants to encode an electric field transmission function consisting of both amplitude and phase information and written as $h(x) = |h(x)|e^{i\phi(x)}$. This situation could represent holographic recording, encoding of a matched spatial filter for optical pattern recognition, or a diffractive optical element.

The accurate encoding of this function has vast implications on the fidelity of the output function and the distribution of energy among the different diffraction components. We can introduce this by means of diffraction gratings.

We begin with a blazed diffraction grating having a linear phase where $h(x) = \exp(i2\pi x/d)) = \exp(i\gamma x)$, where $\gamma = 2\pi/d$ and d is the period of the grating. This ideal grating produces a single $+1$ diffraction order. Now we assume that this function cannot be accurately encoded and we will discuss several examples. In this case, this periodic function will be written as a Fourier series as

$$h(x) = \sum_{n=-\infty}^{\infty} c_n \exp[in\phi(x)] = \sum_{n=-\infty}^{\infty} c_n \exp[in\gamma x] \tag{6.17}$$

where n denotes the diffraction order and

$$c_n = \frac{1}{2\pi} \int_0^d h(x) \exp[-in\gamma x]dx \tag{6.18}$$

Equation 6.17 shows that other diffraction orders appear, each with a relative intensity given by $|c_n|^2$. We will show later that this decomposition is more general, and can also be applied to other elements different than linear gratings. But next we start by analyzing various types of gratings, including binary amplitude/phase gratings, and linear phase gratings with limited phase depth.

6.4.2.1 Binary Amplitude and Binary Phase Gratings

Figure 6.9 shows the simplest transformation of the linear blazed grating $h(x) = \exp(i2\pi x/d) = \exp(i\gamma x)$ onto a binary grating. The period (p) is divided in two regions with complex transmissions t_1 and t_2, respectively. The transition is produced at threshold level (a), typically at mid period. Two typical situations are considered: a binary amplitude grating, with $t_1=1$ and $t_2=0$, and a binary π-phase grating, with $t_1 = 1$ and $t_2 = -1$. Obviously, the intensity pattern for the different diffraction orders is symmetrical due to the symmetry of the grating.

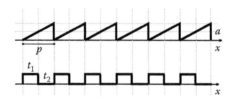

FIGURE 6.9   Transformation of a linear blazed grating into a binary grating.

The Fourier series for the amplitude grating leads to the following coefficients:

$$c_n^{amp} = \frac{1}{2}\operatorname{sinc}\left(\frac{n}{2}\right) \tag{6.19}$$

where the sin c function is defined as $\operatorname{sinc}(x) = \sin(\pi x)/(\pi x)$. Note that the intensity of the zero (DC) order is $|c_{n=0}^{amp}|^2 = 1/4 = 25\%$. All even orders vanish, and the intensity of the odd orders is $\|c_{n\ odd}^{amp}|^2 = 1/(n\pi)^2$, where the most intense orders are for $n = \pm 1$, with $|c_{n=\pm 1}^{amp}|^2 = 1/\pi^2 = 10.1\%$.

The binary π-phase grating is much more efficient, first because transmission through the grating is 100%, and also because the DC term is null, $c_{n=0}^{phase} = 0$. Consequently, all of the energy is distributed into the other orders. If the Fourier coefficients are calculated for this grating, the next result is obtained:

$$c_{n\neq 0}^{phase} = \operatorname{sinc}\left(\frac{n}{2}\right) \tag{6.20}$$

leading to a intensity at the $n = \pm 1$ orders $\left|c_{n=\pm 1}^{amp}\right|^2 = 4/\pi^2 = 40.5\%$.

6.4.2.2 Blazed Gratings with Limited Phase Depth

For a linear phase grating, the efficiency in the first diffraction order reaches 100%. But this happens only when the maximum phase range reaches 2π radians. In the case, where the phase depth is reduced below 2π radians, then the diffraction efficiency into the first order will also be reduced with energy appearing in other orders.

For example, if we assume that the phase depth changes linearly from $-M\pi$ to $M\pi$ over the period, then the energy is mainly diffracted into the zero and -1 orders. The intensity diffracted into the $+1$ order is given by

$$|c_n|^2 = \operatorname{sinc}^2[\pi(n - M)] \tag{6.21}$$

This problem was analyzed in the frame of computer generated holography and diffractive lenses displayed onto LCDs with limited phase range [39,40]. We will make use of this in Section 6.5.

6.4.3 Diffraction Orders of Generalized Phase Functions

It is convenient to apply this Fourier analysis to other kinds of diffractive phase patterns, including lenses, axicons, and holograms. As an example, we consider a converging lens function having a focal length f whose phase can be represented as $h(r) = \exp(i\pi r^2/\lambda f)$. This function is periodic in r^2 and can be represented in a Fourier series as

$$h(r) = \sum_{n=-\infty}^{\infty} c_n \exp[in\pi r^2/\lambda f] \tag{6.22}$$

Now we find that this encoded function creates a number of both converging and diverging lens functions whose focal lengths are given by f/n and whose intensities are given by $|c_n|^2$. The analysis is exactly similar to the case of the diffraction grating. For example, if we use the binary phase encoding, then 40.5% of the light is in the first-order converging beam while 40.5% is also in a first-order diverging beam.

Similar results are obtained by encoding nondiffraction beams and combining vortex patterns onto these axicons or lenses.

6.4.4 Dammann Gratings

In the majority of applications, the goal is 100% efficiency into the desired order. However, there are other applications where the incident energy should be distributed equally into a given number of orders. Such a goal was realized by Dammann where each period of the grating is subdivided into a number of regions having a transmission of $+1$ and other regions having a transmission of -1 [41]. For that purpose, additional transitions points are included in a period. A tabulation of these transition points can be found for a given number of N orders with identical intensity [42]. Recently, this concept has been generalized to other phase patterns, including-vortex producing gratings [43].

6.4.5 Nyquist Limit Encoding Effects

Whenever a continuous sinusoidal function is encoded onto a pixelated array, a Nyquist limit can be reached where the period of the sinusoidal function is shorter than the period between pixels. This Nyquist limit period is usually defined as 2 pixels/period and corresponds to a π phase difference between adjacent pixels.

This concept becomes more interesting when different phase functions, such as a lens function, are to be encoded. Here, the phase increases nonlinearly with the radius of the function as $h(r) = \exp(i\pi r^2/\lambda f)$. We can now define a Nyquist focal length that can be encoded onto an SLM having N pixels where the phase of the $(N/2)$ pixel (at the outer radius) differs by π radians from the phase of the $(N/2) - 1$ pixel. This defines a minimum Nyquist focal length that can be encoded onto an SLM of $f_N = N\Delta^2/\lambda$. Here, λ is the operating wavelength and $(N/2)$ is the pixel size as before.

If we encode lenses having shorter focal lengths, then a pattern begins to resemble a lenslet array with several interesting properties [44]. Similar constraints occur with more complicated phase patterns, including the cubic phase required for accelerating Airy beams [45].

6.5 Encoding Amplitude Information Onto Phase-Only Devices

In order to perform a number of operations, including matched spatial filtering, holograms, and beam-shaping of laser beams, we have to incorporate both amplitude and phase into the diffractive optical element. To obtain this goal, some groups proposed the use of two LCDs, one working in the amplitude-only regime, and the other in the phase-only regime [46]. Other technique involves the use of pseudorandom encoding [47]. However, a much simpler technique to obtain an arbitrary combination of amplitude and phase modulation is based in spatially changing the diffraction efficiency, as we proposed originally in Ref. [48]. This approach has been successfully used in a number of different pattern recognition and optical element design experiments [49–54].

The basic idea is shown in Figure 6.10 and it is based on the spatial modulation of the phase depth encoded on the filter as discussed in Equation 6.21. If a phase grating has a phase depth of 2π radians, then all the light will be diffracted into the first order. However, as the phase depth is reduced, the amount of light diffracted into the first order decreases while other diffraction orders appear, in particular the zero-order corresponding to nondiffracted light. As given in Equation 6.21, the diffracted light intensity at the first diffraction order is given by

$$|c_1|^2 = \text{sinc}^2[\pi(1 - M)] \tag{6.23}$$

If we spatially modulate the phase modulation depth as a function $M(x)$, we can then spatially control the amount of energy diffracted onto the first diffraction order.

In Figure 6.10, the phase depth is highest near the center and decreases near the edges. For this pattern, the diffracted light acts as if it were spatially filtered with a low-pass filter while the zero-order light acts as if it were spatially filtered with a high-pass filter. Note that the idea is fundamentally different from the usual concept of amplitude operation. Normally, an amplitude filter blocks the light from going into a certain direction. Here, we do not block the light, but simply direct it into a different direction. Using this approach, we can generate a very simple encoding technique to include amplitude information onto the phase modulation.

Therefore, this technique can be used for encoding amplitude information onto a phase-only function. Let $F(u)$ denote the complex amplitude to implement, given by

$$F(u) = M(u) \exp[i\phi(u)] \tag{6.24}$$

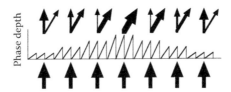

FIGURE 6.10 Technique to encode amplitude information onto a phase-only grating. As the phase depth changes, the fraction of light diffracted into the +1 and 0 orders varies. Here, we show that the modulation creates a low-pass filter in the +1 order and a high-pass filter in the 0 order.

where $M(u)$ is the normalized amplitude distribution, which is in the range [0,1], and $\phi(u)$ is the phase distribution, which we assume in the range $[\pi, +\pi]$. This complex function can be encoded onto a phase-only function. First, a function $M'(u)$ is computed that fulfills the relation

$$\frac{\sin[\pi(1 - M'(u))]}{\pi(1 - M'(u))} = M(u) \tag{6.25}$$

The function $M'(u)$ is also defined in the range [0,+1] and consequently there is one-to-one correspondence between the values of $M(u)$ and $M'(u)$. Then a phase-only function $T(u)$ is calculated as

$$T(u) = \exp[iM'(u)\phi(u)] \tag{6.26}$$

By applying the same considerations as in Equation 6.17, $T(u)$ can be expanded as a Fourier series as

$$T(u) = \sum_{n=-\infty}^{+\infty} T_n(u) \exp[in\phi(u)] \tag{6.27}$$

With the assumption that the phase $\phi(u)$ is in the range $[-\pi + \pi]$, the coefficients $T_n(u)$ are given by

$$T_n(u) = \frac{\sin[\pi(n - M'(u))]}{\pi(n - M'(u))} \tag{6.28}$$

The first diffraction order ($n = 1$) of Equation 6.26 reproduces the original phase function $\exp[i\phi(u)]$ with an amplitude modulation $T_1(u) = M(u)$. In this way, the function $F(u)$ will be exactly reproduced in the first diffraction order ($n=1$). Therefore, this method permits codification of any complex function in a phase-only SLM.

With the proposed technique, it is, therefore, possible to obtain any complex function in the first diffraction order. However, the rest of the orders will be superimposed onto the first order and will cause additive noise. These orders can be spatially separated by adding either a linear phase or a quadratic phase [39]. An alternative technique adapted to binary SLMs was also proposed [55].

6.5.1 Application in the Generation of Laser Modes

By combining a phase-only LCD with the technique for encoding amplitude distributions, a wide variety of applications have been shown where we encode complex amplitude and phase masks for generating desired output patterns.

One such application involved beam shaping where an incident uniform intensity beam was transformed into a variety of Gaussian beam propagating waves. The allowed modes for laser resonators are solutions to the paraxial wave equation given by

$$\left(\nabla_t^2 - 2ik\frac{\partial}{\partial z}\right)\Psi(\vec{r}) = 0 \tag{6.29}$$

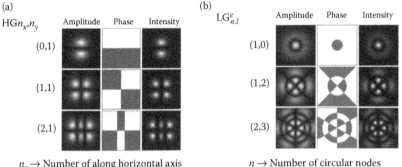

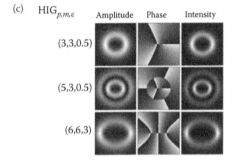

$(p-m)/2 \rightarrow$ Number of elliptical nodes
$m \rightarrow$ Size of central minimum

FIGURE 6.11 Examples of (a) Hermite–Gaussian laser modes, (b) Laguerre–Gaussian laser modes, and (c) Ince–Gaussian laser modes.

The solutions to this equation are the product of a Gaussian beam with a polynomial series that depends on the geometry of the resonator (see Chapters 1 and 8 for more details). The beam is effectively confined by the Gaussian beam.

The most common modes are given in Cartesian coordinates where the polynomial series are Hermite polynomials. Consequently, these modes are the Hermite–Gaussian modes. Figure 6.11a shows the amplitude distributions, the phase maps, and the familiar intensity distributions for three modes.

The second series of modes are given in cylindrical symmetry where now the polynomial series are Laguerre polynomials. Figure 6.11b shows similar amplitude, phase, and intensity maps for some examples of these modes.

In 2004, M. Bandres and J. C. Gutiérrez-Vega at the Tecnológico de Monterrey (Mexico), discovered solutions to this wave equation in elliptical coordinates [56]. In this case, the polynomial series are Ince polynomials and are characterized by elliptical and hyperbolic nodes. Some examples of these modes are shown in Figure 6.11c.

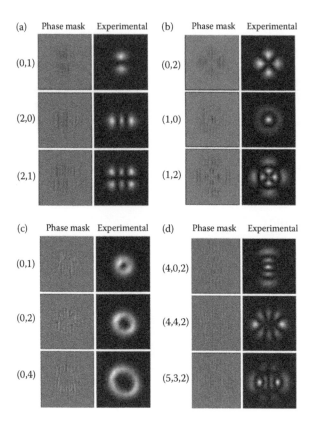

FIGURE 6.12   Phase masks and outputs for (a) Hermite–Gaussian beams, (b) Laguerre–Gaussian beams, (c) circular Laguerre–Gaussian beams, and (d) Ince-Gaussian beams.

A series of experiments were performed using the previously discussed technique for encoding amplitude and phase masks onto a single SLM. The results in Figure 6.12 show the phase masks and experimental results for these three kinds of modes. Experimental results are discussed in more detail in other references [57,58]. However, the approach works exceptionally well.

6.6 Polarization Encoding

The next frontier is the encoding of polarization onto a single SLM. The generation of two-dimensional polarization distributions and polarizing diffractive elements is interesting for many applications, including polarization imaging [59], data encoding, and polarization multiplexing [60]. Other approaches for obtaining spatially structured polarization gratings have been shown, including subwavelength grating structures [61] and liquid-crystal spatial light modulators (LC-SLM) [62]. The use of the LC-SLM is particularly interesting because it is programmable.

A first step was based on the use of the parallel-aligned LCD. When inserted in between two QWPs oriented at ±45° to the liquid-crystal director, it operates as a polarization rotator, where the rotation is directly related to the phase shift introduced by the display. Therefore, spatial polarization patterns were demonstrated [63]. The same technique was later demonstrated with a TNLCD, where the rotated eigenvector configuration described in Section 6.3.2.2 was employed to operate this display approximately as a pure phase modulator among two orthogonal polarization states [64]. However, these methods do not allow a complete polarization control, since not all states can be created.

Recently, some other approaches for spatial polarization control have been reported that use a split screen configuration on an SLM, so different actuation can be performed onto two different polarization components. For instance, we demonstrated to generate polarization-selective computer-generated holograms by using two SLMs [63]. One device displayed two phase-only holograms on different halves, while the other one was used to control the corresponding output polarization. More recently, a parallel aligned reflective liquid-crystal display was employed to display blazed gratings with different orientations in order to generate optical traps in a microscope [64]. For that purpose, the screen of the display was again divided into two halves, and a half-wave-plate had to be placed in the beam reflected in only one half of the screen. An alternative system, again using a reflective SLM, was proposed to generate vectorial laser beams, which used a Wollaston prism to spatially separate two orthogonal linear polarization components of the incoming light beam, and direct them to the two halves of the display screen [65]. Other techniques for producing arbitrary spatial polarization patterns involve selective polarization elements in the Fourier plane of optical 4f processor architectures [66,67].

As mentioned above, it is very easy to encode phase-only patterns onto a parallel-aligned LC-SLM. By adding a spatially modulated carrier, amplitude information can also be encoded. However, the parallel-aligned LC-SLM only diffracts the vertically polarized component of input light (which is parallel to the LC director). The horizontal component of input light is unaffected. In order to achieve full polarization control, the two orthogonal electric field components must be independently controlled. We presented an optical architecture for two-dimensional polarization control that includes amplitude and phase control based on a single parallel-aligned LC-SLM [68]. This architecture requires a double pass configuration where different patterns are encoded onto the two halves of the SLM. Experimental results are shown.

6.6.1 Optical System

The experimental system is shown below in Figure 6.13. A laser beam having both horizontal and vertical polarization components enters from the left through an aperture and a nonpolarizing beamsplitter and illuminates half of the parallel-aligned LC-SLM onto which a diffractive optical element is encoded. The display only diffracts the vertically polarized component of input light (which is parallel to the liquid-crystal director). The horizontal component of input light is unaffected by the LC-SLM. Following the method

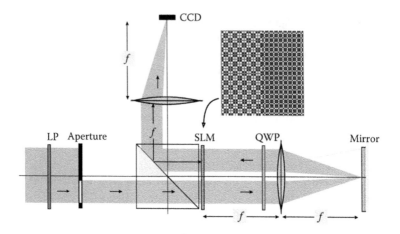

FIGURE 6.13 Double-pass optical architecture to achieve complete polarization control with a parallel aligned LCD.

described in Section 4, we can control the fraction of light that is diffracted by controlling the phase modulation depth M_A of the diffraction grating that is encoded onto the SLM. We can also add a phase bias ϕ_A to the grating, which is applied to both the zero and first diffracted orders, and adds control over the polarization state of the zero-order diffracted order. If the incident light onto the SLM has two polarization states, we can alter the polarization state of the zero-order diffracted light.

The transmitted light passes through a QWP, is reflected, and passes through the QWP again, effectively reversing the two polarization states. The light is then incident onto the second half of the LC-SLM onto which a different diffractive optical element is encoded. The previously vertically polarized light is now horizontally polarized and not affected by the LC-SLM. However, the previously horizontally polarized light is now vertically polarized and will be diffracted. Again, the fraction of the new vertically polarized light can be controlled by the second phase modulation depth M_B and by the second phase bias ϕ_B of the diffractive optical element that is encoded onto this grating on the second half of the LC-SLM.

The transmitted beam is then reflected by the beamsplitter to the detector. We can either image the output or perform a Fourier transform as shown in Figure 6.13. The two patterns can create differently polarized outputs in different areas of the output or they can be combined.

6.6.2 Polarization-Controlled Computer-Generated Holograms

Here, we show an example of the full capability of this system to generate an output image having an arbitrary polarization state.

A first simple example is shown in Figure 6.14. Figure 6.14a shows the phase-only mask, where the two halves correspond to the two generated phase-only holograms.

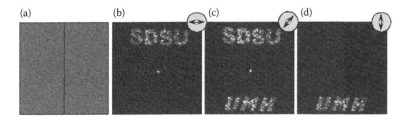

FIGURE 6.14 Dual-phase mask and experimental results showing the reconstruction of computer-generated holograms with different output polarizations.

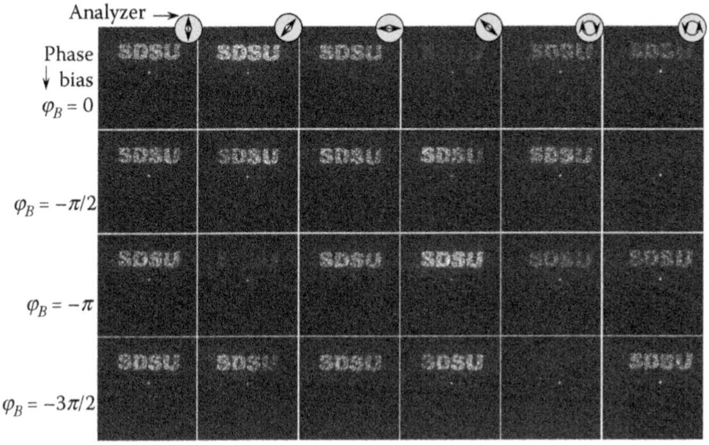

FIGURE 6.15 Experimental results showing different output polarizations for different phase biases encoded onto the two patterns encoded onto the LC-SLM.

Experimental results presented in Figures 6.14b trough 6.14d correspond to an analyzer placed before the CCD camera, oriented horizontally, at 45°, and vertically. These results show that the image of "SDSU" is horizontally polarized, while the diffracted image of "UMH" is vertically polarized.

Figure 6.15 shows additional results that prove the output polarization control. We now use the letters "SDSU" to generate the phase of both holograms. Therefore, we encode the Fourier transform of the same pattern onto both halves of the SLM, and insure that their Fourier transforms will overlap. However, we can also add an additional uniform phase relative between the two halves of the screen (by either φ_A or φ_B). As a result, a single pattern will be produced but the state of polarization is dependent on the phase bias $\varphi_B - \varphi_A$.

We show four cases (the four rows of the figure) where we added phase bias values $\varphi_B = 0$, $-\pi/2$, $-\pi$, and $-3\pi/2$, and we keep $\varphi_A = 0$. These phase bias values will produce the diffraction pattern with four polarization states corresponding to linear polarization at 45°, right circularly polarized, linear polarization at 135°, and left circular polarized, respectively. Each column in Figure 6.15 corresponds to the image captured with the CCD when an analyzer is placed just in front of it. A linear polarizer oriented at 0, 45, 90, and 135°, and two R and L circular polarizer analyzers have been employed. These results prove that the SDSU pattern is being generated with the state of polarization defined by the phase bias φ_B. Experimental results confirm that each image in the different rows is polarized as expected. The system is fully programmable in that different polarization states or output patterns can be obtained.

It must be noted, however, that to achieve these results it has been necessary to very accurately align the system. Otherwise, interference fringes appear over the reconstructed pattern. We have also shown additional results where we encode polarization-selective Dammann gratings as well as amplitude and phase diffractive optical elements [68].

6.7　Conclusions

In conclusion, we have reviewed the use of liquid-crystal SLMs to produce programmable diffractive optical elements. We have described how to operate the usual TN displays, as well as parallel aligned nematic devices. We first reviewed techniques to operate them to produce phase-only modulation. Then we presented techniques to introduce amplitude encoding, thus achieving a total complex (amplitude and phase) spatial programmable control of a light beam. Finally we showed how to add polarization control to a system. Consequently, our methods show how to precisely spatially control the phase, amplitude, and polarization state of a light beam.

Acknowledgments

We have many people to thank for their guidance, collaboration, and support. The whole list would surpass the length of this chapter. However, in particular, we would like to acknowledge Don Cottrell (San Diego State University), Maria Yzuel and Juan Campos (Universidad Autonoma de Barcelona), Julio C. Gutiérrez-Vega (Tecnológico de Monterrey, México), and Tomio Sonehara (Epson Corporation).

References

1. Efron, U. 1995. *Spatial Light Modulators Technology: Materials, Devices and Applications*. New York: Marcel Dekker.
2. Yeh, P. 1999. *Optics of Liquid Crystal Displays*. New York: John Wiley & Sons.
3. Liu, H.K., Davis, J.A., and Lilly, R.A. 1985. Optical data-processing properties of a liquid-crystal television spatial light modulator, *Opt. Lett.* 10: 635–7.
4. Gregory, D.A. 1986. Real-time pattern recognition using a modified liquid crystal television in a coherent optical correlator. *Appl. Opt.* 25: 467–9.
5. Yu, F.T.S., Jutamulia, S., and Huang, X.L. 1986. Experimental application of low-cost liquid crystal TV to white-light signal processing, *Appl. Opt.* 25: 3324–6.

6. Amako, J. and Sonehara, T. 1990. Computer generated hologram using TFT active matrix liquid crystal spatial light modulator (TFT-LCSLM), *Jpn. J. Appl. Phys.* 29: L1533–5.

7. Grier, D.G. 2003. A revolution in optical manipulation, *Nature* 810–6.

8. Bougrenet de la Tocnaye, J.L., Dupont, L. 1997. Complex amplitude modulation by use of liquid crystal spatial light modulators, *Appl. Opt.* 36: 1730–41.

9. Barnes, T.H., Eiju, T., Matsuda, K., and Ooyama, N. 1989. Phase-only modulation using a twisted nematic liquid crystal television, *Appl. Opt.* 28: 4845–52.

10. Amako, J. and Sonehara, T. 1991. Kinoform using an electrically controlled birefringent liquid-crystal spatial light modulator, *Appl. Opt.* 30: 4622–8.

11. Tanone, A., Zhang, Z., Uang, C.-M., Yu, F.T.S., Gregory, D.A. 1993. Phase-modulation depth for a real-time kinoform using a liquid crystal television, *Opt. Eng.* 32: 517–21.

12. Paul-Hus, G. and Sheng, Y. 1993. Optical on-axis real-time phase-dominant correlator using liquid crystal television, *Opt. Eng.* 32: 2165–72.

13. Neto, L.G., Roberge, D., and Sheng, Y. 1995. Programmable optical phase-mostly holograms with coupled-mode modulation liquid crystal television, *Appl. Opt.* 34: 1944–50.

14. Laude, V. and Refregier, P. 1994. Multicriteria characterization of coding domains with optimal Fourier SLM filters, *Appl. Opt.* 33: 4465–71.

15. Scharf, T. 2007. *Polarized Light in Liquid Crystals and Polymers.* New York: John Wiley & Sons.

16. Davis, J.A., Tsai, P., Cottrell, D.M., Sonehara, T., and Amako, J. 1999. Transmission variations in liquid crystal spatial light modulators caused by interference and diffraction effects, *Opt. Eng.* 38: 1051–7.

17. Konforti, N., Marom, E., and Wu, S.-T. 1988. Phase-only modulation with twisted nematic liquid-crystal spatial light modulators, *Opt. Lett.* 13: 251–3.

18. Efron, U., Wu, S.-T., and Bates, T.D. 1986. Nematic liquid crystals for spatial light modulators: recent studies, *J. Opt. Soc. Am. B* 3: 247–52.

19. Soutar, C., Lu, K. 1994. Determination of the physical properties of an arbitrary twisted-nematic liquid crystal cell, *Opt. Eng.* 8: 2704–12.

20. Davis, J.A., Allison, D.B., D'Nelly, K.G., and Moreno, I. 1999. Ambiguities in measuring the physical parameters for twisted nematic liquid crystal spatial light modulators, *Opt. Eng.* 38: 705–9.

21. Davis, J.A., Tsai, P., D'Nelly, K.G., Moreno, I. 1999. Simple technique for determining the extraordinary axis direction for twisted nematic liquid crystal spatial light modulators, *Opt. Eng.* 38: 929–32.

22. Giust, R. and Goedgebuer, J.P. 1998. Determination of the twist angle and the retardation properties of twisted nematic liquid crystal television by spectral measurements, *Opt. Eng.* 37: 629–34.

23. Moreno, I., Bennis, N., Davis, J.A., and Ferreira, C. 1998. Twist angle determination in liquid crystal displays by location of local adiabatic points, *Opt. Comm.* 158: 231–8.

24. Yariv, A. and P. Yeh. 1984. Optical Waves in Crystals. New York: Wiley.

25. Lu, K. and Saleh, B. E. A. 1990. Theory and design of the liquid crystal TV as an optical spatial phase modulator, *Opt. Eng.* 29: 240–6.

26. Coy, J. A., Zaldarriaga, M., Grosz, D. F., and Martínez, O. E. 1996. Characterization of the liquid crystal television as a programmable spatial light modulator, *Opt. Eng.* 35: 15–9.

27. Márquez, A., Campos, J., Yzuel, M.J., Moreno, I., Davis, J.A., Iemmi, C., Moreno, A., and Robert, A. 2000. Characterization of edge effects in twisted nematic liquid crystal displays, *Opt. Eng.* 39: 3301–7.

28. Márquez, A., Iemmi, C., Moreno, I., Davis, J.A., Campos J., Yzuel, M.J. 2001. Quantitative prediction of the modulation behavior of twisted nematic liquid crystal displays, *Opt. Eng.* 40: 2558–64.

29. Pezzaniti, J.L. and Chipman, R.A. 1993. Phase-only modulation of a twisted nematic liquid crystal TV by use of eigenpolarization states, *Opt. Lett.* 18: 1567–9.

30. Davis, J.A., Moreno, I., and Tsai, P. 1998. Polarization eigenstates for twisted nematic liquid crystal displays, *Appl. Opt.* 37: 937–45.

31. Moreno, I., Davis, J.A., D'Nelly, K.G., and Allison, D.B. 1998. Transmission and phase measurements for polarization eigenvectors in twisted-nematic liquid crystal spatial light modulators, *Opt. Eng.* 37: 3048–52.

32. Moreno, I., Fernández-Pousa, C.R., Davis, J.A., and Franich, D.J. 2001. Polarization eigenvectors for reflective twisted nematic liquid crystal displays, *Opt. Eng.* 40: 2220–6.

33. Nicolás, J., Campos, J., and Yzuel, M.J. 2002. Phase and amplitude modulation of elliptic polarization states by nonabsorbing anisotropic elements: Application to liquid-crystal devices, *J. Opt. Soc. Am. A* 19: 1013–20.

34. Durán, V., Lancis, J., Tajahuerce, E., and Fernández-Alonso, M. 2006. Phase-only modulation with a twisted nematic liquid crystal display by means of equi-azimuth polarization states, *Opt. Express* 14: 5607–16.

35. Durán, V., Lancis, J., Tajahuerce, E., and Climent, V. 2007. Poincaré sphere method for optimizing the phase modulation response of a twisted nematic liquid crystal display, *J. Displ. Technol.* 3: 41883.

36. Davis, J.A., Nicolas, J., and Marquez, A., Phasor analysis of eigenvectors generated in liquid crystal displays, *Appl. Opt.* 41: 4579–84.

37. Martínez, J.L., Moreno, I., Davis, J.A., Hernandez, T.J., and McAuley, K.P. 2010. Extended phase modulation depth in twisted nematic liquid crystal displays, *Appl. Opt.* 49: 5929–37.

38. Davis, J.A., Chambers, J.B., Slovick, B.A., and Moreno, I. 2008. Wavelength dependent diffraction patterns from a liquid crystal display, *Appl. Opt.* 47: 4375–80.

39. Moreno, I., Campos, J., Gorecki, C., and Yzuel, M.J. 1995. Effects of amplitude and phase mismatching errors in the generation of a kinoform for pattern recognition, *Jpn. J. Appl. Phys.* 34: 6423–32.

40. Moreno, I., Iemmi, C., Marquez, A., Campos, J., and Yzuel, M.J. 2004. Modulation light efficiency of diffractive lenses displayed in a restricted phase-mostly modulation display, *Appl. Opt.* 43: 6278–84.

41. Dammann, H. and Klotz, E. 1977. Coherent optical generation and inspection of two- dimensional periodic structures, *Opt. Acta* 24: 505–15.

42. Zhou, C. and Liu, L. 1995. Numerical study of Dammann array illuminators. *Appl. Opt.* 34: 5961–9.

43. Moreno, I., Davis, J.A., Cottrell, D.M., Zhang, N., and Yuan, X.-C. 2010. Encoding generalized phase functions on Dammann gratings, *Opt. Lett.* 35: 1536–8.

44. Cottrell, D.M., Davis, J.A., Hedman, T.R., and Lilly, R.A. 1990. Multiple imaging phase-encoded optical elements written on programmable spatial light modulators, *Appl. Opt.* 29: 2505–9.

45. Davis, J.A., Cottrell, D.M., and Sand, D. 2012. Abruptly autofocusing vortex beams, *Opt. Express* 20: 13302–9.

46. Neto, L.G., Roberge, D., and Sheng, Y. 1996. Full range, continuous complex modulation by the use of two coupled-mode liquid crystal television, *Appl. Opt.* 35: 4567–76.

47. Cohn, R.W. and Ling, M. 1994. Approximating fully complex spatial modulation with pseudo-random phase-only modulation, *Appl. Opt.* 33: 4406–15.

48. Davis, J.A., Cottrell, D.M., Campos, J., Yzuel, M.J., and Moreno, I. 1999. Encoding amplitude information onto phase-only filters, *Appl. Opt.* 38: 5004–13.

49. Davis, J.A., McNamara, D.E., Cottrell, D.M., Campos, J., Yzuel, M.J., and Moreno, I. 2001. Encoding complex diffractive optical elements onto a phase-only liquid-crystal spatial light modulator, *Opt. Eng.* 40: 327–9.

50. Davis, J.A., Cottrell, D.M., Campos, J., Yzuel, M.J., and Moreno, I. 1999. Bessel function output from an optical correlator with a phase-only encoded inverse filter, *Appl. Opt.* 38: 6709–13.

51. Campos, J., Yzuel, M.J., Márquez, A., Davis, J.A., Cottrell, D.M., and Moreno, I. 2000. Fully complex synthetic discriminant functions written onto phase-only filters, *Appl. Opt.* 39: 5965–70.

52. Marquez, A., Iemmi, C., Escalera, J.C., Campos, J., Ledesma, S., Davis, J.A., and Yzuel, M.J. 2001. Amplitude apodizers encoded onto Fresnel lenses implemented on a phase-only spatial light modulator, *Appl. Opt.* 40: 2316–22.

53. Yzuel, M.J., Campos, J., Marquez, A., Escalera, J.C., Davis, J.A., Iemmi, C., and Ledesma, S. 2000. Inherent apodization of lenses encoded on liquid crystal devices, *Appl. Opt.* 39: 6034–9.

54. Gualdrón, O., Davis, J.A., Nicolás, J., Campos, J., and Yzuel, M.J., 2003. Complex encoding of rotation invariant filters onto a single phase-only SLM, *Appl. Opt.* 42: 1973–80.

55. Davis, J.A., Valadez, K.O., and Cottrell, D.M. 2003. Encoding amplitude and phase information into a binary phase-only spatial light modulator. *Appl. Opt.* 42: 2003–8.

56. Bandres, M.A. and Gutiérrez-Vega, J.C. 2004. Ince-Gaussian beams, *Opt. Lett.* 29: 144–7.

57. Bentley, J.B., Davis, J.A., Bandres, M.A., and Gutiérrez-Vega, J.C. 2006. Generation of helical Ince-Gaussian beams with a liquid crystal display, *Opt. Lett.* 31: 649–51.

58. Davis, J.A., Bentley, J.B., Bandres, M.A., and Gutiérrez-Vega, J.C. 2006. Generation of helical Ince-Gaussian beams: beam-shaping with a liquid crystal display, *Proc. SPIE* 6290: 62900R1–11.

59. Solomon, J.E. 1981. Polarization imaging, *Appl. Opt.* 20: 1537–44.

60. Davis, J.A., Evans, G.H., Moreno, I. 2005. Polarization-multiplexed diffractive optical elements with liquid-crystal displays, *Appl. Opt.* 44: 4049–52.

61. Bomzon, Z., Kleiner, V., and Hasman, E. 2001. Formation of radially and azimuthally polarized light using space-variant subwavelength metal stripe gratings, *Appl. Phys. Lett.* 79: 1587–9.

62. Davis, J. A., McNamara, D.E., Cottrell, D.M., and Sonehara, T. 2000. Two-dimensional polarization encoding with a phase-only liquid crystal spatial light modulator, *Appl. Opt.* 39: 1549–54.

63. Moreno, I., Martínez, J.L., and Davis, J.A. 2007. Two-dimensional polarization rotator using a twisted-nematic liquid crystal display, *Appl. Opt.* 46: 881–7.

64. Preece D., Keen, S., Botvinick, E., Bowman, R., Padgett, M., and Leach, J. 2008. Independent polarisation control of multiple optical traps, *Opt. Express* 16: 15897–902.

65. Maurer, C., Jesacher, A., Fürhapter, S., Bernet, S., and Ritch-Marte, M. 2007. Tailoring of arbitrary optical vector beams, *New J. Phys.* 9: 78.

66. Wang, X.-L., Ding, J., Ni, W.-J., Guo, C.-S., and Wang, H.-T. 2007. Generation of arbitrary vector beam with a spatial light modulator and a common path interferometric arrangement, *Opt. Lett.* 32: 3549–51.

67. Moreno, I., Iemmi, C., Campos, J., and Yzuel, M.J. 2011. Jones matrix treatment for optical Fourier processors with structured polarization, *Opt. Express* 19: 4583–94.

68. Moreno, I., Davis, J. A., Hernandez, T. M., Cottrell, D.M., and Sand, D. 2011. Complete polarization control of light from a liquid crystal spatial light modulator, *Opt.Express* 20: 364–76.

Novel Laser
Beams

7

Flat-Top Beams

Andrew Forbes
CSIR National Laser Centre

7.1 Introduction

While many lasers will output a Gaussian mode, this is not always the most desirable beam profile for a particular application. As we have seen, Gaussian beams have an $M^2 = 1$ and consequently have a low divergence for a given waist size; this makes them suitable for applications where the laser beam must travel a long distance and still maintain some degree of energy concentration. Gaussian beams are also easy to understand in terms of their propagation since they remain invariant in shape, only changing in size following simple analytical rules (see Chapter 1). However, in many applications a flat-top beam profile, sometimes called a top-hat beam, would be more advantageous, for example, in materials processing, lithography, micromachining, and medical applications. In many of these applications, a beam with a near uniform distribution of energy is desirable: while a Gaussian beam would have a peak energy density of double the average, a flat-top beam would ideally have the same energy density across the entire active area of the beam. In Via drilling this would mean a drill rate that is the same across the area of the hole. In medical applications, say eye surgery, it would mean that the rate of material removal would be equal across the beam and would avoid potential damage due to the large peak energy density at the center of a Gaussian beam. Unfortunately, flat-top beams suffer from some disadvantages: they are not part of the mode set of standard laser resonators and so require custom optics either inside or outside the laser in order to create them; they do not propagate in a shape invariant manner, and so passing them through optical elements (e.g., lenses) would not only change the size but also the profile of the beam. In fact as a general rule, flat-top beams tend not to remain flat-top for very long. However, this can be overcome through a better understanding of the propagation properties of such beams.

In this chapter, we will consider the classes of flat-top beams, discuss how to describe their propagation analytically, use the theory to suggest practical means to propagate flat-top beams over extended distances and finally outline how they might be created in the laboratory.

7.2 Classes of Flat-Top Beams

Flat-top beams, sometimes referred to as top-hat beams, have the property of a near uniform intensity in the central region that falls off very sharply (to zero) near the edges of the beam. In the ideal case a flat-top beam intensity $I_{FT}(r)$ would, therefore, be defined as a step function

$$I_{FT}(r) = \begin{cases} I_0, & r \leq w \\ 0, & r > w \end{cases}, \tag{7.1}$$

where w is the characteristic width of the beam. This intensity profile is not possible to create perfectly in the laboratory because of the infinite spatial frequencies required, and it has a further disadvantage that it cannot be propagated analytically. Instead there are many classes of laser beams that approximate this function when a so-called order parameter is large [1]. Four such laser beams would be: the Fermi–Dirac (FD) beam, the super-Lorentzian (SL) beam, the super-Gaussian (SG) beam, and the flattened-Gaussian (FG) beam, given as

$$I_{FD}(r) = \frac{3\beta^2}{\pi w_{FD}^2} \left(3\beta^2 + 6 \, \mathrm{dilog} \left[1 + \exp(-\beta)\right] + \pi^2\right)^{-1}$$

$$\times \left(1 + \exp\left[\beta\left(\frac{r}{w_{FD}} - 1\right)\right]\right)^{-1}; \tag{7.2}$$

$$I_{SL}(r) = q \sin(2\pi/q)[2\pi^2 w_{SL}^2]^{-1} \left[1 + (r/w_{SL})^q\right]^{-1}; \tag{7.3}$$

$$I_{SG}(r) = \frac{4^{1/p} p}{2\pi w_{SG}^2 \Gamma\left(\frac{2}{p}\right)} \exp\left[-2\left(\frac{r}{w_{SG}}\right)^p\right]; \tag{7.4}$$

$$I_{FG}(r) = \frac{2(N+1)}{\pi w_{FG}^2} \sum_{m,n=0}^{N} \left[\frac{m!n!2^{m+n}}{(m+n)!}\right]$$

$$\times \exp\left[-2(N+1)\left(\frac{r}{w_{FG}}\right)^2\right]^{n+m} \sum_{m,n=0}^{N} \frac{1}{n!m!} \left[(N+1)\left(\frac{r}{w_{FG}}\right)^2\right]. \tag{7.5}$$

Here r is the radial coordinate, w_x represents the flat-top beam size (e.g., w_{SG} is the beam size of the flat-top beam in the super-Gaussian approximation), and all the beams have been normalized to contain an energy of 1. It has already been shown that all these beams can be made to be indistinguishable by selecting the scale and order parameters appropriately, with a conversion table to switch from one beam to another available [1]. It is, therefore, not necessary to discuss all the beams and instead we will focus our attention mostly on the two more common classes of flat-top beams, namely, the SG beam and the FG beam. Although the functional form of Equations 7.2 through 7.5 look very different they do in fact all have similar characteristics. First, they all have an order parameter (p and N in the case of the SG and FG beams, respectively) that determines how closely they resemble a perfect flat-top beam: in the limit of very large values of the order parameter the beams all approach perfect flat-top beams. They also have a characteristic scale parameter (w_{SG} and w_{FG} in the case of the SG and FG beams, respectively) that determines their size. The usefulness of these definitions is that the Gaussian intensity profile is included in the definitions: by setting $p = 2$ for the SG beam, and $N = 0$ for the FG beam, both become Gaussian profiles. In the rest of the chapter we will discuss flat-top beams using the SG and FG beams as examples, and the discussion will assume that the order has been selected to create a good approximation to a flat-top beam. Figure 7.1 shows the intensity profiles of SG beams of orders $p = 2$, $p = 10$, and $p = 100$.

The interest in these beams is evident from the following consideration: the ratio of the peak intensity to the average intensity is approximately 1 (and equal to 1 in the limit of a perfect flat-top beam), whereas for the Gaussian beam this same ratio is always 2; the power density is constant across a flat-top beam but varies continuously across a Gaussian

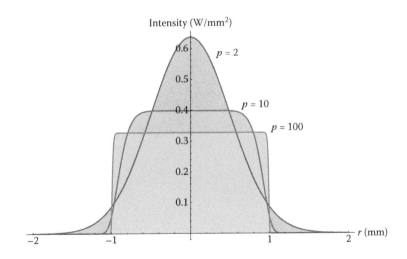

FIGURE 7.1 Plots of the intensity of SG beams for three order parameters: a Gaussian mode ($p = 2$), a super-Gaussian mode of medium order ($p = 10$) and a near perfect flat-top beam ($p = 100$). The energy is normalized to 1 and $w_{SG} = 1$ mm.

beam. This can also be seen from the peak intensity relations for the SG and FG beams:

$$I_{SG}(r) = \frac{4^{1/p}p}{2\pi w_{SG}^2 \Gamma\left(\frac{2}{p}\right)};\tag{7.6}$$

$$I_{FG}(r) = \frac{2(N+1)}{\pi w_{FG}^2} \sum_{m,n=0}^{N} \left[\frac{m!n!2^{m+n}}{(m+n)!}\right].\tag{7.7}$$

In the above equations, the total energy in each beam has been normalized to 1, so that each beam has an average intensity of $1/(\pi w^2)$, where w is the scale of the beam (e.g., $w = w_{SG}$ for the SG beam). Equations 7.6 through 7.7 can be used to find the peak intensity for a chosen order parameter, and it is clear that the peak *decreases* as the order parameter *increases*. This has an implication when using such beams in an application where the peak intensity is important: for example, for the same power content and beam size, the SG beam has a significantly lower peak intensity than an equivalent Gaussian beam. It is easy to show that in order to create an SG beam with the same peak intensity as a Gaussian beam, given that they have the same width parameter, the SG beam would require an increased power content given by

$$P_{SG} = \frac{4\Gamma\left(\frac{2}{p}\right)}{4^{1/4}p}P_G,\tag{7.8}$$

where P_G is the power content of the Gaussian beam and P_{SG} is the power content of the SG beam.

This is shown in Figure 7.2 where it is clear that a substantial power increase is needed to create the same peak intensity as for a Gaussian beam, approaching 2 in the limit of a

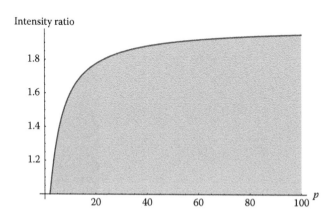

FIGURE 7.2 The required energy increase in order to maintain the same peak intensity as the Gaussian mode as p increases.

perfect flat-top beam. Alternatively if the power is fixed, as it is in many practical cases, then the SG beam size may be adjusted to increase the peak intensity.

7.3 Propagation

7.3.1 Propagation through Free Space

The propagation of flat-top beams cannot, in general, be described by simple analytical expressions and must rather be found by solving the Fresnel diffraction integral given in Chapter 1. It is primarily for this reason that the FG beam has become a popular choice as a descriptor of flat-top beams. In this section, we will use this approximation to discuss the propagation of flat–top beams. The FG beam has become a popular choice as a descriptor of flat-top beams since it allows an analytical expression to be found for the propagation of such beams. The idea is very simple [2]: express the flat-top beam at the plane $z = 0$ as a series of Laguerre–Gaussian beams; since the Laguerre–Gaussian terms have a known (analytical) expression for their propagation, then it follows that the sum of all of them is also known analytically. The steps of this approach will not be reproduced here and instead we simply quote the final result for the FG beam at any plane z:

$$
\begin{aligned}
U_{FG}(r, z) = A_0 \, &\frac{w_N(0)}{w_N(z)} \, \exp\{i[kz - \Phi_N(z)]\} \\
&\times \exp\left[\frac{ikr^2}{2R_N(z)} - \frac{r^2}{w_N^2(z)}\right] \\
&\times \sum_{n=0}^{N} c_n L_n\left[\frac{2r^2}{w_N^2(z)}\right] \exp[-2in\Phi_N(z)],
\end{aligned}
\tag{7.9}
$$

where k is the wavenumber and c_n, w_N, R_N, and Φ_N are given by

$$
c_n = (-1)^n \sum_{m=n}^{N} \frac{1}{2^m} \binom{m}{n};
\tag{7.10}
$$

$$
w_N(z) = w_N(0) \sqrt{1 + \left(\frac{\lambda z}{\pi w_N^2(0)}\right)^2};
\tag{7.11}
$$

$$
R_N(z) = z\left\{1 + \left(\frac{\pi w_N^2(0)}{\lambda z}\right)^2\right\};
\tag{7.12}
$$

$$
\Phi_N(z) = \arctan\left[\frac{\lambda z}{\pi w_N^2(0)}\right].
\tag{7.13}
$$

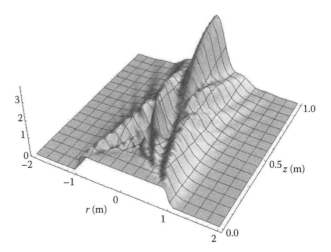

FIGURE 7.3 (**See color insert.**) A 3D plot showing the free-space propagation of a flat-top beam using the FG beam approximation.

Here the parameter $w_N(0)$ is related to the desired beam size (w_{FG}) by

$$w_N(0) = \frac{w_{FG}}{\sqrt{N+1}}.$$

The shape of the FG (flat-top) beam changes during propagation. This can be seen in Figure 7.3 where the cross-section of a FG beam is shown as the initial field at $z = 0$, which then changes shape during propagation toward its far-field profile: Gaussian-like in the center but an airy pattern or sinc function for circular or rectangular flat-top beams, respectively. This is because flat-top beams do not have an angular spectrum that is also a flat-top profile, but rather the Fourier transform of a circular (rectangular) flat-top profile is an airy (sinc) pattern.

The rate of this change depends strongly on the order of the beam; the effective Rayleigh range decreases with increasing order, N. In the near-field region, the shape changes in a rather complicated manner, but converges to an airy pattern in the far field. For a given order, the shape change will be the same for two flat-top beams with the same Fresnel number, $N_F = w_{FG}^2/(\lambda z)$. For example, a $\lambda = 1064$ nm flat-top beam with $w_{FG} = 1$ mm at $z = 1$ m will look the same in shape as another with $w_{FG} = 2$ mm at $z = 4$ m, since they both have the same Fresnel number. When the order parameter is very high, ripples rapidly occur on the beam profile, growing in magnitude during the near-field propagation. This can be delayed by increasing the beam size so that the Fresnel number increases. There is, therefore, a compromise to be made: perfect flat-top beams with very steep edges and flat central regions (high-order parameters) will not propagate very far, whereas larger, near-flat-top beams, with smoother edges will propagate much further. This information can also be captured through the effective Rayleigh range of these beams, given as usual by $z_R = \pi w^2/M^2\lambda$, where once again w is the beam size and

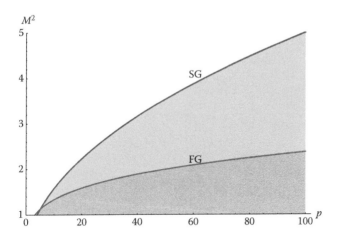

FIGURE 7.4 The change in the beam quality factor for the SG and FG beam.

M^2 is the beam quality factor for the class of flat-top beam chosen. All four of the original flat-top beam classes have simple expressions for the beam quality factor in the limit that the order parameter is very large (tending to infinity), and are given by

$$M_{FG}^2 = \left(\frac{N}{\pi}\right)^{1/4};$$ (7.14)

$$M_{SG}^2 = \frac{1}{2}\sqrt{p};$$ (7.15)

$$M_{FD}^2 = \sqrt{\frac{1+\beta}{8}};$$ (7.16)

$$M_{SL}^2 = \sqrt{\frac{q}{8}}.$$ (7.17)

We see from Equations 7.14 through 7.17 that the beam quality factor increases with the order parameter, approaching infinity, as shown in Figure 7.4. This explains the very rapid divergence increase and shape-invariant distance decrease as we approach better and better approximations to perfect flat-top beams. Why does the M^2 "blow up" so? The reason can be found in the number of spatial frequencies required to create a perfect flat-top beam—an infinite series of harmonics are needed. The result is that the second moment of the intensity in the spatial frequency domain becomes unbounded, and hence the product of the spatial and spatial-frequency moments also become unbounded.

7.3.2 Propagation through ABCD Systems

Since the FG beam approximation to a flat-top beam is made up of a series of Laguerre–Gaussian modes, just as in the case of free-space propagation, so the flat-top beam transformation through an ABCD matrix system may also be described analytically. The result for a beam of initial size w_{FG} at the plane $z = 0$, after passing through an optical system (ABCD) is given by [3]

$$U(r) = \sum_{n=0}^{N} c_n \frac{(A - B/q_0)^n}{(A + B/q_0)^{n+1}} \exp\left(-\frac{ikr^2}{2q}\right) L_n\left[\frac{2r^2}{W^2}\right], \qquad (7.18)$$

where

$$q_0 = \frac{i\pi w_N^2(0)}{\lambda}; \qquad (7.19)$$

$$\frac{1}{q} = \frac{C + D/q_0}{A + B/q_0}; \qquad (7.20)$$

$$W = w_N(0)\sqrt{A^2 + \left(\frac{B\lambda}{\pi w_N^2(0)}\right)^2}, \qquad (7.21)$$

and as before $w_N(0) = w_{FG}/\sqrt{N+1}$. This is a very useful result as usually the flat-top beam is created some distance from the desired application plane and must be transported there through some appropriate optical system. As we have already seen, propagating through free–space will alter the flat-top beam's profile as well as its size. Similarly, the flat-top beam cannot be focussed down by a lens to reduce its size at the target plane since at the focal plane the flat-top beam would appear just as it does when propagating to the far field—we see the Fourier transform of the initial beam, our airy pattern. This is the fundamental reason why Laguerre–Gaussian and Hermite–Gaussian beams retain their shape during propagation and why other beams do not—the former have the property that their Fourier transform returns the same function back, albeit with a scale change. The result is that when transforming such beams through an ABCD system, only the scale and phase of the beam changes, but not the shape. In the case of flat-top beams, the scale, phase, *and* shape of the beam changes.

An application of Equation 7.18 is shown in Figure 7.5 where an FG beam is propagated through a lens of focal length f followed by a distance z. In such a case we have $A = 1 - z/f$, $B = z$, $C = -1/f$, and $D = 1$. The flat-top beam rapidly changes shape as it propagates to its far field at the focal plane of the lens. If we assume that we have a near perfect flat-top beam, as given by Equation 7.1, then the intensity profile at the focal plane of a lens of focal length f will be given by Equation 7.18 with $A = 0$, $B = f$, $C = -1/f$, and $D = 1$. When the order is very high the intensity function (I_f) becomes close to the well-known airy pattern (see Figure 7.6):

$$I_f(r) = I_0 \left(\frac{J_1(kw_{FT}r/f)}{kw_{FT}r/f}\right)^2 \qquad (7.22)$$

with peak intensity $I_0 = A_0 kw_{FT}^2/f$ for a flat-top beam of radius w_{FT} and electric field strength of A_0, taken to be constant across the beam.

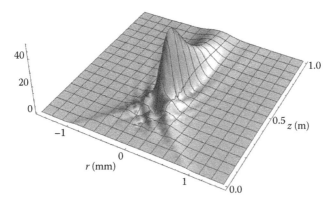

FIGURE 7.5 (**See color insert.**) A 3D plot showing a flat-top beam after passing through a lens, as it propagates towards and through the focus.

FIGURE 7.6 A comparison between an initial flat-top beam and its profile at the focal plane of a lens.

Clearly some care is required in the transport of flat-top beams. The following are possible solutions to deliver a flat-top beam over an extended distance to a target: (i) make sure the effective Rayleigh range is very long by having a suitably large beam radius and/or an appropriately low order; (ii) relay image the beam rather than focus it. A relay imaging system is simply a telescope where the separation of the two lenses is the sum of the focal lengths, as illustrated in Figure 7.7. A beam at plane **A** will be converted to its Fourier transform at **B**, and then inverted back to its original profile at **C**, except for a magnification given by $m = f_2/f_1$. Several relay imaging systems can be used to transfer the flat-top beam from one plane to another without the adverse effects of propagation—that is to say, at plane **C** it appears as if the beam has not propagated at all since both the intensity and phase are unchanged (except for a magnification, which may be set to $m = 1$). Yet it has covered a distance of $2(f_1 + f_2)$. The magnification factor can be used at the final stage to reduce the flat-top beam size to the desired value.

Using Equation 7.18 we can calculate the flat-top profile after any desired optical system.

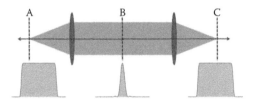

FIGURE 7.7   A schematic of a relay imaging system comprising a two-lens telescope. The phase and intensity of the beam are relayed through such an optical system from plane **A** to plane **C**, passing through the Fourier plane at **B**. The distance from **A** to **C** is $2(f_1 + f_2)$, and half this between **A-B** and **B-C**.

7.4 Generation of Flat-Top Beams

There has been considerable attention paid to the creation of flat-top beams both internal and external to laser cavities. Unfortunately flat-top beams are not eigenmodes of stable cavities with spherical mirrors, and hence custom optics are required to realize a flat-top beam as the output mode. In this section, we will briefly review the standard means to do these transformations and refer the reader to the literature for details.

7.4.1 Intra-Cavity Generation

One of the easiest methods of designing a laser resonator to output a flat-top beam is the reverse propagation method [4]. In this approach, the desired flat-top beam at the output coupler plane is specified. This beam is propagated backward by a distance corresponding to the resonator length (say $z = L$) and the wavefront of the flat-top beam determined. The mirror phase, Φ_m, is then related to this beam phase, Φ_0, by $\Phi_m(r) = \Phi_0(r, L)$. Using Equation 7.9 we can easily determine the required mirror surface as $\Phi_m(r) = \arg[U_{FG}(r, L)]$. This approach results in a graded-phase mirror surface that after reflection creates the conjugate of the incoming beam. Our initial flat-top beam at the output coupler would, therefore, propagate across to the graded-phase mirror, after reflection would become the conjugate of the incoming beam, so that it propagates back to its initial profile. Not only is the flat-top beam now a mode of the cavity (since it keeps repeating after each round trip), but the phase conjugation at the graded-phase mirror has ensured that the unwanted propagation effects do not manifest themselves over the repeated back and forth trips around the resonator. This approach has been used successfully to create and study flat-top beam resonators [4–7].

7.4.2 External Generation

There are several laser beam-shaping techniques avaliable to create a flat-top beam from a Gaussian beam, and the choice of which one to use can be found from a dimensionless parameter β give as [8]

$$\beta = \frac{2\sqrt{2\pi}\, w_G w_{FT}}{\lambda f}, \tag{7.23}$$

where f is the characteristic distance between the shaping element and the plane at which the flat-top beam is formed, usually in the far field at a distance f after a lens of focal length, f, w_G is the $1/e^2$ Gaussian radius and w_{FT} is the flat-top beam size. When β is large (>30) the solution is easy to implement, the flat top will be of a high quality, and the geometric approximations are valid. In such a case any of the solutions would be appropriate, including refractive beam shaping elements [9]. In such a beam-shaping system, the first shaping element converts the Gaussian into a flat-top beam (at the plane of the second element) while leaving the phase as a free parameter, and the second shaping element then corrects the path length of each ray so that the flat-top returns with a flat wavefront. When β is small (<5) the beam-shaping problem is very difficult and few solutions will work. At inbetween values diffractive optics have traditionally been used with success [10]. As discussed in Chapter 6, it is possible to replace the diffractive optical element with an appropriate digital hologram. There are several coding techniques available to convert an arbitrary input beam in to a flat-top beam, but since the most common transformation is from a Gaussian to a flat top, we give the expression for the

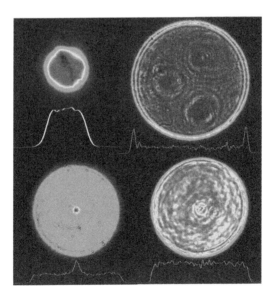

FIGURE 7.8 (**See color insert.**) Examples of flat-top beams created in the laboratory (clockwise from top left): resonator method, two refractive elements, single diffractive element, and phase-only hologram written to a spatial light modulator. The peak in the center of the diffractive optic example is due to an undiffracted zeroth order.

required phase change explicitly:

$$\phi(r) = \beta \frac{\sqrt{\pi}}{2} \int_0^{\frac{\sqrt{2}r}{w_{FT}}} \sqrt{1 - \exp(-\rho^2)}\,d\rho. \tag{7.24}$$

This function is symmetric and thus a function of the radial coordinate (r) only, and converts a circular Gaussian beam of width w_G into a flat-top beam of width w_{FT} at a distance f after a lens of focal length f. In Equation 7.24, the required phase of the diffractive optical element or digital hologram, ϕ, is given as a function of the input and output beams through the parameter β. Some examples of flat-top beams created by these approaches are shown in Figure 7.8.

7.5 Summary

Flat-top beams have many uses but using them requires knowledge of their propagation characteristics. We have seen that while the propagation of ideal flat-top beams may not be described analytically, there are very accurate approximations to such beams that can be described analytically. The result is that the propagation of flat-top beams through any optical system can be modelled. In practice, one usually relays images of the beam from plane to plane in order to cover large distances, or the flat-top should be made large and of low order if free-space propagation is necessary. Finally, at high powers the traditional tools of refractive and diffractive elements can be used to create such beams, while at lower powers the techniques of digital holograms on spatial light modulators are easier to implement.

References

1. Shealy, D.L. and Hoffnagle, J.A. 2006. Laser beam shaping profiles and propagation, *Appl. Opt.* 45: 5118–31.
2. Gori, D.L. 1994. Flattened gaussian beams, *Opt. Commun.* 107: 335–41.
3. Lu, B. and Luo, S. 2001. Approximate propagation equations of flattened Gaussian beams passing through a paraxial ABCD system with hard-edge aperture, *J. Mod. Opt.* 48: 2169–78.
4. Belanger, P.A., Lachance, R.L., and Pare, C. 1992. Super-Gaussian output from a CO_2 laser by using a graded-phase mirror resonator, *Opt. Lett.* 17: 739–41.
5. Leger, J.R., Chen, D., and Wang, W. 1994. Diffractive optical element for mode shaping of a Nd:YAG laser, *Opt. Lett.* 19: 108–10.
6. Litvin, I.A. and Forbes, A. 2009. Gaussian mode selection with intracavity diffractive optics, *Opt. Lett.* 48: 2169–78.
7. Litvin, I.A. and Forbes, A. 2009. Intra-cavity flat-top beam generation, *Opt. Express* 17: 15891–03.

8. Dickey, F.M. and Holswade, S.C. 1996. Gaussian laser beam profile shaping, *Opt. Eng.* 35: 3285–95.
9. Hoffnagle, J.A. and Jefferson, C.M. 2000. Design and performance of a refractive optical system that converts a Gaussian to a flattop beam, *Appl. Opt.* 39: 5488–99.
10. Forbes, A., Dickey, F., DeGama, M., and du Plessis, A. 2012. Wavelength tunable laser beam shaping, *Opt. Lett.* 37: 49–51.

8

Helmholtz–Gauss Beams

Julio C.
Gutiérrez-Vega
Tecnologico de Monterrey

8.1 Introduction

The interest in nondiffracting optical beams is due to the fact that, under ideal conditions, they propagate indefinitely without a change in their transverse shape [1]. Plane waves, Bessel beams [2–4], Mathieu beams [5–7], and parabolic beams (PBs) [8] constitute the four orthogonal and complete families of ideal nondiffracting optical beams in Cartesian, circular cylindrical, elliptic cylindrical, and parabolic cylindrical coordinates, respectively.

The term Helmholtz–Gauss (HzG) beam refers to a field whose disturbance at the plane $z = 0$ reduces to the product of the transverse field of an arbitrary nondiffracting beam (i.e., a solution of the two-dimensional [2D] Helmholtz equation) and a radial Gaussian function. The well-known Bessel–Gauss (BG) beams are a special case of the HzG beams [9,10]. In this chapter, we discuss the propagation characteristics of the HzG beams [11] and provide explicit expressions for the fundamental families: cosine-Gauss (CG) beams in Cartesian coordinates, Mathieu–Gauss (MG) beams in elliptic coordinates, and parabolic-Gauss (PG) beams in parabolic coordinates. Unlike ideal nondiffracting beams, HzG beams carry finite power and can be realized experimentally to a very good approximation [12].

8.2 Scalar HzG Beams in Free Space

We commence the analysis by discussing the classic HzG beam propagating in free space. Suppose that a monochromatic wave $U(\mathbf{r})$ with time dependence $\exp(-i\omega t)$ has a disturbance across the plane $z = 0$ given by

$$U_0(\mathbf{r}_t) = \exp\left(-\frac{r^2}{w_0^2}\right) W(\mathbf{r}_t; \alpha), \tag{8.1}$$

where $\mathbf{r}_t = (x, y) = (r, \theta)$ denotes the transverse coordinates, $W(\mathbf{r}_t; \alpha)$ is the transverse pattern of the ideal nondiffracting beam given by $W(\mathbf{r}_t; \alpha) \exp(ik_z z)$, and w_0 is the waist size of a Gaussian envelope. The transverse α and longitudinal k_z components of the wave vector $\mathbf{k}$ satisfy the relation $k^2 = \alpha^2 + k_z^2$.

The transverse distribution $W(\mathbf{r}_t; \alpha)$ of the ideal nondiffracting beam fulfills the 2D Helmholtz equation

$$\left(\frac{\partial^2}{\partial x^2} + \frac{\partial^2}{\partial y^2} + \alpha^2\right) W(\mathbf{r}_t; \alpha) = 0, \tag{8.2}$$

and can be written as an integral expansion of plane waves either in Cartesian coordinates $\mathbf{r}_t = (x, y)$ or in polar coordinates $\mathbf{r}_t = (r, \theta)$ as follows:

$$W(\mathbf{r}_t; \alpha) = \begin{cases} \displaystyle\int_{-\pi}^{\pi} A(\varphi) \exp\left[i\alpha(x\cos\varphi + y\sin\varphi)\right] d\varphi, \\[2ex] \displaystyle\int_{-\pi}^{\pi} A(\varphi) \exp\left[i\alpha r\cos(\phi - \theta)\right] d\varphi, \end{cases} \tag{8.3}$$

where $A(\varphi)$ is the angular variation of the spectrum of the ideal nondiffracting beam.

The solution of the paraxial wave equation with the boundary condition of transverse field in Equation 8.1 gives the propagated field $U(\mathbf{r})$ [11]

$$U(\mathbf{r}) = \exp\left(-i\frac{\alpha^2}{2k}\frac{z}{\mu}\right) GB(\mathbf{r}) \, W\left(\frac{x}{\mu}, \frac{y}{\mu}; \alpha\right), \tag{8.4}$$

where $GB(\mathbf{r})$ is the fundamental Gaussian beam

$$GB(\mathbf{r}) = \frac{\exp(ikz)}{\mu} \exp\left(-\frac{r^2}{\mu w_0^2}\right), \tag{8.5}$$

and

$$\mu = \mu(z) = 1 + iz/z_R, \tag{8.6}$$

with $z_R = kw_0^2/2$ being the Rayleigh distance of the Gaussian beam.

The angular spectrum of a HzG beam across a plane parallel to the (x, y) plane at a distance z from the origin is given by the 2D Fourier transform

$$\overline{U}(u, v; z) = \frac{1}{2\pi} \iint U(x, y, z) \exp(-ixu - iyv) dx dy, \tag{8.7}$$

where (u, v) are the Cartesian coordinates in frequency space, and the double integral is carried out over the whole plane (x, y). By substituting Equation 8.4 into Equation 8.7, it can be proved that the spectrum of the HzG beams is given by [11]

$$\overline{U}(u, v; z) = D(z) \exp\left(-\frac{w_0^2 \mu}{4} \rho^2\right) W\left(\frac{w_0^2}{2i} u, \frac{w_0^2}{2i} v; \alpha\right), \tag{8.8}$$

where $\rho = (u^2 + v^2)^{1/2}$ is the transverse radius in the frequency space and $D(z) = (w_0^2/2) \exp(ikz - \alpha^2 w_0^2/4)$ is a complex amplitude factor that depends only on z. Equations 8.4 and 8.8 constitute the general expressions of the scalar HzG beam and its spectrum, respectively.

8.2.1 Propagation Properties

From Equation 8.4, it is straightforward to verify that when $w_0 \to \infty$, the HzG beam becomes

$$U(\mathbf{r}) = \exp[i(k - \alpha^2/2k)z] W(\mathbf{r}_t; \alpha), \tag{8.9}$$

which is indeed the equation of an ideal nondiffracting beam with the longitudinal wave vector k_z expressed in the paraxial approximation $k - \alpha^2/2k$. On the other hand, in Equation 8.3 we see that when α tends to zero, the function W becomes a constant; consequently, the HzG beam reduces to a pure Gaussian beam.

Following the physical picture used by Gori et al. [9], to gain a basic understanding on the propagation features of the BG beams, one may imagine that a HzG beam is formed as a coherent superposition of fundamental Gaussian beams that have their waist planes coincident with the plane $z = 0$, whose mean propagation axes lie on the surface of a cone with a half-aperture angle $\theta_0 = \arcsin(\alpha/k) \approx \alpha/k$ and whose amplitudes are modulated angularly by the function $A(\varphi)$. The propagation characteristics are thereby governed by the spreading of the beam due to the conical propagation and the diffraction of the constituent Gaussian beams whose diffraction angle is $\theta_G = 2/kw_0$. The parameter

$$\gamma \equiv \frac{\theta_0}{\theta_G} = \frac{1}{2} \alpha w_0 \tag{8.10}$$

plays an important role in the propagation of the HzG beams. When $\gamma \gg 1$, a significant superposition of all the constituent Gaussian beams will survive up to a distance

$$z_{\max} = \frac{w_0}{\sin \theta_0} \approx \frac{w_0 k}{\alpha} = \frac{z_R}{\gamma}. \tag{8.11}$$

This is a conservative estimate because the spot size of the Gaussian beams actually increases along the propagation axis. For $\gamma \gg 1$, the HzG beam retains the nondiffracting propagation properties of the ideal nondiffracting beam within the range $z \in [-z_{max}, z_{max}]$. Outside this zone, the HzG beam will diverge and acquire wavefront curvature, forming a ring-shaped far-field pattern with mean radius $z \tan \theta_0$, an angular variation given approximately by $A(\varphi)$, and leaving the central region near the z-axis practically obscure.

The case $\gamma \ll 1$ occurs when $\lambda_t = (2\pi/\alpha) \gg \pi w_0$; thus, the outer radial oscillations of the function W at the plane $z = 0$ are strongly damped leaving only a Gaussian-like spot that is angularly modulated by the azimuthal dependence of the ideal nondiffracting beam. The case $\gamma \sim 1$ corresponds to the transition zone between the Gaussian-like behavior ($\gamma \ll 1$) and the nondiffracting-like behavior ($\gamma \gg 1$).

By setting $r = 0$ in Equation 8.4, we can obtain the normalized axial irradiance distribution $I(z) = |U(z)|^2 / |W(0)|^2$ of the HzG beams, namely

$$I(\bar{z}) = \frac{1}{1 + \bar{z}^2} \exp\left(\frac{-2\gamma^2 \bar{z}^2}{1 + \bar{z}^2}\right), \tag{8.12}$$

where $\bar{z} = z/z_R$ is the normalized propagation distance. Equation 8.12 can be applied only to beams for which the field at the origin does not vanish. The irradiance distribution as a function of z is depicted in Figure 8.1 for $\gamma = 1, 3, \ldots, 15$. For each curve, the vertical dashed line is located at the maximum distance $z_{max} = (2k/\alpha^2)\gamma$ (see Equation 8.11), for

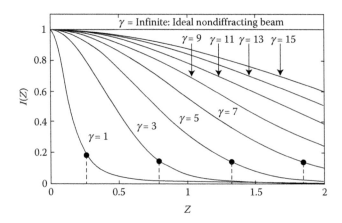

FIGURE 8.1 Normalized axial irradiance distribution of a HzG beam as a function of z for $\gamma = 1, 3, \ldots, 15$. The vertical dashed line is located at the maximum distance $z_{max} = z_R/\gamma$. For numerical purposes, $\lambda = 632.8$ nm and $\theta_0 = 0.05°$.

which the axial irradiance assumes the value

$$I(z_{max}) = \frac{\gamma^2}{1+\gamma^2} \exp\left(\frac{-2\gamma^2}{1+\gamma^2}\right). \tag{8.13}$$

Note that the axial irradiance of the HzG beams is a monotonic decreasing function of the propagation distance z; consequently, it does not present axial irradiance oscillations as it occurs for apertured ideal nondiffracting beams [3].

8.2.2 Basis for HzG Beams

Nondiffracting beam solutions $W(\mathbf{r}_t;\alpha)\exp(ik_z z)$ of the Helmholtz equation can be expressed not only in terms of plane waves, but also in all coordinate systems in which the transverse coordinates and the z-coordinate are separable. In these cases, the mode solutions form a complete basis to construct any nondiffracting beam or HzG beam in the three-dimensional space. We will briefly review these family solutions.

The cosine field

$$W(\mathbf{r}_t;\alpha) = \cos(\alpha y), \tag{8.14}$$

resulting from the superposition of two ideal plane waves $\exp(i\alpha y)/2 + \exp(-i\alpha y)/2$ is one of the simplest examples of a nondiffracting beam in Cartesian coordinates. Its angular spectrum is $A(\varphi) = \delta(\varphi - \pi/2) + \delta(\varphi + \pi/2)$, where $\delta(\cdot)$ is the Dirac delta function. Taking Equation 8.14 as the functional form of W in Equation 8.4, the resulting field is the CG beam

$$CG(\mathbf{r}) = \exp\left(-i\frac{\alpha^2}{2k}\frac{z}{\mu}\right) GB(\mathbf{r}) \cos\left(\frac{\alpha y}{\mu}\right). \tag{8.15}$$

For circular cylindrical coordinates $\mathbf{r}_t = (x,y) = (r\cos\theta, r\sin\theta)$, the separable solutions of Equation 8.2 are given by the well-known Bessel beams [3,4,9]

$$W(\mathbf{r}_t;\alpha) = B_m(r,\theta) = J_{|m|}(\alpha r)\exp(im\theta), \tag{8.16}$$

where J_m is the mth-order Bessel function of the first kind. The angular spectrum of the Bessel beams is located on a single ring of radius $\rho = \alpha$ in the frequency space and its angular dependence is $A(\varphi) \propto \exp(im\varphi)$. From Equation 8.4, the BG beams are given by

$$BG_m(\mathbf{r}) = \exp\left(-i\frac{\alpha^2}{2k}\frac{z}{\mu}\right) GB(\mathbf{r}) J_m\left(\frac{\alpha r}{\mu}\right)\exp(im\phi). \tag{8.17}$$

BG waves are often called vortex beams and carry a well-defined intrinsic orbital angular momentum content equal to $m\hbar$ per photon.

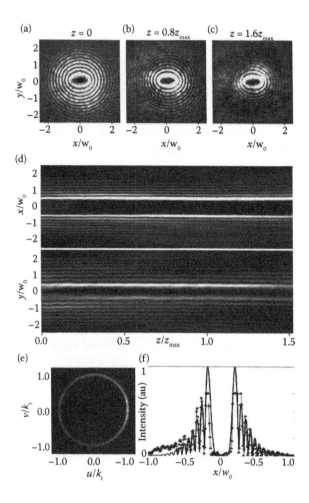

FIGURE 8.2 (a)–(c) Experimental transverse intensity distribution of a seventh-order MG beam at different z-planes; (d) propagation of the intensity along the (x, z) and the (y, z) planes; (e) intensity distribution of the power spectrum in the Fourier plane; and (f) transverse cut at $y = 0$ of the intensity profile of the beam at the waist. The dashed line is the theoretical intensity distribution.

In elliptic cylindrical coordinates, the mth-order even and odd MG beams are given by

$$\mathrm{MG}_m^e \left(\mathbf{r}, q \right) = \exp \left(-i \frac{\alpha^2}{2k} \frac{z}{\mu} \right) \mathrm{GB} \left(\mathbf{r} \right) \mathrm{Je}_m \left(\bar{\xi}, q \right) \mathrm{ce}_m \left(\bar{\eta}, q \right), \qquad (8.18a)$$

$$\mathrm{MG}_m^o \left(\mathbf{r}, q \right) = \exp \left(-i \frac{\alpha^2}{2k} \frac{z}{\mu} \right) \mathrm{GB} \left(\mathbf{r} \right) \mathrm{Jo}_m \left(\bar{\xi}, q \right) \mathrm{se}_m \left(\bar{\eta}, q \right), \qquad (8.18b)$$

where $\mathrm{Je}_m(\cdot)$ and $\mathrm{Jo}_m(\cdot)$ are the mth-order even and odd radial Mathieu functions, $\mathrm{ce}_m(\cdot)$ and $\mathrm{se}_m(\cdot)$ are the mth-order even and odd angular Mathieu functions, respectively. The parameter $q = f_0^2\alpha^2/4$ carries information about the transverse spatial frequency α and the ellipticity of the coordinate system through f_0. In a transverse z-plane, the complex elliptic variables $(\bar{\xi}, \bar{\eta})$ are determined by the following relations

$$x = f_0(1 + iz/z_R) \cosh\bar{\xi} \cos\bar{\eta}, \tag{8.19a}$$

$$y = f_0(1 + iz/z_R) \sinh\bar{\xi} \sin\bar{\eta}, \tag{8.19b}$$

with f_0 being the semifocal separation at the waist plane $z = 0$. While the elliptic variables $(\bar{\xi}, \bar{\eta})$ at the plane $z = 0$ are real, outside this plane, they become complex to satisfy the requirement that the Cartesian coordinates (x, y) keep real in the entire space. As shown in Figure 8.2, from the stationary mode solutions described by Equations 8.18a and 8.18b, it is possible to construct helical MG (HMG) beams of the form

$$\mathrm{HMG}_m^{\pm}(\mathbf{r}, q) = C_m(q)\mathrm{MG}_m^e(\mathbf{r}, q) \pm i S_m(q)\mathrm{MG}_m^o(\mathbf{r}, q), \tag{8.20}$$

but whose phase now rotates elliptically around a strip defined by $(|x| \leq f, 0, z)$. In Equation 8.20, $C_m(q)$ and $S_m(q)$ are normalization factors to ensure that the even and odd beam components carry the same power. It can be shown that for $m \geq 2$ and $q \lesssim m^2/2 - 1$, the normalization factors are numerically very similar, that is, $C_m(q) \approx S_m(q)$ and hence, they can be factorized out. Mathieu fields (8.20) can be written as a superposition of plane waves of the form (8.3) by setting $A_m^{\pm}(\varphi) = \mathrm{ce}_m(\varphi; q) \pm i\, \mathrm{se}_m(\varphi; q)$.

The parabolic cylindrical coordinates (ξ, η) are defined by $x = (\eta^2 - \xi^2)/2$ and $y = \xi\eta$, where the variables range in $\xi \in [0, \infty)$, $\eta \in (-\infty, \infty)$. In terms of the parabolic coordinates, the expressions for the even and odd PG beams are written as

$$\mathrm{PG}^e(\mathbf{r}; a) = \exp\left(-i\frac{\alpha^2}{2k}\frac{z}{\mu}\right)\mathrm{GB}(\mathbf{r})\,\frac{|\Gamma_1|^2}{\pi\sqrt{2}}P_e\left(\sqrt{2\alpha/\mu}\xi; a\right)P_e\left(\sqrt{2\alpha/\mu}\eta; -a\right), \tag{8.21a}$$

$$\mathrm{PG}^o(\mathbf{r}; a) = \exp\left(-i\frac{\alpha^2}{2k}\frac{z}{\mu}\right)\mathrm{GB}(\mathbf{r})\,\frac{2|\Gamma_3|^2}{\pi\sqrt{2}}P_o\left(\sqrt{2\alpha/\mu}\xi; a\right)P_o\left(\sqrt{2\alpha/\mu}\eta; -a\right), \tag{8.21b}$$

where $\Gamma_1 = \Gamma(1/4 + ia/2)$, $\Gamma_3 = \Gamma(3/4 + ia/2)$, and the parameter a represents the continuous order of the beam and can assume any real value in the range $(-\infty, \infty)$. The functions $P_e(\cdot)$ and $P_o(\cdot)$ are the even and odd solutions to the parabolic cylinder differential equation $\left[d^2/dx^2 + (x^2/4 - a)\right]P(x; a) = 0$. The angular spectra for the PBs (8.21)

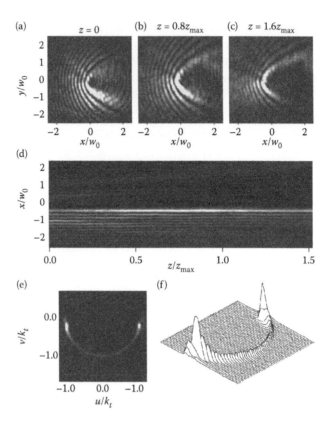

FIGURE 8.3 (a)–(c) Experimental transverse intensity distribution of a TPG beam of order $a = 3$ at different z-planes; (d) propagation of the intensity along the (x, z) plane; (e) distribution of the power spectrum in the Fourier plane; and (f) surface contour plot of the power spectrum. Note the semiannular profile with high intensity at $v = 0$.

are given by

$$A_e(\varphi; a) = \frac{1}{2(\pi \, |\sin \varphi|)^{1/2}} \exp\left(ia \ln \left| \tan \frac{\varphi}{2} \right| \right), \tag{8.22}$$

$$A_o(\varphi; a) = i \begin{cases} A_e(\varphi; a), & \varphi \in (-\pi, 0) \\ -A_e(\varphi; a), & \varphi \in (0, \pi) \end{cases}, \tag{8.23}$$

respectively.

From the stationary beam solutions described by Equations 8.21, it is possible to construct traveling parabolic-Gauss (TPG) beams of the form

$$\text{TPG}^{\pm}(\mathbf{r}; a) = \text{PG}^e(\mathbf{r}; a) \pm i\text{PG}^o(\mathbf{r}; a), \tag{8.24}$$

whose overall phase travels around the semiplane $(x \geq 0, z)$ for $a > 0$. When observed at fixed transverse planes, the phase seems to follow confocal parabolic trajectories. The sign in Equation 8.24 defines the traveling direction. For $a > 0$, the transverse intensity pattern consists of well-defined nondiffracting parabolic fringes with a dark parabolic region around the positive x-axis. From Equations 8.22 and 8.23, the angular spectra of the traveling PBs are given by $A^{\pm}(\varphi; a) = A_e(\varphi; a) \pm i A_o(\varphi; a)$.

We show in Figures 8.3a through c the transverse experimental shapes of the traveling beams TPG^{-} $(\eta, \xi; a = 3)$ at planes $z = 0, 0.8z_{\max}$ and $1.6z_{\max}$. Parabolic nodal lines and fringes along constant ξ are clearly defined. The propagation sequence along the plane (x, z) shown in Figure 8.3d was obtained by recording the transverse field at 30 transverse planes evenly spaced in the range $0 \leq z \leq 1.5z_{\max}$.

8.3 Transformation of HzG Beams by Paraxial Optical Systems

We now discuss a useful generalized form of the scalar HzG beams. This form will be referred to as *generalized* HzG (gHzG) beams and can handle propagation not only in free space, but also in the more general types of paraxial optical systems characterized by complex ABCD matrices. It is found that the gHzG beams are a class of fields that exhibit the property of *form invariance* under paraxial optical transformations. This form-invariance property does not have to be confused with the shape-invariance property of the Hermite–Gauss or Laguerre–Gauss beams that preserve, except by a scaling factor, the same transverse shape under paraxial transformations.

Let us write the complex field amplitude of a gHzG beam at the input plane $z = z_1$ of a radially symmetrical first-order optical ABCD system

$$U_1(\mathbf{r}_1) = \exp\left(\frac{ikr_1^2}{2q_1}\right) W(\mathbf{r}_1; \alpha_1), \quad q_1'' < 0, \tag{8.25}$$

where $\mathbf{r}_1 = (x_1, y_1) = (r_1, \theta_1)$ denotes the transverse coordinates, k is the wave number, and single (′) and double (″) apostrophe denote the real and the imaginary parts of a complex quantity, respectively.

Equation 8.25 results from the product of two functions, each depending on one parameter. The Gaussian apodization is characterized by a complex beam parameter $q_1 = q_1' + iq_1''$. In assuming a complex q_1, we are allowing for the possibility that the Gaussian modulation has a converging ($q_1' < 0$) or diverging ($q_1' > 0$) spherical wavefront. For simplicity in dealing with the ABCD system, the parameter q_1, instead of the width $w_1 = \sqrt{i2q_1/k}$, is used throughout the chapter. The function $W(\mathbf{r}_1; \alpha_1)$ in Equation 8.25 is a solution of the 2D Helmholtz equation $[\partial_{xx} + \partial_{yy} + \alpha_1^2] W = 0$ and describes the transverse field of an ideal nondiffracting beam. It can be expanded as [1]

$$W(\mathbf{r}_1; \alpha_1) = \int_{-\pi}^{\pi} g(\phi) \exp\left[i\alpha_1 r_1 \cos(\phi - \theta_1)\right] d\phi, \tag{8.26}$$

where α_1 and $g(\phi)$ are the transverse wave number and the angular spectrum of the ideal nondiffracting beam, respectively.

In the traditional approach to nondiffracting beams, the wave number α_1 in Equation 8.26 is customarily assumed to be real and positive [1]. For the sake of generality, we will let $\alpha_1 = \alpha_1' + i\alpha_1''$ be arbitrarily complex allowing the possibility of having three main cases: (a) real $\alpha_1 = \alpha_1'$ leads to the *ordinary* HzG (oHzG) beams for which $W(\mathbf{r}_1; \alpha_1)$ is a purely oscillatory function, (b) purely imaginary $\alpha_1 = i\alpha_1''$ leads to evanescent functions $W(\mathbf{r}_1, \alpha_1'')$ that satisfy the modified Helmholtz equation $\left[\partial_{xx} + \partial_{yy} - \alpha_1''^2\right] W = 0$. Owing to this, we shall refer to this case as *modified* HzG (mHzG) beams. Particular examples are given by the cosh-Gaussian and modified BG beams, and (c) complex α_1 leads to our gHzG beams, which, as we will see, can be interpreted as intermediate beam solutions between oHzG and mHzG beams.

To find the transverse field of the gHzG beams at the output plane $z = z_2$ of the ABCD system, we proceed by writing the Huygens diffraction integral

$$U_2(\mathbf{r}_2) = \frac{k \exp(ik\zeta)}{i2\pi B} \iint_{-\infty}^{\infty} U_1(\mathbf{r}_1) \exp\left[\frac{ik}{2B}\left(Ar_1^2 - 2\mathbf{r}_1 \cdot \mathbf{r}_2 + Dr_2^2\right)\right] d^2\mathbf{r}_1, \tag{8.27}$$

where $\mathbf{r}_2 = (x_2, y_2) = (r_2, \theta_2)$ denotes the transverse coordinates at the output plane and ζ is the optical path length from the input to the output plane measured along the optical axis. After substituting Equation 8.25 into Equation 8.27 and using Equation 8.26, the integration can be performed, applying the changes of variables $x_j = u_j \cos\phi - v_j \sin\phi$ and $y_j = u_j \sin\phi + v_j \cos\phi$, for $j = 1, 2$. Upon returning to the original variables, we obtain

$$U_2(\mathbf{r}_2) = \exp\left(\frac{\alpha_1 \alpha_2 B}{i2k}\right) GB(\mathbf{r}_2, q_2) W(\mathbf{r}_2; \alpha_2), \tag{8.28}$$

where

$$GB(\mathbf{r}_2, q_2) = \frac{\exp(ik\zeta)}{A + B/q_1} \exp\left(\frac{ikr_2^2}{2q_2}\right), \tag{8.29}$$

is the output field of a Gaussian beam with input parameter q_1 traveling axially through the ABCD system and the transformation laws for q_1 and α_1 from the plane z_1 to the plane z_2 are

$$q_2 = \frac{Aq_1 + B}{Cq_1 + D}, \quad \alpha_2 = \frac{\alpha_1}{A + B/q_1}. \tag{8.30}$$

Equation 8.28 permits an arbitrary gHzG beam to be propagated in closed form through a real or complex ABCD optical system. Apart from a complex amplitude factor, the output field has the same mathematical structure as the input field; thus, the

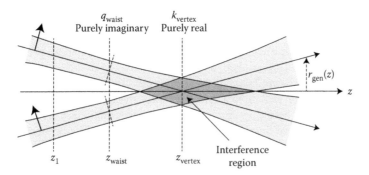

FIGURE 8.4 Physical picture of the decomposition of a gVHzG beam propagating in free space in terms of the fundamental vector Gaussian beams whose mean propagation axes lie on the surface of a double cone.

gHzG beams constitute a class of fields whose form is invariant under paraxial optical transformations. This form-invariance property does not have to be confused with the shape-invariance property of the Hermite–Gauss or Laguerre–Gauss beams that preserve, except by a scaling factor, the same transverse shape under paraxial transformations. The shape of the gHzG beams will change because α_1 and α_2 are not proportional to each other through a real factor leading to different profiles of the function W, and moreover because the parameters q_1 and α_1 are transformed according to different laws.

A physical insight of the gHzG beams is gained by studying some particular cases. First, we consider the free space propagation along a distance $L = z_2 - z_1$, which is represented by the matrix $[A, B; C, D] = [1, L; 0, 1]$. The input and output fields are given by Equations 8.25 and 8.28, where from Equation 8.30, the transformation laws become $q_2 = q_1 + L$ and $\alpha_2 = \alpha_1 q_1/(q_1 + L)$. Note that in this case, the product $q_2 \alpha_2 = q_1 \alpha_1$ remains constant under propagation. A useful physical picture of the gHzG beams in free space can be built up from Equations 8.25 and 8.26 by noting that each constituent plane wave of the function W is multiplied by the Gaussian apodization. In this way, gHzG beams can be viewed as a superposition of tilted Gaussian beams whose mean propagation axes lie on the surface of a double cone whose generatrix is found to be

$$r_{\text{gen}} = \frac{|q_1|^2 \alpha_1''}{kq_1''} + \left(\frac{\alpha_1' q_1'' + \alpha_1'' q_1'}{kq_1''}\right)(z - z_1), \qquad (8.31)$$

and whose amplitudes are modulated angularly by $g(\phi)$ (see Figure 8.4). The position of the waist planes of the constituent Gaussian beams coincides with the plane where the radial factor $\exp\left(ikr_2^2/2q_2\right)$ becomes a real Gaussian envelope, that is, $z_{\text{waist}} = z_1 - q_1'$. At z_{waist}, the parameter q becomes purely imaginary $q_{\text{waist}} = iq_1''$, whereas the wave number reduces to $\alpha_{\text{waist}} = \alpha_1\left(1 - iq_1'/q_1''\right)$.

The extent of the region where the constituent Gaussian beams interfere is maximized where their main propagation axes intersect, that is, the vertex of the double cone. The evaluation of the condition $r_{gen} = 0$ yields $z_{vertex} = z_1 - |q_1|^2 \alpha_1'' / (\alpha_1' q_1'' + \alpha_1'' q_1')$. Now, at the vertex plane, the transverse wave number α becomes purely real $\alpha_{vertex} = \alpha_1' + \alpha_1'' q_1'/q_1''$, with the consequence that at this plane, the transverse amplitude profile belongs to the oHzG kind with $q_{vertex} = \alpha_1 q_1 / \alpha_{vertex}$.

In particular, oHzG beams Equation 8.1 are a special case of Equation 8.25 when $\alpha_1'' = 0$, $q_1' = 0$, and $z_1 = 0$, leading to a cone generatrix given by $r_{gen} = (\alpha_1'/k) z$. On the other side, the mHzG beams occur when $\alpha_1' = 0$; if we additionally set $q_1' = 0$, then $r_{gen} = r_0 = q_1'' \alpha_1''/k$ and thus, the mHzG beams may be viewed as a superposition of Gaussian beams whose axes are parallel to the z-axis and lie on the surface of a circular cylinder of radius r_0.

Further understanding of the gHzG beams can be gained by determining their fractional Fourier transform (FrFT). To do this, we recall that the propagation over a distance $L = z_2 - z_1 = p\pi a/2$ in a graded refractive-index (GRIN) medium with index variation $n(r) = n_0(1 - r^2/2a^2)$ results in the pth FrFT [13]. The ABCD matrix from plane z_1 to plane z_2 is given by

$$\begin{bmatrix} A & B \\ C & D \end{bmatrix} = \begin{bmatrix} \cos(p\pi/2) & a\sin(p\pi/2) \\ -\sin(p\pi/2)/a & \cos(p\pi/2) \end{bmatrix}. \tag{8.32}$$

The output field is described in Equation 8.28 provided that

$$q_2 = a\frac{q_1 \cos(p\pi/2) + a\sin(p\pi/2)}{-q_1 \sin(p\pi/2) + a\cos(p\pi/2)}, \tag{8.33a}$$

$$\alpha_2 = \frac{\alpha_1 q_1}{q_1 \cos(p\pi/2) + a\sin(p\pi/2)}. \tag{8.33b}$$

By setting $p = 1$ in Equation 8.32, we get the ABCD matrix $[0, a; -1/a, 0]$ corresponding to the conventional Fourier transform. This is indeed the matrix transformation from the first to the second focal plane of a converging thin lens of focal length a. Let us assume that the input field of this Fourier transformer is an oHzG profile (i.e., $\alpha_1'' = 0$) with real Gaussian apodization (i.e., $q_1' = 0$). From Equation 8.33, we see that both parameters $q_2 = ia^2/q_1''$ and $\alpha_2 = i\alpha_1' q_1''/a$ become purely imaginary. It is now evident that if an oHzG profile is Fourier transformed, an mHzG profile will be obtained and vice versa. In particular, the propagating field will belong to the oHzG kind when $p = 0, \pm2, \pm4, \cdots$, and to the mHzG kind when $p = \pm1, \pm3, \pm5, \cdots$. For intermediate values of p, the gHzG profiles can be regarded as a continuous transition between oHzG and mHzG profiles.

Equation 8.28 can be applied to propagate gHzG beams through complex ABCD systems. To illustrate this point, we show in Figure 8.5 the propagation of a generalized fourth-order MG profile through two free space regions of thickness 1 m separated by a thin lens with complex power $1/f = 1.5 - i0.03$ m^{-1}. Physically, this element represents a thin lens with focal length 2/3 m apodized by a Gaussian transmittance $\exp(-0.03kr^2/2)$.

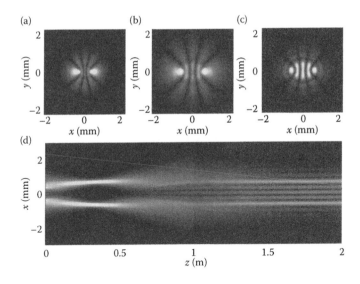

FIGURE 8.5 Propagation of an $MG_4(\mathbf{r}, \varepsilon = 8)$ ($\kappa_1 = -18,394 + i4751 \text{ m}^{-1}$, $q_1 = -0.331 - i0.102$ m) through two free space regions separated by a complex thin lens. The transverse field is at (a) $z = 0$, (b) after the complex lens $z = 1$ m, (c) at $z = 2$ m, and (d) evolution along the plane (z, x).

8.4 Generation of HzG Beams

Since the original works by Durnin et al. [3,4], several methods to generate nondiffracting and HzG beams have been reported. Essentially, all methods consist of superposing a conical arrangement of plane waves for ideal nondiffracting beams or Gaussian beams for HzG beams. Two approaches are possible: (a) passive optical systems fed by laser light using ring apertures [3,4,6], axicons [14–16], holographic plates [17,18], diffractive optical elements [19], spatial light modulators (SLMs) [20,21], anisotropic nonlinear crystals [22,23], and so on. (b) Active optical systems using laser resonators with reflective axicons [24–28], output mirrors with special transmittances [29,30], and intracavity elements [31–35].

8.4.1 Passive Methods

The original experimental setup used by Durnin et al. [3,4] to produce Bessel beams is based on the fact that an annulus is the Fourier transform of a perfect ideal Bessel beam. Therefore, an annular aperture can be positioned in the back focal plane of a lens creating a Bessel mode. By modulating the amplitude and phase of the annulus, we can generate any other nondiffracting beam. For example, the Durnin setup was successfully implemented to generate fundamental and high-order Mathieu fields [6]

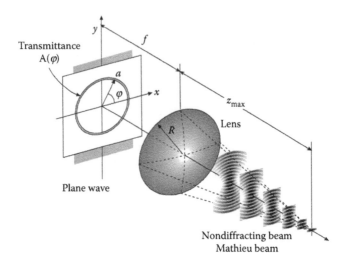

FIGURE 8.6 Durnin experimental setup to generate nondiffracting beams.

(see Figure 8.6) and stationary even and odd PBs [36]. The Durnin setup also permits the generation of vector HzG beams if the field at the ring is polarized properly with, for example, space-variant diffractive polarization components. Unfortunately, the Durnin scheme suffers from low light efficiency, which excludes its use in most of the applications of HzG beams. Therefore, high-efficiency methods are often preferable, even if they have limitations such as a non-uniform longitudinal intensity profile.

In practical terms, HzG beams may be more efficiently generated by the use of an axicon. This is an optical element with a conical refracting or reflecting surface, which transforms an incident plane wave into a converging conical wave (see Figure 8.7a). If the illumination is uniform, the constituent plane waves of the conical waves superpose in the interference region, building up a Bessel beam. Careful illumination of such a device

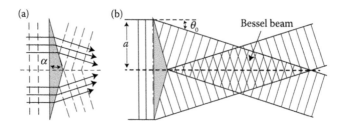

FIGURE 8.7 (a) An axicon transforms an incident plane wave into a converging conical wave. (b) The constituent plane waves of the conical waves superpose in the interference region, building up a Bessel beam.

is required to avoid any aberrations. If the axicon is illuminated with a properly modulated plane wave, one can get different nondiffracting profiles. Microfabricated axicons have also been generated that are important for applications in microfluidics and optical trapping [37].

The holographic techniques are widely used to generate HzG beams in an efficient and versatile way. The holograms can be either static etched glass elements [17] or reconfigurable SLMs [18]. Holography also allows one to generate superpositions of nondiffracting beams for several purposes. Higher-order MG beams have been generated holographically [7]. The experimental generation of all families of HzG beams using holographic techniques has also been reported by López-Mariscal et al. [12]. For some applications, it is desirable to alter the shape of the HzG beam in real time. Davis et al. [20] were the first to demonstrate this by writing diffractive axicons in SLMs. The techniques to generate HzG beams with SLMs are covered extensively in Chapter 6.

8.4.2 Active Methods

Active sources of HzG beams can be realized using laser resonators with HzG natural modes. The first evidence can be traced back to 1989 when Durnin and Eberly proposed to place a thin annular aperture at one of the mirrors of a stable cavity. This scheme was later demonstrated by Uehara and Kikuchi in 1989 with limited results [29]. In 1994, Friberg and Turunen [38] showed that apodized Bessel modes are the exact eigenmodes of an infinite-aperture (passive) Fabry–Perot resonator. Unfortunately, the Gaussian-to-Bessel conversion efficiency in a Fabry–Perot resonator is low. Trying to solve this problem, in 1998, Pääkkönen and Turunen [30] applied the concept of graded-phase mirrors, introduced by Bélanger and Paré [32,33], to suggest resonators with aspheric mirrors, which conjugate the phase of the incident field to produce BG modes.

Axicon-based resonators supporting Bessel beams were proposed independently by Rogel et al. [24] and Khilo et al. [25] in 2001 based on the fact that an axicon transforms an incident plane wave into a Bessel beam. This configuration has the technical advantage that it does not require intracavity optics or special shapes of the active medium and employs conventional mirrors (not complex-shaped mirrors) and commercially available axicons. In 2003, Gutiérrez-Vega et al. [26] continued the exploration of the axicon-based resonator properties, extending the analysis to concave spherical mirrors and developing a formal geometrical and wave analysis of the performance of the bare cavity. The axicon-based resonator with convex output coupler operating in an unstable regime was proposed by Tsangaris et al. [39] in 2003 and was later studied by Hernandez-Aranda et al. [27] in 2005.

The configuration of the axion-based resonator (ABR) is shown in Figure 8.8. The resonator consists of a refractive conical mirror and a conventional output mirror separated at a distance $L = a/(2 \tan \theta_0)$, where a and θ_0 are the radius and characteristic angle of the axicon mirror. This resonator has been employed for generating both BG and MG modes using different active media. For example, Figure 8.9 shows the measured intensity distributions and their theoretical predictions of several even MG modes for different

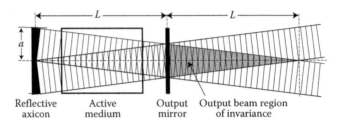

FIGURE 8.8 Configuration of the ABR. The conical wave fronts represent the geometric field distribution inside and outside the cavity. The output beam preserves the nondiffracting behavior inside the conical green region.

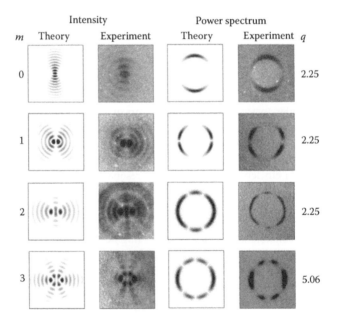

FIGURE 8.9 Theoretical and experimental intensity patterns of even-parity MG beams and their power spectra for the orders $m = \{0, 1, 2, 3\}$ and ellipticities q.

orders and ellipticities generated in a 10-W cw CO_2 laser resonator composed of a copper axicon and a plane output coupler operating at 10.6 μm [28]. Figure 8.9 also shows the power spectra of the output beams recorded at the focal plane of a converging lens of focal distance $f = 25$ cm. As expected, the patterns correspond to circular thin rings whose angular intensity variation is described by the angular Mathieu function, that is $|\text{ce}_m(\theta, q)|^2$.

8.5 Conclusion

In this chapter, the basic theory of HzG beams has been covered, considering monochromatic scalar fields propagating in paraxial ABCD systems. However, it is important to remark that the theory and applications of HzG beams go beyond this domain. The model of the scalar HzG beams has been extended to the vector domain as formal solutions of the Maxwell equations in free space and also in ABCD paraxial optical systems [40]. In this case, the electric and magnetic components of the vector HzG beams are determined starting from scalar solutions of the 2D Helmholtz and Laplace equations [41]. The propagation of scalar HzG beams in dielectric media exhibiting loss or gain has also been studied by Guizar et al. [42] obtaining and discussing general expressions for the field propagation, the time-averaged power on propagation, the trajectory of the beam centroid, the beam spreading, the nondiffracting distance, and the far field of HzG beams. The stability and propagation of HzG beams in turbulent media has also been investigated by Noriega et al. [43] who derived expressions for the first- and second-order normalized Born approximations, the second-order moments, and the transverse intensity pattern of the HzG beams at arbitrary propagation distances under the Rytov approximation. Finally we stress that the theory and applications of the HzG beams are still under active study. In particular, the subject of partially coherent and partially polarized HzG beams has been barely touched [44–47].

The particular properties of HzG beams immediately suggest a wide range of applications, some of which were already recognized by McGloin et al. in a very good review of application published in 2005 [48]. The obvious applications of HzG beams may be found in optical alignment, pointing, and tracking. Guiding, collimation, focusing, trapping, and cooling of neutral atoms are possible through the dipole force exerted on the atom by near-resonant BG beams [49,50].

References

1. Bouchal, Z. 2003. Nondiffracting optical beams: Physical properties, experiments, and applications, Czech, *J. Phys.* 53: 537–624.
2. Sheppard, C.J.R. and Wilson T. 1978. Gaussian-beam theory of lenses with annular aperture, *Micro. Opt. Acoust.* 2: 105–12.
3. Durnin, J. 1987. Exact solutions for nondiffracting beams. I. The scalar theory, *J. Opt. Soc. Am. A* 4: 651–4.
4. Durnin, J., Micely, J.J., and Eberly, J.H. 1987. Diffraction-free beams, *Phys. Rev. Lett.* 58: 1499–501.
5. Gutiérrez-Vega, J.C., Iturbe-Castillo, M.D., and Chávez-Cerda, S. 2000. Alternative formulation for invariant optical fields: Mathieu beams, *Opt. Lett.* 25: 1493–5.
6. Gutiérrez-Vega, J.C., Iturbe-Castillo, M.D., Ramírez, G.A., Tepichín, E., Rodríguez-Dagnino, R.M., Chávez-Cerda, S., and New, G.H.C. 2001. Experimental demonstration of optical Mathieu beams, *Opt. Comm.* 195: 35–40.

7. Chávez-Cerda, S., Padgett, M.J., Allison I., New, G.H.C., Gutiérrez-Vega, J. C., O'Neil, A.T., MacVicar, I., and Courtial, J. 2002. Holographic generation and orbital angular momentum of high-order Mathieu beams, *J. Opt. B: Quantum Semiclass. Opt.* 4: S52–57.

8. Bandres, M.A., Gutiérrez-Vega, J.C., and Chávez-Cerda, S. 2004. Parabolic nondiffracting optical wave fields, *Opt. Lett.* 29: 44–6.

9. Gori, F., Guattari G., and Padovani, C. 1987. Bessel-Gauss beams, *Opt. Comm.* 64: 491–5.

10. Kiselev, A.P. 2004. New structures in paraxial Gaussian beams, *Opt. Spectros.* 96: 479–81.

11. Gutiérrez-Vega, J.C. and Bandres, M.A. 2005. Helmholtz–Gauss waves, *J. Opt. Soc. Am. A* 22: 289–98.

12. López-Mariscal, C., Bandres, M.A., and Gutiérrez-Vega, J.C. 2006. Observation of the experimental propagation properties of Helmholtz–Gauss beams, *Opt. Eng.* 45: 068001.

13. Ozaktas, H.M., Zalevsky, Z., and Kutay, M.A. 2001. *The Fractional Fourier Transform*, Chichester: Wiley.

14. McLeod, J.H. 1954. The axicon: A new type of optical element, *J. Opt. Soc. Am.* 44: 592–7.

15. Indebetouw, G. 1989. Nondiffracting optical fields: Some remarks on their analysis and synthesis, *J. Opt. Soc. Am. A* 6: 150–2.

16. Scott, G. and McArdle, N. 1992. Efficient generation of nearly diffraction-free beams using an axicon, *Opt. Eng.* 31: 2640–3.

17. Turunen, J., Vasara, A., and Friberg, A.T. 1998. Holographic generation of diffraction-free beams, *Appl. Opt.* 27: 3959–62.

18. Vasara, A., Turunen, J., and Friberg, A.T. 1989. Realization of general nondiffracting beams with computer-generated holograms, *J. Opt. Soc. Am. A* 6: 1748–54.

19. Wen-Xiang, C., Nan-Xian, C., and Ben-Yuan, G., 1998. Generation of nondiffracting beams by diffractive phase elements, *J. Opt. Soc. Am. A* 15: 2362–4.

20. Davis, J.A., Guertin, J., and Cottrell, D.M. 1993 Diffraction-free beams generated with programmable spatial light modulators, *Appl. Opt.* 32: 6368–70.

21. Davis, J.A., Carcole, E., and Cottrell, D.M. 1996. Nondiffracting interference patterns generated with programmable spatial light modulators, *Appl. Opt.* 35: 599–602.

22. Khilo, N.A., Petrova, E.S., and Ryzhevich, A.A. 2001. Transformation of the order of Bessel beams in uniaxial crystals, *Quantum Electron.* 31: 85–9.

23. Belyi, V.N., Kazak, N.S., and Khilo, N.A. 2000. Frequency conversion of Bessel light beams in nonlinear crystals, *Quantum Electron.* 30: 753–66.

24. Rogel-Salazar J., New G.H.C., and Chávez-Cerda, S. 2001. Bessel–Gauss beam optical resonator, *Opt. Commun.* 190: 117–22.

25. Khilo, N.A., Katranji, E.G., and Ryzhevich, A.A. 2001. Axicon-based Bessel resonator: Analytical description and experiment, *J. Opt. Soc. Am. A* 18: 1986–92.

26. Gutiérrez-Vega, J.C., Rodríguez-Masegosa, R., and Chávez-Cerda, S. 2003. Bessel–Gauss resonator with spherical output mirror: Geometrical- and wave-optics analysis, *J. Opt. Soc. Am. A* 20: 2113–22.

27. Hernández-Aranda, R.I., Chávez-Cerda, S., and Gutiérrez-Vega, J.C. 2005. Theory of the unstable Bessel resonator, *J. Opt. Soc. Am. A* 22: 1909–17.

28. Alvarez-Elizondo, M.B., Rodríguez-Masegosa, R., and Gutiérrez-Vega, J.C. 2008. Generation of Mathieu–Gauss modes with an axicon-based laser resonator, *Opt. Express* 16: 18770–5.

29. Uehara, K. and Kikuchi, H. 1989. Generation of nearly diffraction-free laser beams, *Appl. Phys. B* 48: 125–9.

30. Pääkkönen, P. and Turunen, J. 1998. Resonators with Bessel–Gauss modes, *Opt. Commun.* 156: 359–66.

31. Durnin, J. and Eberly, J. H. 1998. Diffraction free arrangement, U.S. Patent 4887885, December 19.

32. Bélanger, P.A. and Paré, C. 1991. Optical resonators using graded-phase mirrors, *Opt. Lett.* 16: 1057–9.

33. Bélanger, P. A., Lachance, R.L., and Paré, C. 1992. Super-Gaussian output from a CO_2 laser by using a graded-phase mirror resonator, *Opt. Lett.* 17: 739–41.

34. Litvin, I.A. and Forbes, A. 2008. Bessel–Gauss resonator with internal amplitude filter, *Opt. Commun.* 281: 2385–92.

35. Litvin, I.A., Khilo, N.A., Forbes, A., and Belyi, V.N. 2010. Intra-cavity generation of Bessel-like beams with longitudinally dependent cone angles, *Opt. Express* 18: 4701–8.

36. López-Mariscal, C., Bandres, M.A., Chávez-Cerda, S., and Gutiérrez-Vega, J.C. 2005. Observation of parabolic nondiffracting optical fields, *Opt. Express* 13: 2364–9.

37. Cheong, W.C., Ahluwalia, B.P.S., Yuan, X.-C., Zhang, L.-S., Wang, H., Niu, H. B., and Peng, X. 2005. Fabrication of efficient microaxicon by direct electron-beam lithography for long nondiffracting distance of Bessel beams for optical manipulation, *Appl. Phys. Lett.* 87: 024104.

38. Friberg A. T. and Turunen J. 1994. Spatially partially coherent Fabry–Perot modes, *J. Opt. Soc. Am. A* 11: 227–35.

39. Tsangaris C. L., New G. H. C., and Rogel-Salazar J. 2003. Unstable Bessel beam resonator, *Opt. Commun.* 223: 233–8.

40. Bandres M. A. and Gutiérrez-Vega J. C. 2005. Vector Helmholtz–Gauss and vector Laplace–Gauss beams, *Opt. Lett.* 30: 2155–7.

41. Hernandez-Aranda, R.I., Gutiérrez-Vega, J.C., Guizar-Sicairos, M., and Bandres M.A. 2006. Propagation of generalized vector Helmholtz–Gauss beams through paraxial optical systems, *Opt. Express* 14: 8974–88.

42. Guizar-Sicairos, M. and Gutiérrez-Vega, J.C. 2006. Propagation of Helmholtz–Gauss beams in absorbing and gain media, *J. Opt. Soc. Am. A* 23: 1994–2001.

43. Noriega-Manez, R.J. and Gutiérrez-Vega, J.C. 2007. Rytov theory for Helmholtz–Gauss beams in turbulent atmosphere, *Opt. Express* 15: 16328–41.

44. Davila-Rodriguez, J. and Gutiérrez-Vega, J.C. 2007. Helical Mathieu and parabolic localized pulses, *J. Opt. Soc. Am. A* 24: 3449–55.

45. Kiselev, A.P. 2007. Localized light waves, *Opt. Spectros.* 102: 603–22.

46. Hernández-Figueroa, H.E., Zamboni-Rached, M., and Recami E. (eds.) 2008. *Localized Waves.* New York: Wiley.

47. Fischer, P., Brown, C., Morris, J., López-Mariscal, C., Wright, E., Sibbett, W., and Dholakia, K. 2005. White light propagation invariant beams, *Opt. Express* 13: 6657–66.

48. McGloin, D. and Dholakia, K. 2005. Bessel beams: Diffraction in a new light, *Contemp. Phys.* 46: 15–28.

49. Arlt, J., Dholakia, K., Soneson, J., and Wright, E.M. 2001. Optical dipole traps and atomic waveguides based on Bessel light beams, *Phys. Rev. A* 63: 063602.

50. López-Mariscal, C., Gutiérrez-Vega, J.C., Milne, G., and Dholakia, K. 2006. Orbital angular momentum transfer in helical Mathieu beams, *Opt. Express* 14: 4182–7.

9

Vector Beams

Qiwen Zhan
University of Dayton

9.1 Introduction

Polarization is an important property of light. This vector nature of light and its interactions with matters have been exploited in many areas of optical sciences and optical engineering. Most past research dealt with optical beams with spatially homogeneous states of polarization (SOP), such as linear, elliptical, and circular polarizations. For these cases, the spatial dependence of SOP in the beam cross section has been largely ignored. This simple approach has worked very well in most of the cases in the past. With the recent rapid advances in high-power computing and micro- and nanofabrication, optical beams with spatially variant SOPs within the beam cross section are increasingly available and the study of their properties both numerically and experimentally has become more accessible. By engineering the spatial SOP distribution of a light beam purposefully and carefully, new effects and phenomena that can expand the functionality and enhance the capability of optical systems are expected.

These unconventional polarization states of light have received rapidly increasing interests in the past decade. Some of the developments in optical vector beams, particularly in the special class of cylindrical vector (CV) beams, have been summarized in several extensive review volumes [1–3]. To capture the rapid development in this important emerging field of optics, an *optics express* focus issue on the unconventional polarization states of light was successfully organized and published in 2010 [4]. This chapter is intended to offer a general overview of the field and some of the latest developments in the past couple of years. In the first section, the mathematical description, recently developed generation and manipulation methods, the focusing properties, and applications of CV beams will be reviewed. In the next section, a more generalized and complicated type of vector beams—full Poincaré beams—will be introduced. The generation of this type of vector beams and their application in focus shaping will be presented. In the last section, I will present the concept of generalized vector beams, the methods to generate these type of beams, and discuss the potential applications of these most generalized vector beams.

9.2 CV Beams

One particular class of vector beams that has been intensively studied is optical beams with cylindrical symmetry in polarization, the so-called CV beams. Owing to their interesting properties and potential applications, there has been a rapid increase of the number of publications on CV beams. In this section, we introduce the commonly used mathematical descriptions, present some of the recently developed generation and manipulation methods, examine the focusing properties, and review some of the applications of these CV beams.

9.2.1 Mathematical Description

The CV beams are vector beam solutions of Maxwell's equations that obey axial symmetry in both amplitude and phase. In free space, the paraxial harmonic beam-like solutions are obtained by solving the scalar Helmholtz equation. As we have already seen (Chapter 1, Section 1.7) for a beam-like paraxial solution in Cartesian coordinates, with slow-varying envelop approximation, the well-known Hermite–Gauss solution HG_{mn} mode can be derived as

$$u(x, y, z) = AH_m\left(\sqrt{2}\frac{x}{w(z)}\right) H_n\left(\sqrt{2}\frac{y}{w(z)}\right) \frac{w_0}{w(z)}$$

$$\exp[-i\phi_{mn}(z)] \exp\left[-\frac{k}{2q(z)}r^2\right], \tag{9.1}$$

with $H_m(x)$ being the Hermite polynomials that satisfy the differential equation

$$\frac{d^2 H_m}{dx^2} - 2x\frac{dH_m}{dx} + 2mH_m = 0,$$

and $w(z)$ is the beam size, w_0 is the beam size at beam waist, $z_0 = \frac{\pi w_0^2}{\lambda}$ is the Rayleigh range, $q(z) = z + jz_0$ is the complex beam parameter, and $\phi_{mn}(z) = (m + n + 1)$

$\tan^{-1}(z/z_0)$ is the Gouy phase shift. For $m = n = 0$, this solution becomes the fundamental Gaussian beam:

$$u(r,z) = A\frac{w_0}{w(z)}\exp[-i\phi(z)]\exp\left[-\frac{k}{2q(z)}r^2\right]. \tag{9.2}$$

For a beam-like paraxial solution in cylindrical coordinates, the Laguerre–Gauss (LG) solution LG_{pl} modes can be obtained as

$$u(r,\phi,z) = A\left(\sqrt{2}\frac{r}{w(z)}\right)^l L_p^l\left(2\frac{r^2}{w^2(z)}\right)\frac{w_0}{w(z)}\exp\left[-\frac{kr^2}{2q(z)}\right]\exp(-jl\phi), \tag{9.3}$$

where $L_p^l(x)$ is the associated Laguerre polynomials that satisfy the differential equation

$$x\frac{d^2L_p^l}{dx^2} - (l+1-x)\frac{dL_p^l}{dx} + pL_p^l = 0,$$

and $\phi_{pl}(z) = (2p+l+1)\tan^{-1}(z/z_0)$ is the Gouy phase shift. For $l = p = 0$, the solution also reduces to the fundamental Gaussian beam solution.

These commonly known solutions are the paraxial beam-like solutions to the scalar Helmholtz equation that correspond to spatially homogeneous polarization or scalar beams. For these beams, the electric field oscillation trajectory (i.e., SOP) does not depend on the location of observation points within the beam cross section. Thus, a single SOP is sufficient to describe the polarization property of the beam. However, if we consider the full vector wave equation for the electric field

$$\nabla \times \nabla \times \vec{E} - k^2\vec{E} = 0, \tag{9.4}$$

there are beam-like solutions that have spatially variant SOP distributions across the beam cross section. For example, if we impose axial symmetry to the beam-like vector solution of Equation 9.4, one particular solution would have the electric field aligned in the azimuthal direction with the following form:

$$\vec{E}(r,z) = U(r,z)\exp[i(kz - \omega t)]\vec{e}_\varphi, \tag{9.5}$$

where $U(r,z)$ satisfies the following equation under paraxial and slow-varying envelop approximations:

$$\frac{1}{r}\frac{\partial}{\partial r}\left(r\frac{\partial U}{\partial r}\right) - \frac{U}{r^2} + 2ik\frac{\partial U}{\partial z} = 0. \tag{9.6}$$

The solution that obeys azimuthal polarization symmetry has a trial solution as

$$U(r,z) = AJ_1\left(\frac{\beta r}{1 + iz/z_0}\right)\exp\left[-\frac{i\beta^2 z/(2k)}{1 + iz/z_0}\right]u(r,z), \tag{9.7}$$

where $u(r, z)$ is the fundamental Gaussian solution given in Equation 9.2, $J_1(x)$ is the first order Bessel function of the first kind, and β is a constant scale parameter. This solution corresponds to an azimuthally polarized vector Bessel–Gauss beam solution. In a similar way, we can derive a transverse magnetic field solution that is given by

$$\vec{H}(r, z) = -BJ_1\left(\frac{\beta r}{1 + iz/z_0}\right) \exp\left[-\frac{i\beta^2 z/(2k)}{1 + iz/z_0}\right] u(r, z) \exp[i(kz - \omega t)]\vec{h}_\varphi. \qquad (9.8)$$

For this solution, the corresponding electric field in the transverse plane is aligned in the radial direction. Hence, it represents the radial polarization for the electric field. It should be noted that the electric field is not purely transverse and there should be a z-component of the electric field as well. However, this z-component is typically very weak and can be ignored under paraxial conditions.

Examples of the spatial distributions of the instantaneous electric field vector for several linearly polarized Hermite–Gauss, LG modes, and the CV modes are illustrated in Figure 9.1. The SOPs of the modes shown in Figure 9.1a through f are considered spatially homogeneous even if the electric field may have an opposite instantaneous direction caused by the inhomogeneous phase distribution across the beam. The polarization patterns of the radial polarization and azimuthal polarization are illustrated in Figure 9.1g and h, respectively. A linear superposition of these two polarizations can be used to form the generalized CV beam as shown in Figure 9.1i. One characteristic of these CV modes feature is a null or a singularity point of the beam cross section at the beam center. This is dictated by the transverse field continuity of the electrical field. Taking the radial polarization (Figure 9.1g) as an example, the electrical vector fields on either side of the beam center point to the opposite directions. As one approaches the beam center, to ensure the transverse electrical field to be continuous, the amplitude of the electrical field has to vanish, leading to a dark core of the beam cross section.

In many applications that use CV beams with a large cross section, other simplified distribution can be used. For very small β, the vector Bessel–Gauss beam at the beam waist can be approximated as

$$\vec{E}(r, z) = Ar \exp(-\frac{r^2}{w^2})\vec{e}_i, i = r, \varphi. \qquad (9.9)$$

This is the LG_{01} modes without the vortex phase term $\exp(-i\varphi)$. Consequently, using Equations 9.1 and 9.2, it is fairly straightforward to show that CV beams can also be obtained with a superposition of orthogonally polarized Hermite–Gauss HG_{01} and HG_{10} modes:

$$\vec{E}_r = HG_{10}\vec{e}_x + HG_{01}\vec{e}_y, \qquad (9.10)$$

$$\vec{E}_\varphi = HG_{01}\vec{e}_x + HG_{10}\vec{e}_y, \qquad (9.11)$$

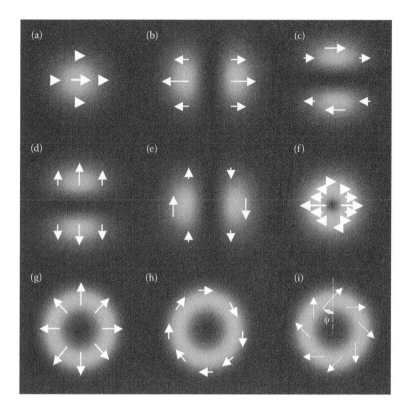

FIGURE 9.1 Illustration of the spatial distribution of the instantaneous electric field for different modes. (a) x-Polarized fundamental Gaussian mode, (b) x-polarized HG_{10} mode, (c) x-polarized HG_{01} mode, (d) y-polarized HG_{01} mode, (e) y-polarized HG_{01} mode, (f) x-polarized LG_{01} mode, (g) radially polarized mode, (h) azimuthally polarized mode, and (i) generalized CV beams. (From Zhan, Q. 2009. Cylindrical vector beams: From mathematical concepts to applications. *Adv. Opt. Photon.* 1: 1–57. With permission of Optical Society of America.)

where E_r and E_φ denote the radial and azimuthal polarization, respectively. This linear superposition is illustrated in Figure 9.2. Such an observation has led to some of the interferometric techniques to generate CV beams.

9.2.2 Generation and Manipulation

In general, the techniques used to generate vector beams can be divided into two categories: active and passive, depending on whether a gain medium is involved. Since 1972 [5,6], and particularly the late 1990s, various active and passive methods have been developed to generate CV beams. An overview of a variety of these methods can be found in a previous review article [2]. Among the passive methods, a very popular and powerful method involves the use of a liquid crystal spatial light modulator (LC-SLM). Despite

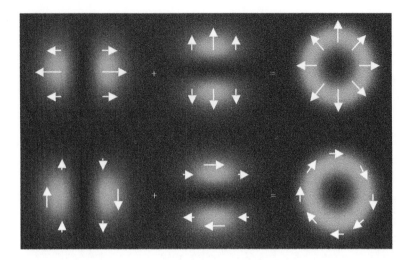

FIGURE 9.2 Formation of radial and azimuthal polarizations using the linear superposition of orthogonally polarized HG CV beams. (From Zhan, Q. 2009. Cylindrical vector beams: From mathematical concepts to applications. *Adv. Opt. Photon.* 1: 1–57. With permission of Optical Society of America.)

its relative high cost, the LC-SLM offers the flexibility and capability to generate almost arbitrary complex field distributions (see, e.g., Chapter 6). LC-SLM-based devices that are specifically designed for CV beam generation with lower cost have also been developed and are commercially available nowadays [7].

The other important type of passive CV beam generation technique utilizes the polarization sensitivity of subwavelength dielectric and metallic structures. For subwavelength dielectric structures, the form birefringent effect is exploited. For example, a direct laser writing technique produced radial polarization converters (known as z-polarizer, z-plate, q-plate, or s-waveplate) that convert linearly polarized input into CV beams are available commercially [8]. A subwavelength metallic structure such as concentric rings (also known as bulls-eye pattern) etched into a metallic thin film is another popular choice for creating CV beams. Utilizing the high differences (extinction ratio) in the transmittance for transverse electric (TE) and transverse magnetic (TM) polarization, metallic bulls-eye structures can be used to convert circular polarization into radial or azimuthal polarizations. However, it is important to pay attention to a spiral geometric phase term when these types of structures are illuminated with a circularly polarized beam [9]. This additional phase needs to be compensated to achieve true CV beams.

The generation of CV beams with few-mode fibers is another technique that deserves special attention. If the parameters are chosen correctly, a multimode step index optical fiber can support the TE_{01} and TM_{01} annular modes possessing cylindrical polarization symmetry, with the TE_{01} mode being azimuthally polarized and the TM_{01} mode being radially polarized (Figure 9.3). Under weakly guiding approximation, these modes have the same cutoff parameter that is lower than all the other modes except the HE_{11}

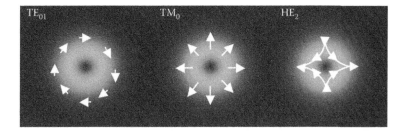

FIGURE 9.3 Illustrations of the polarization patterns of LP_{11} modes of the optic fiber. The TE_{01} and TM_{01} modes have azimuthal and radial polarization symmetry, respectively.

fundamental mode. CV mode excitation in the fiber can be achieved with careful misalignment with an effort to minimize the excitation of the fundamental mode and higher-order modes [10]. A few-mode fiber with metallic bulls-eye pattern fabricated on the fiber end (Figure 9.4) has also been recently demonstrated to facilitate the launching of CV beams into an optical fiber [11].

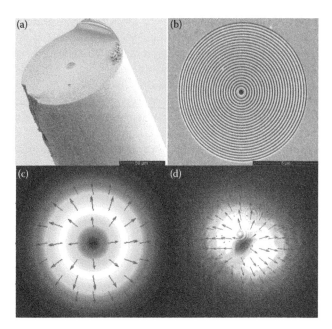

FIGURE 9.4 (a) Scanning electron microscopy (SEM) image of a few-mode fiber with bulls-eye pattern integrated at the core region, (b) zoom-in of the bulls-eye structure, (c) calculated polarization pattern transmitted through the bulls-eye structure and (d) experimental measurement of the polarization pattern. (From Chen, W. et al., 2011. Generating cylindrical vector beams with subwavelength concentric metallic gratings fabricated on optical fibers. *J. Opt.* 13: 015003. With permission.)

Particular interests were shown in developing compact CV beam sources that may facilitate some of the applications in the past several years. For example, GaAs vertical cavity surface emitting laser (VCSEL) operated at 850 nm that can generate a radially polarized output has been demonstrated [12]. In this work, a gold bulls-eye structure is integrated within the output aperture on top of the VCSEL. A similar structure has been used on an InGaAsP semiconductor laser (termed as radial Bragg laser [RBL]) operated at 1.6 μm [13]. Depending on the metallic bulls-eye design, the laser could operate either in the radial polarization mode or in the azimuthal polarization mode. In another effort, a dye-doped nematic liquid crystal (LC) microlaser has been demonstrated to produce an azimuthally polarized output [14]. The polarization selection was achieved through a concentric alignment of LC molecules by patterned rubbing on one of the high-reflectivity surface of the cavity. A very interesting feature of this simple structure is that it also allows the lasing wavelength to be tuned electrically without losing the polarization properties. Over 20 nm tuning range centered around 630 nm has been reported at a 1.6-v driving voltage. However, this dye-doped LC laser requires a pump laser that makes it bulky and cumbersome. Integrating it with a monolithic solid-state pump laser would be a possible route to shrink the size.

The fact that a few-mode optical fiber can support both radially and azimuthally polarized modes enables the design of CV fiber lasers. Fiber lasers have attracted increasing interest owing to their relatively high gain, compactness, and flexibility. Various types of fiber laser designs have been explored to produce CV beams. The basic underlying principle is to use appropriate elements in the cavity, such as dual conical prism, axicon, spatially variable retarder, and so on, to select the desired polarization. In most of the cases, only one type of the CV beams (either radial or azimuthal polarization) can be generated. Recently, a simple design that can generate both radially and azimuthally polarized beams using a *c*-cut calcite crystal with a three-lens telescope in an erbium-doped fiber laser cavity operated near 1.6 μm has been demonstrated

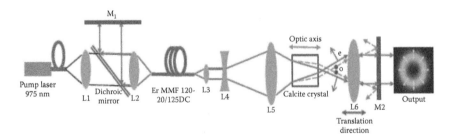

FIGURE 9.5 Experiment setup for a fiber laser capable of generating radially and azimuthally polarized modes. The solid rays indicate extraordinary beams (radial polarization) and the dashed rays indicate ordinary rays (azimuthal polarization). The output mode can be switched between the radial and azimuthal modes through a translating lens L6. (Reprinted with permission from Zhou, R. et al., Fiber laser generating switchable radially and azimuthally polarized beams with 140 mW output power at 1.6 μm wavelength. *Appl. Phys. Lett.* 95: 191111. Copyright 2009, American Institute of Physics.)

(Figure 9.5) [15]. Maximum power of 140 mW was reported, with the potential of much higher output power with optimization. Detailed studies of the mode polarization evolution also show that the modes inside the laser cavity can be spatially homogeneously polarized in one section of the cavity while being spatially inhomogeneously polarized in another section of the cavity, and opens the opportunities for many potential new fiber laser design possibilities, including frequency selection and intracavity frequency conversion [16]. Furthermore, more complicated vectorial vortex output modes other than radial and azimuthal polarizations can be generated by translating one lens of the telescope setup within the laser cavity. Higher-order vectorial vortex modes are also observed by purposefully introducing angular misalignments of the calcite crystal (Figure 9.6) [16].

In the fiber laser design above, there is still a portion of the cavity using bulk optics in free space. It is much desirable to have an all-fiber vectorial laser without the use of bulk optical elements. A bulls-eye pattern integrated on the fiber end offers a viable route to achieve an all-fiber vectorial laser design [11]. In another recent work, low-threshold single-wavelength all-fiber laser-generating CV beams using a few-mode fiber Bragg grating (FBG) was demonstrated (Figure 9.7) [17]. Both radially and azimuthally polarized beams have been generated with very good modal symmetry and polarization purity higher than 94%. The output can be switched between radially and azimuthally polarized modes by adjusting fiber polarization controllers built in the fiber laser cavity. This fiber laser operates at a single wavelength of 1053 nm with a 3-dB linewidth of <0.02 nm and a threshold as low as 16 mW.

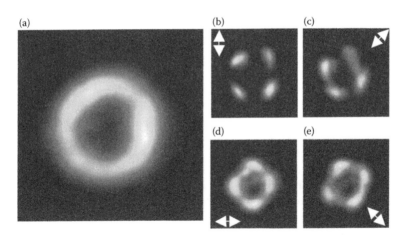

FIGURE 9.6 Experimental results of higher-order vectorial vortex output mode with angular misalignment. (a) Intensity profile of the output mode, (b)–(e) beam profiles after a linear polarizer whose transmission axis is indicated by the white arrow. The experimental results agree well with numerical simulation results of linear superposition of radial and azimuthal polarization with opposite topological charges. (From Zhou, R. et al., 2010. Vectorial fiber laser using intracavity axial birefringence. *Opt. Express* 18: 10839–10847. With permission of Optical Society of America.)

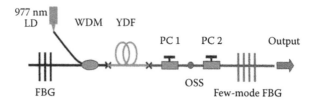

FIGURE 9.7 Schematic illustration of the all-fiber vectorial laser design using a few-mode FBG. OSS, offset splicing spot; PC, polarization controller. A single-mode fiber is used to the left side of the Yb-doped fiber (YDF) and a few-mode fiber is used to the right of the YDF. (From Sun, B., Wang, A., Xu, L. et al. 2012. Low-threshold single-wavelength all-fiber laser generating cylindrical vector beams using a few-mode fiber Bragg grating. *Opt. Lett.* 37: 464–466. With permission of Optical Society of America.)

For a conventional few-mode fiber, the three second-order modes (TM_{01}, TE_{01}, and HE_{21}) are nearly degenerate, which could lead to instability that limits the length of propagation within the fiber. Recently, a vortex fiber with specially tailored index profile has been developed to increase the effective index difference between TM_{01} from TE_{01} and HE_{21} such that the radially polarized mode can be excited more efficiently and propagate in longer fiber length [18]. It is reported that the conversion efficiency can be as high as 99%. The high conversion efficiency combined with the stability and longer propagation length within the optical vortex fiber makes this a very good platform for producing a broadband radially polarized mode with high purity through a nonlinear optical process such as stimulated Raman scattering. A broadband radially polarized source with 250-nm bandwidth (from 1064 to 1310 nm) over a 100-m-long fiber has been reported [18]. The radially polarized mode is able to stably propagate through the fiber even with fiber-bend radii <7 cm.

For practical applications of the CV beams, it is necessary to be able to perform basic manipulations such as reflection, polarization rotation, and retardation of these beams. In most cases, it is critical to maintain the polarization symmetry during or after these manipulations. When CV beams are reflected and steered, polarization symmetry could be easily broken due to different reflection coefficients for *s*- and *p*-polarizations. Even if the magnitudes of these reflection coefficients are close, the phase difference can still destroy the polarization symmetry and requires special attention. In general, a metallic mirror should preserve the polarization symmetry better than dielectric mirrors. However, metallic mirrors typically have protective coatings that could give rise to drastically different reflection coefficients for *s*- and *p*-polarizations, especially in terms of the phases. The combination of two identical beam splitters with twisted orientation has been implemented to maintain polarization symmetry while they provides the steering function for CV beams [19]. A similar arrangement with two identically coated metallic mirrors can also be used to provide higher throughput.

Polarization rotation can be achieved with an optically active material (such as quartz) or Faraday rotators. These types of rotators typically have a fixed rotation angle and it is difficult to manufacture a Faraday rotator with large-enough clear aperture to accommodate the needs of some applications. A simple polarization rotator (PR) using two cascaded $\lambda/2$-plates has been proposed and demonstrated [20]. If the angle between the fast axes of the two $\lambda/2$-plates is $\Delta\phi$, the Jones matrix of this device can be shown to be:

$$T = \begin{pmatrix} \cos(2\Delta\phi) & -\sin(2\Delta\phi) \\ \sin(2\Delta\phi) & \cos(2\Delta\phi) \end{pmatrix} = R(2\Delta\phi). \tag{9.12}$$

This is a pure polarization rotation function that is independent of the incident polarization. The polarization rotation angle can be readily tuned by adjusting the angle between the fast axes of the two plates. This device can be used to conveniently adjust the polarization pattern of a CV beam into the desired generalized CV beam polarization patterns. A nonmechanical PR using electro-optics (EO) variable retarders sandwiched between two orthogonally oriented $\lambda/4$-plates (Figure 9.8) has also been designed and demonstrated [21]. The Jones matrix of these devices can be derived as

$$T = R\left(-\frac{\pi}{2}\right)\begin{pmatrix} 1 & 0 \\ 0 & -j \end{pmatrix} R\left(\frac{\pi}{2}\right) R\left(-\frac{\pi}{4}\right)\begin{pmatrix} 1 & 0 \\ 0 & e^{-j\delta} \end{pmatrix} R\left(\frac{\pi}{4}\right)\begin{pmatrix} 1 & 0 \\ 0 & -j \end{pmatrix}$$

$$= -je^{-j\frac{\delta}{2}} R\left(\frac{\delta}{2}\right), \tag{9.13}$$

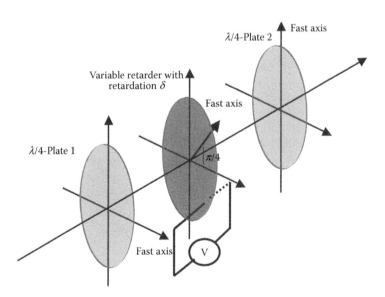

FIGURE 9.8 Diagram of a nonmechanical PR using variable retarders sandwiched between two orthogonally oriented $\lambda/4$-plates.

where δ is the retardation of the EO retarder. Other types of variable retarder such as LC variable retarder can also be used. These devices can be made with a large clear aperture and the amount of rotation can be adjusted and modulated.

Fast EO radial polarization retarder with its local fast axis aligned along the radial direction has also been reported [22]. These types of devices can be used to provide retardation between radial and azimuthal polarization components. Beams with cylindrical symmetrically distributed elliptical polarization states can be generated in this way. For fiber optic devices, the torsion in the optical fiber induces a birefringence in the fiber. It can be shown that the uniformly twisted fiber behaves like a medium exhibiting rotatory power. This provides a convenient means of polarization control. However, careful control is necessary due to the sensitivity.

9.2.3 Focusing Properties

The increasing interests in CV beams are largely due to their unique properties under high-numerical-aperture (NA) focusing. Numerical and experimental studies have shown that a tighter focal spot can be obtained using radial polarization owing to the existence of a strong and localized longitudinal field component [23]. This unique focusing capability of radial polarization can be intuitively understood by examining the radiation characteristics of an oscillating electric dipole. Imagine an electric dipole located at the focal point of a high-NA aplanatic objective lens with the dipole oscillating along the optical axis of the lens (Figure 9.9); the well-known angular radiation pattern of the electric dipole is illustrated with the local polarization. A high-NA objective lens collects the dipole radiation into the top half-space and collimates the radiation. From the illustration, clearly, the polarization pattern at the lens exit pupil plane will be aligned along the radial directions. If we time reverse the optical path and start with the radial polarization pattern at the pupil plane, the corresponding focal field should recover the propagating components in the upper half-plane. From this simple argument, qualitatively, we can see that radial polarization should provide optimal focusing compared with other polarization distributions.

The focusing property of highly focused polarized beams can be numerically analyzed with the vectorial diffraction method. The coordinates for calculation are shown in Figure 9.10. The illumination is a generalized CV beam, which assumes a planar wave front over the pupil. The incident field can be written in the pupil plane cylindrical coordinates (ρ, ϕ, z) as

$$\vec{E}_i(\rho, \phi) = l_0 P(\rho)[\cos\phi_0 \vec{e}_\rho + \sin\phi_0 \vec{e}_\phi], \qquad (9.14)$$

with l_0 being the peak field amplitude at the pupil plane and $P(\rho)$ being the axially symmetric pupil plane amplitude distribution normalized to l_0. An aplanatic lens produces a spherical wave converging to the focal point. The vector fields in the pupil plane are mapped onto this spherical surface as shown in Figure 9.10. By taking a surface integral of the fields on the spherical surface, the focal field can be expressed in the cylindrical

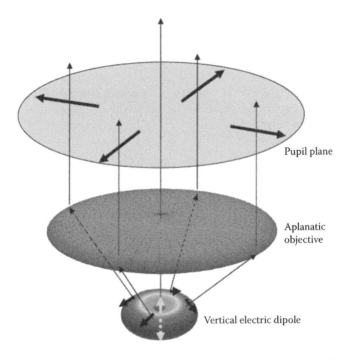

FIGURE 9.9 Illustration of the polarization pattern at the pupil plane for the radiation from a vertical electric dipole collected by a high-NA objective. (From Zhan, Q. 2009. Cylindrical vector beams: From mathematical concepts to applications. *Adv. Opt. Photon.* 1: 1–57. With permission of Optical Society of America.)

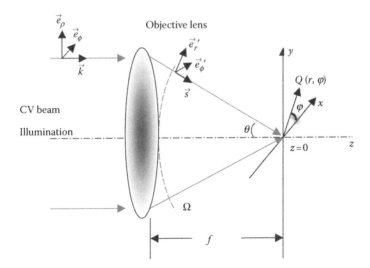

FIGURE 9.10 Focusing of a CV beam. In the diagram, f is the focal length of the objective lens. $Q(r, \varphi)$ is an observation point in the focal plane.

coordinates in the focal space as

$$\vec{E}(r,\varphi,z) = E_r \vec{e}_r + E_z \vec{e}_z + E_\varphi \vec{e}_\varphi, \tag{9.15}$$

where $\vec{e}_r, \vec{e}_z$, and $\vec{e}_\varphi$ are the unit vectors in the radial, longitudinal, and azimuthal directions, respectively. E_r, E_z, and E_φ are the amplitudes of the three orthogonal components that can be written as

$$E_r(r,\varphi,z) = 2A \cos\phi_0 \int_0^{\theta_{max}} P(\theta) \sin\theta \cos\theta J_1(kr\sin\theta)e^{ikz\cos\theta} d\theta, \tag{9.16}$$

$$E_z(r,\varphi,z) = i2A \cos\phi_0 \int_0^{\theta_{max}} P(\theta) \sin^2\theta J_0(kr\sin\theta)e^{ikz\cos\theta} d\theta, \tag{9.17}$$

$$E_\varphi(r,\varphi,z) = 2A \sin\phi_0 \int_0^{\theta_{max}} P(\theta) \sin\theta J_1(kr\sin\theta)e^{ikz\cos\theta} d\theta, \tag{9.18}$$

where θ_{max} is the maximal angle determined by the NA of the objective lens, k is the wave number, and $J_n(x)$ is the Bessel function of the first kind with order n. A is a constant that is given by the objective lens focal length f and the wavelength λ as

$$A = \frac{\pi f l_0}{\lambda}. \tag{9.19}$$

$P(\theta)$ is the pupil apodization function that strongly depends on the objective lens design. Assuming the incident field at the pupil plane has an amplitude distribution of $P(\rho)$ normalized to l_0, with ρ being the radial position in the pupil plane, this amplitude distribution is mapped to the wave front after the objective lens through the ray projection function $g(\theta)$ as

$$\rho/f = g(\theta). \tag{9.20}$$

With this mapping relationship, the pupil apodization function is given by [24]:

$$P(\theta) = P(\rho)\sqrt{\frac{g(\theta)g'(\theta)}{\sin\theta}} = P(fg(\theta))\sqrt{\frac{g(\theta)g'(\theta)}{\sin\theta}}. \tag{9.21}$$

The ray projection function for most of the commonly used objective lens is given by the sine condition

$$g(\theta) = \sin\theta, \tag{9.22}$$

$$\rho = f\sin\theta, \tag{9.23}$$

which leads to the pupil apodization function as

$$P(\theta) = P(fg(\theta))\sqrt{\cos\theta}. \tag{9.24}$$

Other types of ray projection functions are also used, such as Herschel condition, Lagrange condition, Helmholtz condition, and so on. However, these types of objective lens are not as common as objective lenses that obey sine conditions.

Note that all components given in Equations 9.16 through 9.18 are independent of φ, which means the field maintains cylindrical symmetry. From Equation 9.17, one can see that a key feature of the focal field distribution is the existence of a strong longitudinal component (z-component) arising from the radial component of the illumination. One example of the field calculation is given in Figure 9.11, which corresponds to the radial polarization focused by an objective lens with NA = 0.95. For radial polarization incident, the z-component experiences an apodization function of $\sin \theta$. This apodization function for radial polarization focusing places more weight toward the high-spatial-frequency components, consequently leading to a smaller spot size. It also indicates that high NA is required and annular illumination is preferred in order to have prominent effects of the longitudinal component.

9.2.4 Applications

The spatial separation between the longitudinal and transversal components of the focal field for highly focused CV beams (Figure 9.11) leads to a variety of applications in optical focus engineering. Through phase modulation at the pupil plane with diffractive optical elements (DOEs) with multiple concentric zones, optical focal fields with different desirable characteristics have been demonstrated with the use of CV beams. For example, an optical focal field with two-dimension (2D) flattop [20], three-dimensional (3D) flattop, and optical "bubbles" [25] can be realized through modulating the wave front of a generalized CV beams with binary DOE; an optical needle [26] and an optical chain [27] have been realized with radial polarization modulated by DOE; an optical tube has been realized using spatially modulated azimuthal polarization [28], and so on. Among these unusual focal field distributions, an optical needle is particularly interesting due to its unusual energy and vector field distribution along the optical focus and its potential applications in polarization-sensitive orientation imaging, particle manipulation and acceleration, and light–matter interaction on the nanoscale. The electrical field of an optical needle field is substantially polarized in the longitudinal direction with a long depth of focus (DOF) created through the focusing of pupil-modulated radial polarization.

Recently, a systematic approach has been reported to produce an extremely long and uniform optical needle field through the reversal of the radiation pattern of a linear antenna array [29]. This idea is an extension of the argument previously given that links the radial polarization to the emission of oscillating electric dipole (Figure 9.9). Instead of a single electrical dipole, one can calculate the vectorial radiation fields of a linear dipole array collected by an aplanatic objective lens at its pupil plane (Figure 9.12). When this pupil plane vector field is used as the illumination, then a focal field with certain prescribed characteristics (such as an extremely long optical needle field) can be realized. The schematic of this approach is illustrated in Figure 9.12. In Figure 9.12a, identical infinitesimal electric dipoles are located along the optical axis to form a dipole array in the focal volume of the objective lens. A zoom-in diagram of the dipole array with $2N$

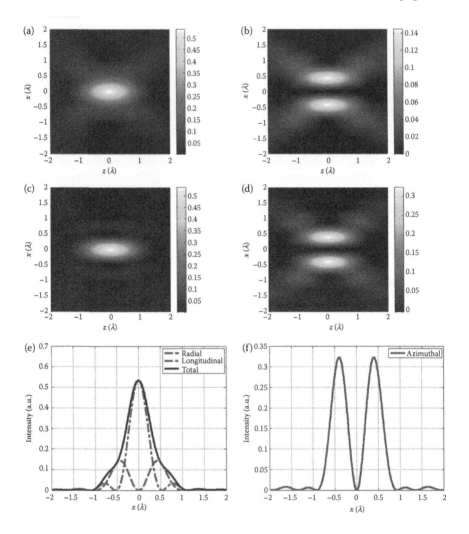

FIGURE 9.11 Calculated focal field distributions for CV beams focused by an objective lens with NA = 0.95. (a) Total intensity distribution for focused radial polarization, (b) radial components of (a), (c) longitudinal components of (a), (d) total intensity distribution of focused azimuthal polarization, (e) line scan of the distribution in the focal plane for (a), and (f) line scan of the distribution in the focal plane for (d); the focal field is solely made of an azimuthal component. For this example, it can be seen that the longitudinal component is spatially separated from the transversal (radial and azimuthal) components.

number of electric dipoles is shown in Figure 9.12b. The electrical field radiated from the dipole array at point A on the spherical surface Ω right after the objective lens can be written as

$$\vec{E}_0(\theta) = E_{DA}(\theta)\vec{a}_\theta = C\sin\theta AF_N\vec{a}_\theta, \qquad (9.25)$$

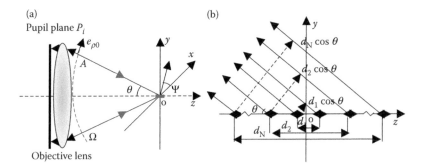

FIGURE 9.12 Schematic of the method of pupil plane field synthesis to achieve specific focal field characteristics through the reversing of the radiation pattern of a dipole array. (a) System layout with a dipole array oscillating along the optical axis in the focal volume and the coordinates used in the calculation and (b) far-field geometry and conceptual diagram of the dipole array with 2N dipole elements. (From Wang, J., Chen, W., and Zhan, Q. 2010. Engineering of high purity ultra-long optical needle field through reversing the electric dipole array radiation. *Opt. Express* 18: 21965–21972. With permission of Optical Society of America.)

with $C = jz_{air}I_0l_0 \exp(-jkf)/4\pi f$ with z_{air} being the intrinsic impedance of air, I_0 and l_0 being the standard current and dipole length, and f being the focal length of the objective lens. Without loss of generality, C can be normalized to 1 in the calculation. The AF_N is an array factor related to the phase delay caused by the spacing distance d_n and the initial phase difference β_n of each pair of the dipole that are mirror symmetric with respect to the focal plane:

$$AF_N = \sum_{n=1}^{N} A_n[e^{j(kd_n \cos\theta + \beta_n)/2} + e^{-j(kd_n \cos\theta + \beta_n)/2}], \qquad (9.26)$$

where A_n is the ratio of the radiation amplitude between the nth dipole pair and the standard dipole pair with the normalized amplitude. In this case, the radiation only has a radial component on Ω due to the specific oscillating direction chosen for the electric dipoles such that it can also be written as $\vec{E}_0(\theta) = E_{DA}(\theta)\vec{a}_{\rho_0}$. For an objective lens that obeys the sine condition, the radiation pattern collected at the pupil plane P_i can be expressed as

$$\vec{E}_i(\rho_i, \theta) = E_{DA}(\theta)(\cos\phi \vec{x}_i + \sin\phi \vec{y}_i)/\sqrt{\cos\theta}, \qquad (9.27)$$

where $\rho_i = f\sin\theta$ is the ray projection function of a sine condition lens given by Equation 9.27. If one starts with this field as the illumination in the pupil plane and reverses its propagation, the electrical field reconstructed in the focal volume can be computed with the vectorial Debye integral as [24]:

$$\vec{E}(r, \psi, z) = \frac{i}{\lambda} \int_0^{\theta_{max}} \int_0^{2\pi} E_{DA}(\theta)(\cos\theta \cos\phi\, \vec{i} + \cos\theta \sin\phi\, \vec{j} + \sin\theta\, \vec{k})$$

$$\exp[-ikr\sin\theta\cos(\phi - \psi) - ikz\cos\theta]\sin\theta\, d\theta\, d\phi. \tag{9.28}$$

Applying this method, numerical examples demonstrated an optical needle field of a DOF up to 8λ with a diffraction-limited transverse spot size ($<0.43\lambda$ FWHM) and longitudinal polarization purity higher the 80% throughout the superlong DOF (Figure 9.13) [29].

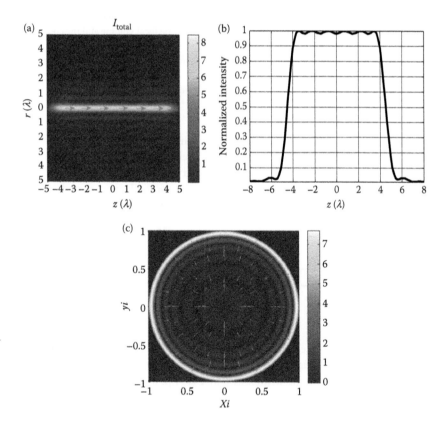

FIGURE 9.13 (**See color insert.**) (a) Total intensity distribution through the focus region and (b) axial intensity distribution for an optical needle field with extended DOF. The corresponding required incident field in the pupil plane is illustrated in (c). (From Wang, J., Chen, W., and Zhan, Q. 2010. Engineering of high purity ultra-long optical needle field through reversing the electric dipole array radiation. *Opt. Express* 18: 21965–21972. With permission of Optical Society of America.)

The possibility of producing a unique focal field distribution with CV beams naturally finds many applications that require the use of tightly focused light beams, such as laser micromachining, laser direct writing, optical imaging and so on. The applications of CV beams in laser micromachining have been reviewed in a previous monograph [2]. Radial polarization has been applied to directly write 3D photonics crystal structure through two-photon polymerization [30]. It has been found that a 27.5% reduction of the lateral size of the fabricated polymer features can be achieved when a radially polarized beam is employed, leading to improved performances of the fabricated 3D photonic crystal device. On imaging, CV beams have been utilized in the determination of single molecular orientation by inspecting the back pupil fluorescence radiation pattern from an excited single molecule [31]. Radial polarization has also been employed in the coherent anti-Stokes Raman scattering (CARS) microscopy and an improvement of 10% in the lateral resolution was demonstrated compared to CARS imaging with linear polarization [32].

The unique polarization symmetry and focal field distribution also lead to interesting phenomena when highly focused CV beams interact with nanostructures, particularly metallic nanostructures. Through a polarization mode matching, radially polarized light has been extensively explored to offer optimal plasmonic excitation in combination with metallic structures with rotational symmetry, such as planar film [33], bulls eye (Figure 9.14) [34], conical tip [35], conical tip integrated within a bulls eye (Figure 9.15) [36], and so on. The strongly localized plasmonic fields produced by these structures have broad potential applications in imaging, sensing, data storage, and so

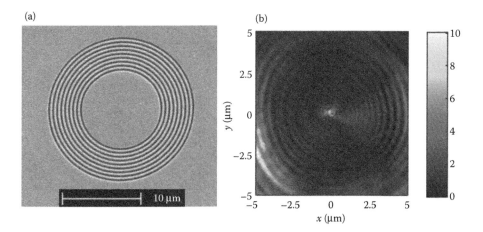

FIGURE 9.14 (a) A bulls-eye plasmonic lens made of nine concentric rings. (b) Near-field scanning optical microscopic image showing the focusing of surface plasmon wave from all directions toward the center when illuminated with radial polarization. (From Chen, W. et al., 2009. Plasmonic lens made of multiple concentric metallic rings under radially polarized illumination. *Nano Lett.* 9: 4320–4325. With permission.)

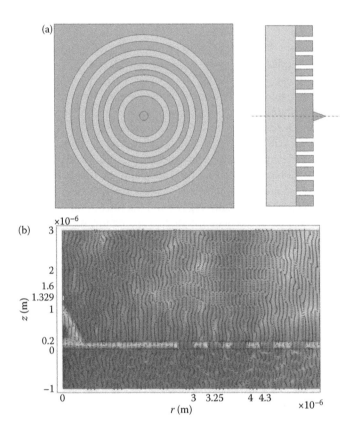

FIGURE 9.15 (**See color insert.**) (a) Diagram of a combined structure of bulls-eye and a conical tip. (b) Logarithmic plot of the 2D electric energy density and the electric field (show a by the arrow) distribution when illuminated by radial polarization. A hot spot with extremely high field enhancement is observed at the tip apex. (From Rui, G. et al., 2010. Plasmonic nearfield probe using the combination of concentric rings and conical tip under radial polarization illumination. *J. Opt.* 12: 035004. With permission.)

on. While radial polarization has attracted lots of attention in plasmonic focusing and localization, azimuthal polarization is generally considered not suitable for plasmonic excitation due to its polarization matching to a magnetic dipole. However, recently, plasmonic focusing with azimuthal polarization has been demonstrated through the use of spatially arranged triangular subapertures (Figure 9.16) [37]. Its application in the design of high-efficiency miniature circular polarization analyzer has also been reported [38].

The focusing properties of CV beams have found important applications in optical trapping. It has been shown that radial polarization can be used to trap particles with refractive index higher than the ambient while the azimuthal polarization is suitable for the trapping of particles with refractive index lower than the ambient (such as hollow particles or air bubbles) [39]. The switching between radial and azimuthal polarization

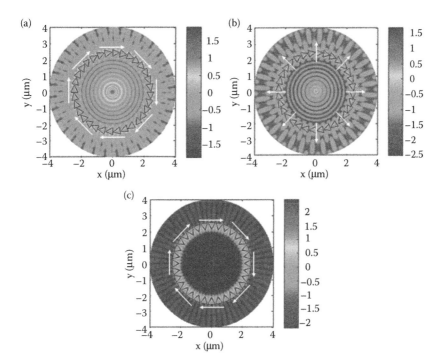

FIGURE 9.16 (**See color insert.**) (a) Logarithmic electric energy density distribution at the air/silver interface with 32 isosceles triangles arranged in an antisymmetric mode under azimuthally polarized illumination. A highly confined focal spot is obtained at the center. (b) Logarithmic electric energy density distribution of the same structure under radially polarized illumination. The solid spot with lower field enhancement is observed. (c) Logarithmic electric energy density distribution with 32 isosceles triangles arranged in a symmetric mode under azimuthally polarized illumination. A completely dark center is observed. (From Chen, W., Nelson, R. L., and Zhan, Q. 2012. Geometrical phase and surface plasmon focusing with azimuthal polarization. *Opt. Lett.* 37: 581–583. With permission of Optical Society of America.)

can be readily achieved with the PR described in Ref. 20, making versatile optical trapping tools with the use of CV beam illumination. One particularly important finding is to improve the trapping efficiency of metallic nanoparticles with radial polarization. Using a dipole approximation for the optical force on Rayleigh particles, the spatial separation between the power density and energy density indicates a spatial separation of the trapping force (proportional to the gradient of electric energy density) and the destabilizing radiation pressure forces (proportional to the power density), leading to the expectation of significantly improved trapping stability of metallic nanoparticles using radial polarization [40]. The advantages offered by radial polarization over linear polarization were confirmed by other computational methods [41,42] and some of the latest experimental works [43,44]. However, it is worthy of pointing out that a recently discovered additional scattering force term (the so-called light spin force) [45], also known as an "unnamed third term" reported previously [46], may reduce the advantages that radial polarization would offer. This term is normally negligible compared to the gradient force and the radiation pressure force and little attention has been paid to it. However, owing to the unique vector field distributions of the highly focused radial polarization, this term becomes comparable to the radiation pressure force unexpectedly. With the consideration of the light spin force, the benefits of using radial polarization to minimize the scattering force appear to be reduced. That being said, the use of vector beams could eventually allow us to shape the vector focal field distribution and tailor the optical force distributions to suit the specific needs of different applications.

The polarization diversity within the beam cross section of CV beams itself offers interesting opportunities in optical measurement. The polarization symmetry was first exploited in a radially symmetric microellipsometer design to boost the signal-to-noise ratio in a spatially resolved ellipsometer to achieve submicron-level spatial resolution [19]. A radial polarization interferometer has been demonstrated with improved accuracy compared with a traditional Michelson interferometer, where the phase difference between the interferometer's arms is manifested as spatially varying intensity distribution through the interfering radially and azimuthally polarized beams [47]. A rapid Mueller matrix polarimetry that can extract 12-Mueller matrix elements from a single-intensity image enabled by the parallel polarization measurement with the use of vector beams was demonstrated and applied to the characterization of anisotropic samples, including metamaterials [48]. Recently, CV beams have also been used as atomic spin analyzers and applied in rapid, single-shot atomic spin visualizations in a vapor cell [49]. In addition to these, it has also been shown that CV beams are more resilient to turbulence propagating through the atmosphere [50]. This could potentially lead to the applications of CV beams in remote sensing and free space optical communications.

9.3 Full Poincaré Sphere Beams

For the CV beams, the local SOPs within the beam cross section are linearly polarized with different orientations. On a Poincaré sphere (PS), these local SOPs will populate the equator. Recently, a more general type of vector beams called full poincaré (FP) beam has been proposed and studied [51]. Unlike the CV beam whose SOPs only populate the equator, the SOPs within the beam cross section of an FP beam span the entire surface of the PS. In this section, we briefly introduce the mathematical description and generation methods of FP beams and the application of fiber bragg (FB) beams in flattop beam shaping under low-NA condition.

9.3.1 Mathematical Description

FP beams are typically generated by superimposing orthogonally polarized beams with spatially different intensity distributions. Most often, a polarized fundamental Gaussian mode and an orthogonally polarized high-order LG mode are used in the generation of FP beams. The spiral vortex phase embedded in a high-order LG mode is necessary to generate these general elliptical SOPs that cover the surface of the entire PS. For example, the generation of the first-order FP beam has been reported using the fundamental Gaussian (LG_{00}) and the first-order LG (LG_{01}) beams of right- and left-hand circular polarization, respectively [51].

The fundamental Gaussian beam can be represented as

$$LG_{00}\,(r,z) = A_0 \frac{w_0}{w\,(z)} \exp\left(-j\left(kz - \varphi\,(z)\right) - r^2\left(\frac{1}{w\,(z)^2} + \frac{jk}{2R\,(z)}\right)\right), \qquad (9.29)$$

where A_0 is the amplitude at the origin, w_0 is the beam waist, and $\varphi(z)$ is the Gouy phase. Similarly, an LG_{0n} beam can be written as

$$LG_{0n}\,(r,z) = A_0 2^{n/2} \frac{r^n}{w\,(z)^n} \frac{w_0}{w\,(z)} \exp\left(-j\left(kz - (n+1)\varphi\,(z)\right)\right.$$
$$\left. -r^2\left(\frac{1}{w\,(z)^2} + \frac{jk}{2R\,(z)}\right) - jn\phi\right), \qquad (9.30)$$

where ϕ is the azimuthal angle.

A superposition of orthogonally linearly polarized LG_{00} and LG_{0n} can be expressed as follows:

$$E_{FP}\,(r,z) = \cos\gamma\,LG_{00}\vec{e}_x + \sin\gamma\,LG_{0n}\vec{e}_y = C\begin{pmatrix}1\\\rho_0 e^{j\delta}\end{pmatrix}. \qquad (9.31)$$

Here, we denote the second term in Jones vector as $\rho_0 e^{j\delta}$, where ρ_0 is the ratio between the amplitude of y- and x-components and δ is the phase difference between them. Angle γ determines the weighting between the x- and y-components. It is straightforward to show that

$$\rho_0 = 2^{n/2}\tan\gamma\,\frac{r^n}{w^n(z)}, \qquad (9.32)$$

$$\delta = n\varphi(z) - n\phi. \qquad (9.33)$$

Simple algebra shows that the polarization is mainly along the x-axis for points close to the origin and the polarization gradually evolves to y-polarization as the observation point moves away from the center. The phase delay is n times of the difference between the azimuthal angle and Gouy phase shift. Therefore, at any cross section of the superimposed beam along the propagation, the phase delay between two components will range from 0 to $2n\pi$. Thus, the SOPs within the superimposed beam cross section will span the entire surface of PS n times. Hence, we call this the nth-order FP beam. Along the propagation, the Gouy phase shift will further introduce an additional phase delay, which will cause the SOP pattern to rotate as it propagates.

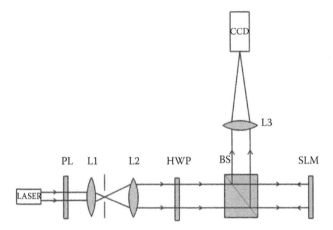

FIGURE 9.17 Diagram of the experimental setup for FP beam generation with an LC SLM. PL, linear polarizer; L1, lens for spatial filter; L2, collimation lens; HWP, half-wave plate; BS, 50/50 nonpolarizing beam splitter; SLM, spatial light modulator; L3, imaging lens. (From Dai, J. and Zhan, Q. 2008. Beam shaping with vectorial vortex beams under low numerical aperture illumination condition. *SPIE Proc.*, 7062: 70620D. With permission.)

9.3.2 Methods of Generation

There are a variety of ways to generate FP beams. Stress-engineered glass has been demonstrated to generate the first-order FP beams [51]. In this technique, the appropriate amount and distribution of stress is applied to a piece of glass window to create stress-induced birefringence. When the device is illuminated with linear polarization, one of the circularly polarized components remains a Gaussian profile while the orthogonally circularly polarized component obtains a topological charge 1 spiral phase. The superposition of these orthogonally polarized components renders a first-order FP beam. LC–SLMs remain to be the most popular and versatile methods to generate FP beams [52,53]. A typical configuration is shown in Figure 9.17. Utilizing the fact that these LC-SLMs only respond to one linear polarization (y-polarization) and leave the orthogonal linear polarization (x-polarization) unaffected, by loading a vortex phase with topological charge of n onto the SLM, the y-polarized component after reflection of the LC-SLM approximates LG_{0n} mode, while the x-polarized component remains to be the fundamental Gaussian LG_{00} mode. After the beam splitter, the two components superimpose together to form an nth-order FP beam. A half-wave plate can be used to conveniently adjust the strength ratio between the two orthogonally polarized components.

9.3.3 Applications

When FP beams are focused by low-NA lens, the LG_{0n} mode component contributes to a donut distribution at the focal plane and the fundamental Gaussian mode component contributes to a Gaussian distribution with a peak at the center. Owing to the orthogonal

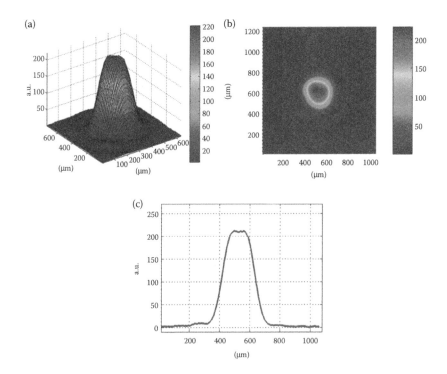

FIGURE 9.18 Flattop generated by focusing the first-order FP beam with low-NA lens. (a) 3D intensity distribution, (b) 2D view of the flattop, and (c) a line scan of (b). A gentle edge roll-off can be observed in this case. (From Dai, J. and Zhan, Q. 2008. Beam shaping with vectorial vortex beams under low numerical aperture illumination condition. *SPIE Proc.,* 7062: 70620D. With permission.)

polarization of these two components, their distributions will linearly superimpose in intensity. By adjusting the incident polarization angle γ, we can change the relative weighting of these two constituting components. Thus, it is conceivable that for an appropriate amplitude ratio, a flattop beam profile at the focal plane or a far field can be obtained. This idea' has been demonstrated with the first-order FP beam (Figure 9.18) [52]. A smooth flattop profile has been obtained with appropriately chosen parameters. However, the edge roll-off is not steep due to the gentle roll-off of the transverse profile for the focused LG_{01} component. This can be improved by using the LG modes with higher topological charges. For example, the flattop focusing with steep edge roll-off using the second-order FP beam has been demonstrated experimentally (Figure 9.19) [53].

9.4 Generalized Vector Beams

Compared with the CV beams, the polarization diversity within the beam cross section of FP beams is dramatically increased. However, the optical field distributions of FP beams

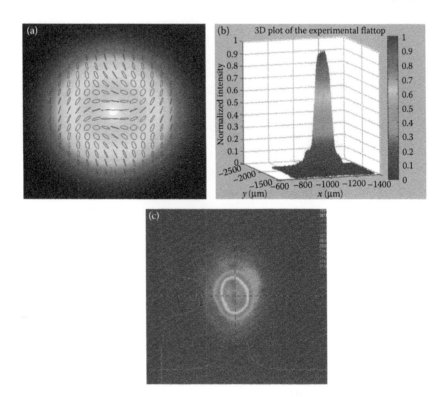

FIGURE 9.19 (a) Polarization pattern of a second-order FP beam, (b) 3D view of the flattop generated by focusing the second-order FP beam with low-NA lens, and (c) 2D view of the flattop with line scans. A steeper edge roll-off can be observed. (From Han, W., Cheng, W., and Zhan, Q. 2011. Flattop focusing with full Poincaré beams under low numerical aperture illumination. *Opt. Lett.* 36: 1605–1607. With permission of Optical Society of America.)

still maintain rotational symmetry in the beam amplitude and the polarization pattern is not arbitrarily addressed. A natural further extension of the concept is to have an optical field whose intensity, phase, and polarization (orientation and polarization ellipticity) can be engineered arbitrarily. It is conceivable that more interesting phenomenon and applications are expected with the availability of such generalized optical vector fields.

9.4.1 Concepts

In the most general format, the transverse electrical field distribution of a generalized optical vector beam can be mathematically described as an extension of Equation 9.31

$$\overrightarrow{E}(x,y) = E_0(x,y)e^{j\varphi(x,y)} \begin{pmatrix} 1 \\ \rho(x,y)e^{j\delta(x,y)} \end{pmatrix}, \tag{9.34}$$

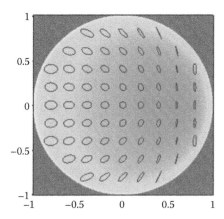

FIGURE 9.20 Illustration of an example of a generalized vector field with amplitude, phase, and polarization spatial variations across the aperture.

where $E_0(x, y)$ is the amplitude distribution, $\varphi(x, y)$ is the wave front, $\rho(x, y)$ is the distribution of the polarization ratio between the y-component and the x-component, and $\delta(x, y)$ is the distribution of the phase difference between the y-component and the x-component. If all these distributions can be controlled and engineered, we can control the intensity, phase, polarization orientation, and polarization ellipticity. Then we have the most generalized optical vector fields as far as a continuous wave is concerned. An example of one such generalized vector beam is illustrated in Figure 9.20, in which the amplitude, phase, polarization orientation, and ellipticity of the optical field are all functions of the transverse coordinates (x, y).

9.4.2 Methods of Generation

Obviously, it is technically challenging to achieve such a comprehensive engineering of the optical beam cross section. However, the latest advances made in LC-SLM manufacturing and the research in nanophotonics, plasmonics, and metamaterials offer the opportunities to manipulate the properties of optical fields on unprecedented levels. For example, one method using LC-SLMs has been proposed to generate generalized vector beams [54]. This method is based on the PR previously described in Figure 9.8. Illustrated in Figure 9.21a, a pixelated pure PR can be realized by sandwiching an LC-SLM between two orthogonally oriented $\lambda/4$-waveplates. The fast axis of the LC-SLM has an angle of $\pi/4$ with respect to the two waveplates. Arbitrarily complex generalized vector beams with amplitude distribution can be obtained using four LC-SLMs (Figure 9.21(b)). The first SLM generates a pure phase pattern for the whole aperture. The combination of the first pure PR and a linear polarizer provides the necessary amplitude modulation distribution. A linearly polarized beam with the desired amplitude distribution is obtained after the linear polarizer. The second pure PR then rotates the

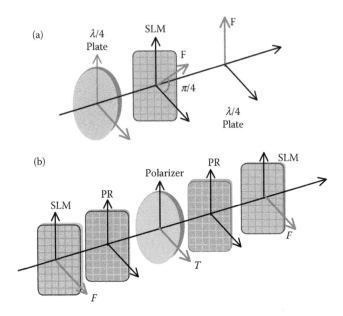

FIGURE 9.21 (a) Illustration of a pixelated PR by sandwiching an LC-SLM between two quarter-wave plates. (b) Proposed setup that is capable of generating arbitrarily complex generalized vector fields.

local polarization to the desired polarization ellipse orientations. The last SLM adds the necessary retardation distributions to realize the designed ellipticity distribution.

Clearly, the approach described above is bulky and careful alignment is necessary. It is also expensive to use four sets of LC-SLM. A very promising alternative approach comes from the latest developments in optical metamaterials and plasmonics. Optical metamaterials are man-made materials that exhibit extraordinary properties that may not exist in natural materials. The development in these fields offers the capabilities to control light–matter interactions on the subwavelength scale, which could be exploited for the full engineering of optical vector fields. One such example is the metasurface using optical nanoantennas [55]. By adjusting the size, orientation, and configuration of these antennas, one can engineer the intensity, phase, and polarization of the scattered light almost arbitrarily. Compared with the approaches using LC-SLMs, this approach potentially offers a much more compact and economical solution. The stringent alignment requirement is also eliminated. Research to realize such a full optical vector beam engineering method is underway and it is anticipated that the results will be reported in the near future.

9.4.3 Applications

The availability of such generalized vector beams opens the door to lots of novel applications that were beyond imagination before. One such example is to fully control the state of polarization within the focal volume while maintaining the optical spot to be diffraction limited. By extending the dipole argument for the focusing properties of radially polarized beams, a novel method that can generate arbitrary 3D state of polarization within an optimal focal spot has been proposed and demonstrated numerically [54]. By combining the electric dipole radiation and Richards–Wolf vectorial diffraction method, the input field at the pupil plane of a high-NA objective lens for generating arbitrary 3D-oriented linear polarization at the focal point with an optimal spot size can be found analytically by solving the inverse problem. The corresponding input field to generate an arbitrary 3D elliptical state of polarization can be found using two electric dipoles situated at the focus and polarized at orthogonal planes with different phase and amplitude. The calculated pupil fields are generalized vector fields that allow the generation of arbitrary 3D polarization states within the diffraction-limited focus. It is believed that this focal field 3D polarization control may find important applications in areas such as single molecular imaging, tip-enhanced Raman spectroscopy, high-resolution optical microscopy, and particle trapping and manipulation.

9.5 Summary and Acknowledgments

This chapter summarizes some of the recent developments in optical beams with spatially variant polarization distributions. Among these optical vector beams, CV beams attracted most of the interests in the past due to their unique focusing properties arising from the polarization symmetry. Various generation methods of CV beams have been developed and several products are available commercially. The applications of cylindrical beams in diverse areas such as laser micromachining, lithography, imaging, optical trapping and manipulation of nanoparticles, optical measurements, and metrology have been pursued. Beyond CV beams, optical vector beams with more complicated spatial polarization distributions attracted rapidly increasing recent interests. The spatial polarization distributions of these vector beams often time have intrinsic connections to the emission patterns of quantum and nanoemitters. Further research into these generalized vector beams is expected to have impacts on nanophotonics, biophotonics, and related areas. In-depth discussions of these very interesting topics are beyond the scope of this chapter due to the limitation of space. Nevertheless, it is safe to conclude that with simpler and more efficient generation methods becoming increasingly accessible, this field will grow further and the list of applications of vector beams is expected to continue to expand rapidly.

The author is grateful to his current and former students and colleagues who have contributed to the understanding of the subject, participated in stimulating discussions, conducted difficult experiments and collected excellent experimental data, and coauthored papers in the past: Dr. Weibin Chen, Dr. Guanghao Rui, Mr Shuangyang Yang, Dr. Alain Tschimwangan, Mr Renjie Zhou, Dr. Wen Cheng, Dr. Wei Han, Professor Jiming Wang, Dr. Yiqiong Zhao, Dr. Jianing Dai, Dr. Liangcheng Zhou, Professor James R. Leger, Professor Daniel Ou-yang, Professor Yongping Li, Professor Hai Ming, Professor Pei Wang, Professor Yonghua Lu, Professor Lixin Xu, Dr. Don

C. Abeysinghe, and Dr. Robert L. Nelson. The author also acknowledges financial support from the Fraunhofer Society, the State of Ohio, the U.S. Air Force Research Laboratory, and the National Science Foundation.

References

1. Hasman, E., Biener, G., Niv, A., and Kleiner, V. 2005. Space-variant polarization manipulation. In *Progress in Optics*, ed. E. Wolf, 47: 215– 89. Oxford, UK: Elsevier.
2. Zhan, Q. 2009. Cylindrical vector beams: From mathematical concepts to applications, *Adv. Opt. Photon.* 1: 1–57.
3. Brown, T. G. 2011. Unconventional polarization states: Beam propagation, focusing and imaging. In *Progress in Optics*, ed. E. Wolf, 56: 81–129. Oxford, UK: Elsevier.
4. Brown, T.G. and Zhan, Q. 2010. Introduction: Unconventional polarization states of light focus issue, *Opt. Express* 18: 10775–6.
5. Pohl, D. 1972. Operation of a ruby laser in the purely transverse electric mode TE_{01}, *Appl. Phys. Lett.* 20: 266–67.
6. Mushiake, Y., Matzumurra K., and Nakajima, N. 1972. Generation of radially polarized optical beam mode by laser oscillation, *Proc. IEEE* 60: 1107–1109.
7. For example, see the radial polarizer from ARCoptix, http://www.arcoptix.com/.
8. For example, see the s-waveplate from Altechna, http://www.altechna.com/.
9. Zhan, Q. and Leger, J.R. 2002. Interferometric measurement of Berry's phase in space-variant polarization manipulations, *Opt. Commun.* 213: 241–45.
10. Grosjean, T., Courjon, D., and Spajer, M. 2002. An all-fiber device for generating radially and other polarized light beams, *Opt. Commun.* 203: 1–5.
11. Chen, W., Han, W., Abeysinghe, D.C., Nelson, R.L., and Zhan, Q. 2011. Generating cylindrical vector beams with subwavelength concentric metallic gratings fabricated on optical fibers, *J. Opt.* 13: 015003.
12. Cai, L., Zhang, J., Bai, W., Wang, Q., Wei, X., and Song, G. 2010. Generation of compact radially polarized beam at 850 nm in vertical-cavity surface-emitting laser via plasmonic modulation, *Appl. Phys. Lett.* 97: 201101.
13. Weiss, O. and Scheuer, J. 2010. Emission of cylindrical and elliptical vector beams from radial Bragg lasers, *Appl. Phys. Lett.* 97: 251108.
14. Yoshida, H., Tagashira, K., Kumagai, T., Fujii, A., and Ozaki, M. 2010. Alignment-to-polarization projection in dye-doped nematic liquid crystal microlasers, *Opt. Express* 18: 12562–12568.
15. Zhou, R., Ibarra-Escamilla, B., Powers, P.E., Haus, J.W., and Zhan, Q. 2009. Fiber laser generating switchable radially and azimuthally polarized beams with 140 mW output power at 1.6 μm wavelength, *Appl. Phys. Lett.* 95: 191111.
16. Zhou, R., Haus, J.W., Powers, P.E., and Zhan, Q. 2010. Vectorial fiber laser using intracavity axial birefringence, *Opt. Express* 18: 10839–47.

17. Sun, B., Wang, A., Xu, L. et al. 2012. Low-threshold single-wavelength all-fiber laser generating cylindrical vector beams using a few-mode fiber Bragg grating, *Opt. Lett.* 37: 464–66.

18. Ramachandran, S., Smith, C., Kristensen, P., and Balling, P. 2010. Nonlinear generation of broadband polarisation vortices, *Opt. Express* 18: 23212–7.

19. Zhan, Q. and Leger, J.R. 2002. Microellipsometer with radial symmetry, *Appl. Opt,* 41: 4630–37.

20. Zhan, Q. and Leger, J. R. 2002. Focus shaping using cylindrical vector beams, *Opt. Express* 10: 324–31.

21. Tschimwang, A. and Zhan, Q. 2010. High spatial-resolution nulling microellipsometer using rotational polarization symmetry, *Appl. Opt.* 49: 1574–80.

22. Lim, B.C., Phua, P.B., Lai, W.J., and Hong, M.H. 2008. Fast switchable electro-optic radial polarization retarder, *Opt. Lett.* 33: 950–52.

23. Dorn, R., Quabis, S., and Leuchs, G. 2003. Sharper focus for a radially polarized light beam, *Phys. Rev. Lett.* 91: 233901.

24. Gu, M. 1999. *Advanced Optical Imaging Theory.* New York: Springer-Verlag.

25. Chen, W. and Zhan, Q. 2006. Three-dimensional focus shaping with cylindrical vector beams, *Opt. Commun.* 265: 411–17.

26. Wang, H.F., Shi, L.P., Lukyanchuk, B., Sheppard, C., and Chong, C.T. 2008. Creation of a needle of longitudinally polarized light in vacuum using binary optics, *Nat. Photon.* 2: 501–505.

27. Zhao, Y. Q., Zhan, Q., Zhang, Y. L., and Li, Y. P. 2005. Creation of a three-dimensional optical chain for controllable particle delivery, *Opt. Lett.* 30: 848–50.

28. Wang, J., Chen, W., and Zhan, Q. 2011. Three-dimensional focus engineering using dipole array radiation pattern, *Opt. Commun.* 284: 2668–71.

29. Wang, J., Chen W., and Zhan, Q. 2010. Engineering of high purity ultra-long optical needle field through reversing the electric dipole array radiation, *Opt. Express* 18: 21965–72.

30. Jia, B., Kang, H., Li, J., and Gu, M. 2009. Use of radially polarized beams in three-dimensional photonic crystal fabrication with the two-photon polymerization method, *Opt. Lett.* 34: 1918–20.

31. Novotny, L., Beversluis, M.R., Youngworth, K.S., and Brown, T.G. 2001. Longitudinal field modes probed by single molecules, *Phys. Rev. Lett.* 86: 5251–54.

32. Lu, F., Zheng, W., and Huang, Z. 2009. Coherent anti-Stokes Raman scattering microscopy using tightly focused radially polarized light, *Opt. Lett.* 34: 1870–72.

33. Chen W. and Zhan, Q. 2009. Realization of an evanescent Bessel beam via surface plasmon interference excited by a radially polarized beam, *Opt. Lett.* 34: 722–24.

34. Chen, W., Nelson, R.L., Abeysinghe, D.C., and Zhan, Q. 2009. Plasmonic lens made of multiple concentric metallic rings under radially polarized illumination, *Nano Lett.* 9: 4320–25.

35. Chen, W. and Zhan, Q. 2007. Numerical study of an apertureless near field scanning optical microscope probe under radial polarization illumination, *Opt. Express* 15: 4106–11.

36. Rui, G., Chen, W., Lu, Y., Wang, P., Ming H., and Zhan, Q. 2010. Plasmonic near-field probe using the combination of concentric rings and conical tip under radial polarization illumination, *J. Opt.* 12: 035004.

37. Chen, W., Nelson, R.L., and Zhan, Q. 2012. Geometrical phase and surface plasmon focusing with azimuthal polarization, *Opt. Lett.* 37: 581–583.

38. Chen, W., Nelson, R. L., and Zhan, Q. 2012. Efficient miniature circular polarization analyzer design using hybrid spiral plasmonic lens, *Opt. Lett.* 37: 1442–44.

39. Zhan, Q. 2003. Radiation forces on a dielectric sphere produced by highly focused cylindrical vector beams, *J. Opt. A: Pure Appl. Opt.* 5: 229–32.

40. Zhan, Q. 2004. Trapping metallic Rayleigh particles with radial polarization, *Opt. Express* 12: 3377–82.

41. Nieminen, T.A., Heckenberg, N.R., and Rubinsztein-Dunlop, H. 2008. Forces in optical tweezers with radially and azimuthally polarized trapping beams, *Opt. Lett.* 33: 122–24.

42. Qin, J.Q., Wang, X.L., Jia, D. et al. 2009. FDTD approach to optical forces of tightly focused vector beams on metal particles, *Opt. Express* 17: 8407–16.

43. Huss, A., Chizhik, A.M., Jäger, R., Chizhik, A.I., and Meixner, A.J. 2011. Optical trapping of gold nanoparticles using a radially polarized laser beam, *SPIE Proc.* 8097: 809720.

44. Huang, L., Guo, H., Li, J., Ling, L., Feng, B., and Li, Z. 2012. Optical trapping of gold nanoparticles by cylindrical vector beam, *Opt. Lett.* 37: 1694–96.

45. Albaladejo, S., Marqués, M.I., Laroche, M., and Sáenz, J.J. 2009. Scattering forces from the curl of the spin angular momentum of a light field, *Phys. Rev. Lett.* 102: 113602.

46. Wong, V. and Ratner, M.A. 2006. Explicit computation of gradient and nongradient contributions to optical forces in the discrete-dipole approximation, *J. Opt. Soc. Am. B* 23: 1801–14.

47. Lerman, G. M. and Levy, U. 2009. Radial polarization interferometer, *Opt. Express* 17: 23234–246.

48. Tripathi, S. and Toussaint, K.C. 2009. Rapid Mueller matrix polarimetry based on parallelized polarization state generation and detection, *Opt. Express* 17: 21396–407.

49. Fatemi, F.K. 2011. Cylindrical vector beams for rapid polarization-dependent measurements in atomic systems, *Opt. Express* 19: 25143–50.

50. Cheng, W., Haus, J.W., and Zhan, Q. 2009. Propagation of vector vortex beams through a turbulent atmosphere, *Opt. Express* 17: 17829–36.

51. Beckley, A.M., Brown, T.G., and Alonso, M.A. 2010. Full Poincaré beams, *Opt. Express* 18: 10777–85.

52. Dai, J. and Zhan, Q. 2008. Beam shaping with vectorial vortex beams under low numerical aperture illumination condition, *SPIE Proc.*, 7062: 70620D.

53. Han, W., Cheng, W., and Zhan, Q. 2011. Flattop focusing with full Poincaré beams under low numerical aperture illumination, *Opt. Lett.* 36: 1605–7.

54. Chen, W. and Zhan, Q. 2010. Diffraction limited focus with controllable arbitrary three dimensional polarization, *J. Opt.* 12: 045707.

55. Yu, N., Genevet, P., Kats, M.A. et al. 2011. Light propagation with phase discontinuities: Generalized laws of reflection and refraction, *Science* 334: 333–7.

10

Low-Coherence Laser Beams

Jari Turunen
*University of Eastern
Finland*

10.1 Introduction

Many lasers exhibit partial spatial or temporal coherence properties, which can have a substantial effect on the propagation of the light beams they generate. Some examples of laser sources with reduced spatial coherence properties are multi-spatial-mode solid-state, gas or semiconductor lasers, excimer lasers, free-electron lasers (FELs), as well as many random lasers and vertical-cavity surface-emitting laser (VCSEL) arrays. On the other hand, excimer lasers and multi-longitudinal-mode continuous-wave lasers also have rather limited temporal coherence properties, which restrict their use in, for example, interference experiments. Finally, there are pulsed sources that are spatially nearly coherent but can have quite arbitrary temporal coherence; the prime examples of these are sources of supercontinuum (SC) light. In this chapter, we apply the second-order coherence theory of both stationary [1,2] and nonstationary [3–5] light to model continuous-wave and pulsed low-coherence laser sources. We further look at some aspects related to focusing, imaging, and controlling the spatial shape of beams generated by such lasers.

This chapter is organized as follows: in Section 10.2, we review the principles of second-order coherence theory of optical fields, introducing the statistical description of partially coherent light. We apply these concepts in Section 10.3 to describe the propagation of partially coherent fields in free space. We will arrive, in general, at four-dimensional (4D) integral formulas, which are not suitable for numerical evaluation. Therefore, in Section 10.4, we introduce two methods for expressing partially coherent fields as superpositions of fully coherent fields, which reduce the 4D integral formulas to superpositions of two-dimensional (2D) integral formulas and are in addition quite natural representations of many practical laser beams. In Section 10.5, we examine the effects of partial coherence in problems related to focusing, imaging, and shaping partially coherent fields using refractive and diffractive optics. Finally, in Section 10.6, we consider the modeling of several practical laser sources using the techniques introduced earlier in this chapter.

10.2 Basic Concepts of Coherence Theory

Let us denote the position vector by $\mathbf{r} = (x, y, z)$, the time by t, and the scalar components of the electric field vector of an arbitrarily fluctuating optical field by $E_i(\mathbf{r}; t)$ with $i = x, y, z$ (we employ the complex-analytic-signal field representation [1] throughout this chapter). Generally, the exact form of $E_j(\mathbf{r}; t)$ associated with a more or less random field is too complicated to be of much practical use even if it was known, and, therefore, statistical concepts are used to describe such fields mathematically. Comparing the fields at two positions and two instants of time, one then considers ensembles of all possible field realizations and defines (second-order) correlation functions of the type $\langle E_j^*(\mathbf{r}_1; t_1) E_i(\mathbf{r}; t_2) \rangle$, where the brackets indicate averaging over the entire ensemble of realizations. If $i = j$, these functions measure the correlation of one field component with itself. If $i \neq j$, they measure cross-correlation between two different field components. For simplicity, we drop the subscripts i and j from now on and define [3–5]

$$\Gamma(\mathbf{r}_1, \mathbf{r}_2; t_1, t_2) = \langle E^*(\mathbf{r}_1; t_1) E(\mathbf{r}_2; t_2) \rangle. \tag{10.1}$$

This so-called (two-time) mutual coherence function (MCF) is understood as one component of a 3×3 mutual coherence matrix formed by all different ij combinations [1]. In the typical case of a linearly polarized, highly directional laser beam or pulse, it is indeed sufficient to consider only one component of the coherence matrix, that is, to use the scalar formulation of coherence theory. The quantity

$$I(\mathbf{r};t) = \Gamma(\mathbf{r},\mathbf{r};t,t) = \langle |E(\mathbf{r};t)|^2 \rangle \tag{10.2}$$

then defines the space–time optical intensity of the random field and one can further introduce a normalized form of the MCF as

$$\gamma(\mathbf{r}_1,\mathbf{r}_2;t_1,t_2) = \frac{\Gamma(\mathbf{r}_1,\mathbf{r}_2;t_1,t_2)}{\sqrt{I(\mathbf{r}_1;t_1)I(\mathbf{r}_2;t_2)}}. \tag{10.3}$$

This function is known as the complex degree of coherence between field fluctuations at two points in space and two instants of time. One can show that the inequalities $0 \leq |\gamma(\mathbf{r}_1,\mathbf{r}_2;t_1,t_2)| \leq 1$ hold; the lower limit indicates complete incoherence and the upper limit implies full coherence at the appropriate space–time points. In the latter case, the MCF takes on the factorable form $\Gamma(\mathbf{r}_1,\mathbf{r}_2;t_1,t_2) = E^*(\mathbf{r}_1;t_1)E(\mathbf{r}_2;t_2)$ and the intensity distribution is simply given by $I(\mathbf{r};t) = |E(\mathbf{r};t)|^2$.

In particular, when considering the propagation of light either in homogeneous media or in optical systems, it is usually far more convenient to consider the space–frequency representation of optical fields rather than the space–time representation introduced in Equations 10.1 through 10.3. To this end, we define the temporal Fourier transform of the electric field as

$$E(\mathbf{r};\omega) = \frac{1}{2\pi} \int_{-\infty}^{\infty} E(\mathbf{r};t)\exp(i\omega t)\,dt \tag{10.4a}$$

so that, by Fourier inversion

$$E(\mathbf{r};t) = \int_{0}^{\infty} E(\mathbf{r};\omega)\exp(-i\omega t)\,d\omega. \tag{10.4b}$$

In analogy with Equation 10.1, we define the cross-spectral density function (CSD) as an ensemble average over the space–frequency representations $E(\mathbf{r};\omega)$ of the field realizations:

$$W(\mathbf{r}_1,\mathbf{r}_2;\omega_1,\omega_2) = \langle E^*(\mathbf{r}_1;\omega_1)E(\mathbf{r}_2;\omega_2)\rangle. \tag{10.5}$$

Thus the CSD measures correlations between fields at two spatial points and two frequencies. The power spectrum of the field at position $\mathbf{r}$ is defined as

$$S(\mathbf{r};\omega) = W(\mathbf{r},\mathbf{r};\omega,\omega) = \langle |E(\mathbf{r};\omega)|^2 \rangle \tag{10.6}$$

and we can again introduce a normalized function

$$\mu(\mathbf{r}_1,\mathbf{r}_2;\omega_1,\omega_2) = \frac{W(\mathbf{r}_1,\mathbf{r}_2;\omega_1,\omega_2)}{\sqrt{S(\mathbf{r}_1;\omega_1)S(\mathbf{r}_2;\omega_2)}}, \tag{10.7}$$

which satisfies the inequalities $0 \leq |\mu(\mathbf{r}_1,\mathbf{r}_2;\omega_1,\omega_2)| \leq 1$. Now, the lower and upper limits indicate complete incoherence and full coherence, respectively, between field fluctuations at two spatial points and two temporal (angular) frequencies. In the case of full coherence in the space–frequency domain, we have $W(\mathbf{r}_1, \mathbf{r}_2; \omega_1, \omega_2) = E^*(\mathbf{r}_1; \omega_1)E(\mathbf{r}_2; \omega_2)$ and $S(\mathbf{r}; \omega) = |E(\mathbf{r}; \omega)|^2$.

Substituting from Equation 10.4a into Equation 10.1 and using Equation 10.5, we see that there is a simple connection between the MCF and the CSD:

$$W(\mathbf{r}_1, \mathbf{r}_2; \omega_1, \omega_2) = \frac{1}{(2\pi)^2} \iint_{-\infty}^{\infty} \Gamma(\mathbf{r}_1, \mathbf{r}_2; t_1, t_2) \exp\left[-i\left(\omega_1 t_1 - \omega_2 t_2\right)\right] dt_1 dt_2 \quad (10.8a)$$

and, by inversion

$$\Gamma(\mathbf{r}_1, \mathbf{r}_2; t_1, t_2) = \iint_{0}^{\infty} W(\mathbf{r}_1, \mathbf{r}_2; \omega_1, \omega_2) \exp\left[i\left(\omega_1 t_1 - \omega_2 t_2\right)\right] d\omega_1 d\omega_2. \quad (10.8b)$$

These results are known as the generalized Wiener–Khintchine theorem. In the case of full coherence, the factorability of the coherence function in one domain (space–frequency or space–time) clearly implies factorability and, therefore, full coherence in the other domain as well, and in this case, the transformation between the two domains is fully described by Equations 10.4a and 10.4b.

Fields that are completely spectrally incoherent at each spatial point (the correlations vanish if $\omega_1 \neq \omega_2$) can be described by writing the CSD in the form

$$W(\mathbf{r}_1, \mathbf{r}_2; \omega_1, \omega_2) = W(\mathbf{r}_1, \mathbf{r}_2; \omega_1)\delta\left(\omega_1 - \omega_2\right), \quad (10.9)$$

where $\delta(\omega)$ is the Dirac delta function. Here, $W(\mathbf{r}_1, \mathbf{r}_2; \omega_1)$ is still an arbitrary function of the two spatial points under consideration. This function generally depends on the angular frequency ω: instead of Equation 10.5 it is defined as

$$W(\mathbf{r}_1, \mathbf{r}_2; \omega) = \langle E^*(\mathbf{r}_1; \omega)E(\mathbf{r}_2; \omega)\rangle \quad (10.10)$$

and the definitions of the spectrum and the complex degree of spectral coherence are modified in an analogous way. By inserting from Equation 10.9 into Equation 10.8b, we see that the MCF takes the form

$$\Gamma(\mathbf{r}_1, \mathbf{r}_2; \Delta t) = \int_{0}^{\infty} W(\mathbf{r}_1, \mathbf{r}_2; \omega) \exp\left(-i\omega\Delta t\right) d\omega, \quad (10.11a)$$

where $\Delta t = t_2 - t_1$. By Fourier inversion

$$W(\mathbf{r}_1, \mathbf{r}_2; \omega) = \frac{1}{2\pi} \int_{-\infty}^{\infty} \Gamma(\mathbf{r}_1, \mathbf{r}_2; \Delta t) \exp\left(i\omega\Delta t\right) d\Delta t. \quad (10.11b)$$

Hence the MCF (and also its normalized version γ) for each pair of spatial points is a function of the time difference Δt only, instead of the two separate instants of time. By writing $\mathbf{r}_1 = \mathbf{r}_2$ in Equations 10.11a and 10.11b, we have the results

$$\Gamma(\mathbf{r}, \mathbf{r}; \Delta t) = \int_0^\infty S(\mathbf{r}; \omega) \exp(-i\omega \Delta t) \, d\omega \qquad (10.12a)$$

and

$$S(\mathbf{r}; \omega) = \frac{1}{2\pi} \int_{-\infty}^\infty \Gamma(\mathbf{r}, \mathbf{r}; \Delta t) \exp(i\omega \Delta t) \, d\Delta t, \qquad (10.12b)$$

which constitute the standard Wiener–Khintchine theorem. It states that the spectral density and the temporal coherence function of a spectrally incoherent field form a Fourier-transform pair at any point in space. Finally, if we write $\Delta t = 0$ in Equation 10.12a, we find that the temporal intensity distribution of a spectrally incoherent field is independent of time:

$$I(\mathbf{r}) = \int_0^\infty S(\mathbf{r}; \omega) d\omega. \qquad (10.13)$$

Thus, the field is temporally stationary, which is a good approximation for continuous-wave lasers.

10.3 Propagation of Partially Coherent Fields

Let us now consider free-space propagation of light toward the positive z-direction, starting from the space–frequency-domain description and assuming that we know the CSD at some reference plane $z = 0$. Denoting the transverse spatial coordinate by $\rho = (x, y)$ and indicating the field at this reference plane by the superscript (0), we thus assume that $W^{(0)}(\rho_1, \rho_2; \omega_1, \omega_2)$ is known for nonstationary fields and that $W^{(0)}(\rho_1, \rho_2; \omega)$ is known for stationary fields.

In general, the propagation of the electric field $E(\mathbf{r}; \omega)$ can be governed by rigorous techniques such as the angular spectrum representation or one of the Rayleigh diffraction formulas [1]. However, in the case of reasonably directional laser beams, it is adequate to assume that the paraxial approximation is valid and to apply the Fresnel diffraction formula (see Chapter 1 for further details)

$$E(\rho, z; \omega) = \frac{\omega \exp(i\omega z/c)}{i2\pi cz} \int_{-\infty}^\infty E^{(0)}(\rho'; \omega) \exp\left[\frac{i\omega}{2cz}(\rho - \rho')^2\right] d^2\rho', \qquad (10.14)$$

where c is the speed of light in vacuum. If we expand the square in the exponential inside the integral and neglect a term that becomes insignificant at sufficiently large propagation distances, we arrive at the Fraunhofer diffraction formula

$$E^{(\infty)}(\rho, z; \omega) = \frac{\omega n(\omega) \exp(iz/c)}{i 2\pi cz} \exp\left(\frac{i\omega}{2cz} |\rho|^2\right)$$

$$\times \int_{-\infty}^{\infty} E^{(0)}(\rho'; \omega) \exp\left(-\frac{i\omega}{cz} \rho \cdot \rho'\right) d^2\rho', \qquad (10.15)$$

where we have denoted the electric field at a large distance from the reference plane by the superscript (∞).

The propagation laws for the CSD in the paraxial approximation can be readily obtained by applying Equations 10.5 and 10.10. For example, the Fresnel formula for nonstationary fields takes the form

$$W(\rho_1, z_1, \rho_2, z_2; \omega_1, \omega_2) = \frac{\omega_1 \omega_2 \exp\left[-i\left(\omega_1 z_1 - \omega_2 z_2\right)/c\right]}{(2\pi c)^2 z_1 z_2} \iint_{-\infty}^{\infty} W^{(0)}(\rho_1', \rho_2'; \omega_1, \omega_2)$$

$$\times \exp\left[-\frac{i\omega_1}{2cz_1} (\rho_1 - \rho_1')^2\right] \exp\left[\frac{i\omega_2}{2cz_2} (\rho_2 - \rho_2')^2\right] d^2\rho_1' d^2\rho_2' \qquad (10.16a)$$

and, for stationary fields, we have

$$W(\rho_1, z_1, \rho_2, z_2; \omega) = \frac{\omega^2 \exp(i\omega\Delta z/c)}{(2\pi c)^2 z_1 z_2} \iint_{-\infty}^{\infty} W^{(0)}(\rho_1', \rho_2'; \omega)$$

$$\times \exp\left\{-\frac{i\omega}{2c}\left[\frac{(\rho_1 - \rho_1')^2}{z_1} - \frac{(\rho_2 - \rho_2')^2}{z_2}\right]\right\} d^2\rho_1' d^2\rho_2', \qquad (10.16b)$$

where $\Delta z = z_2 - z_1$. Thus, in both cases, we generally have 4D integrals and the same applies to the Fraunhofer formulas. This is unpleasant from the viewpoint of analytical calculations. Also, numerical evaluation of the propagation integrals at sufficient accuracy becomes virtually impossible since even the storage of the discrete version of the CSD requires a rather prohibitive amount of computer memory. If, however, the CSD at $z = 0$ is separable in x' and y', the propagation problem becomes numerically manageable since the 4D integrals reduce to products of two 2D integrals.

The transitions between temporal and spectral domains can be carried out with the aid of the generalized Wiener–Khintchine theorems, that is, Equations 10.8a and 10.8b in the nonstationary case and Equations 10.11a and 10.8b in the stationary case.

10.4 Modal Representations of Partially Coherent Fields

Since the direct evaluation of the Fresnel and Fraunhofer propagation integrals in the partially coherent case was found to be difficult compared to the case of fully coherent fields, a question arises: is it possible to represent partially coherent fields as superpositions of a reasonable number of mutually uncorrelated and fully coherent fields? Then one could apply Equations 10.14 and 10.15 to each such "coherent mode" and get the final result for the original partially coherent field by combining the propagated coherent modes. We will first (in Sections 10.4.1 through 10.4.3) present a general solution to this problem and then (in Sections 10.4.4 and 10.4.5) consider another solution that applies to a somewhat restricted class of partially coherent fields but is computationally convenient and provides a valuable physical insight into the properties of partially coherent light. Both representations will prove useful in modeling realistic low-coherence laser beams (see Section 10.6):

10.4.1 Spatial Mercer-Type Coherent-Mode Representations

The general solution to the problem of representing partially coherent fields as superpositions of fully coherent modes has been known for a long time for stationary fields [6–8]. In this case, we may always write the CSD at the plane $z = 0$ in the form of a Mercer representation

$$W^{(0)}(\boldsymbol{\rho}_1, \boldsymbol{\rho}_2; \omega) = \sum_{m=0}^{\infty} \alpha_m(\omega) e_m^{(0)*}(\boldsymbol{\rho}_1; \omega) e_m^{(0)}(\boldsymbol{\rho}_2; \omega), \qquad (10.17)$$

where the coefficients $\alpha_m(\omega)$ are the real, nonnegative eigenvalues and $e_m(\boldsymbol{\rho}; \omega)$ are the orthonormal eigenfunctions of the Fredholm integral equation

$$\int_{-\infty}^{\infty} W^{(0)}(\boldsymbol{\rho}_1, \boldsymbol{\rho}_2; \omega) e_m^{(0)}(\boldsymbol{\rho}_1; \omega) \mathrm{d}^2 \rho_1 = \alpha_m(\omega) e_m^{(0)}(\boldsymbol{\rho}_2; \omega). \qquad (10.18)$$

Obviously, since all the individual terms inside the summation are of factorable form, they represent fully coherent fields and the functions $e_m(\boldsymbol{\rho}; \omega)$ may be called the coherent modes associated with the CSD. Since the summation in Equation 10.17 is a linear combination of fully coherent contributions, it is indeed possible to propagate each individual coherent mode separately and then combine the results to express the propagated CSD in the form [9]

$$W(\boldsymbol{\rho}_1, z_1, \boldsymbol{\rho}_2, z_2; \omega) = \sum_{m=0}^{\infty} \alpha_m(\omega) e_m^*(\boldsymbol{\rho}_1, z_1; \omega) e_m(\boldsymbol{\rho}_2, z_2; \omega), \qquad (10.19)$$

where the propagation of the modes $e_m(\boldsymbol{\rho}, z; \omega)$ are governed by expressions analogous to Equations 10.14 and 10.15.

The coherent-mode representation allows one to express the random electric vector associated with a stationary field in the form [7]

$$E(\mathbf{r}; \omega) = \sum_{m=0}^{\infty} c_m(\omega) e_m(\mathbf{r}; \omega), \tag{10.20}$$

where $c_m(\omega)$ are random variables that satisfy $\langle c_m^*(\omega) c_n(\omega) \rangle = \alpha_m(\omega) \delta_{mn}$; inserting these expressions into the definition 10.5 immediately leads to Equation 10.19.

Recently, the coherent-mode expansion has been extended to nonstationary fields [10–12]. In this case, it reads as

$$W^{(0)}(\boldsymbol{\rho}_1, \boldsymbol{\rho}_2; \omega_1, \omega_2) = \sum_{m=0}^{\infty} \alpha_m e_m^{(0)*}(\boldsymbol{\rho}_1; \omega_1) e_m^{(0)}(\boldsymbol{\rho}_2; \omega_2) \tag{10.21}$$

and the integral equation 10.18 is replaced by

$$\int_{-\infty}^{\infty} \int_0^{\infty} W^{(0)}(\boldsymbol{\rho}_1, \boldsymbol{\rho}_2; \omega_1, \omega_2) e_m^{(0)}(\boldsymbol{\rho}_1; \omega_1) d^2 \rho_1 d\omega_1 = \alpha_m e_m^{(0)}(\boldsymbol{\rho}_2; \omega_2). \tag{10.22}$$

Note that now the eigenvalues α_m no longer depend on the frequency. The propagated form of the coherent-mode expansion for nonstationary fields is

$$W(\boldsymbol{\rho}_1, z_1, \boldsymbol{\rho}_2, z_2; \omega_1, \omega_2) = \sum_{m=0}^{\infty} \alpha_m e_m^*(\boldsymbol{\rho}_1, z_1; \omega_1) e_m(\boldsymbol{\rho}_2, z_2; \omega_2). \tag{10.23}$$

The application of Equation 10.8b gives the time-domain representation

$$\Gamma(\boldsymbol{\rho}_1, z_1, \boldsymbol{\rho}_2, z_2; t_1, t_2) = \sum_{m=0}^{\infty} \alpha_m e_m^*(\boldsymbol{\rho}_1, z_1; t_1) e_m(\boldsymbol{\rho}_2, z_2; t_2), \tag{10.24}$$

where the space–time-domain coherent modes $e_m(\boldsymbol{\rho}, z; t)$ are related to their space–frequency-domain counterparts by expressions strictly analogous to Equations 10.4a and 10.4b. The random electric fields in the space–frequency and space–time domains are now expressible in the forms [11]

$$E(\mathbf{r}; \omega) = \sum_{m=0}^{\infty} c_m e_m(\mathbf{r}; \omega) \tag{10.25}$$

and

$$E(\mathbf{r}; t) = \sum_{m=0}^{\infty} c_m e_m(\mathbf{r}; t), \tag{10.26}$$

where the random variables c_m satisfy $\langle c_m^* c_n \rangle = \alpha_m \delta_{mn}$.

The importance of the coherent-mode representation in laser beam propagation arises from our ability to associate the modes $e_m(\rho, z; \omega)$ with the known cavity modes of lasers. Care should be exercised, though, in this interpretation since some of the modes may be frequency-degenerate [13]. In practice it is likely that cavity imperfections lift the degeneracy and the resulting modal spacings exceed the modal linewidth. In this case, the (stationary) field is fully spatially coherent at each frequency, that is, $\left|\mu^{(0)}(\rho_1, \rho_2; \omega)\right| = 1$ for all ω, but to verify this, one would need high-resolution spectral measurements. In practice, one would typically measure frequency integrals over the entire laser spectrum, that is, the quantity

$$J^{(0)}(\rho_1, \rho_2) = \int_0^\infty W^{(0)}(\rho_1, \rho_2; \omega)d\omega = \Gamma^{(0)}(\rho_1, \rho_2; 0). \tag{10.27}$$

Here, the last equality follows directly from Equation 10.11a. This zero-time-delay value of the MCF is called the mutual intensity. It is often used to characterize the spatial coherence of the field [2].

Under the above-mentioned conditions, we may express the spectral coefficients in the coherent-mode expansion approximately as

$$\alpha_m(\omega) = \alpha_m(\omega_m)\delta\,(\omega - \omega_m)\,, \tag{10.28}$$

where the constants $a_m(\omega_m)$ are proportional to the relative weights of the different transverse modes at their center frequencies ω_m. Then, by using Equation 10.17, we may cast the mutual intensity in the form

$$J^{(0)}(\rho_1, \rho_2) = \sum_{m=0}^\infty \alpha_m(\omega_m)e_m^{(0)*}(\rho_1; \omega_m)e_m^{(0)}(\rho_2; \omega_m). \tag{10.29}$$

Typically, the radiation from continuous-wave lasers is *quasimonochromatic*, that is, the effective spectral width is much smaller than the center frequency (say, ω_0) of the entire laser spectrum. Then the spatial variation of the modes is essentially the same for all ω_m and we can write $\omega_m = \omega_0$ in Equation 10.29, that is,

$$J^{(0)}(\rho_1, \rho_2) = \sum_{m=0}^\infty \alpha_m(\omega_0)e_m^{(0)*}(\rho_1; \omega_0)e_m^{(0)}(\rho_2; \omega_0). \tag{10.30}$$

It should be noted that although the mutual intensity does not obey the exact propagation laws, we may in practice (at least in the quasimonochromatic case) propagate the modal fields in Equation 10.30 using expressions analogous to Equations 10.14 and 10.15, thus essentially treating the mutual intensity as the CSD at the center frequency ω_0.

In the case of nonstationary fields, the instantaneous spatial coherence is characterized by the quantity $\Gamma^{(0)}(\rho_1, \rho_2; t, t)$. However, when dealing with pulsed lasers, one often measures the time average of this quantity over the entire pulse, that is, the (integrated) mutual intensity

$$J^{(0)}(\boldsymbol{\rho}_1, \boldsymbol{\rho}_2) = \int_{-\infty}^{\infty} \Gamma^{(0)}(\boldsymbol{\rho}_1, \boldsymbol{\rho}_2; t, t) \mathrm{d}t = 2\pi \int_{0}^{\infty} W^{(0)}(\boldsymbol{\rho}_1, \boldsymbol{\rho}_2; \omega, \omega) \mathrm{d}\omega, \qquad (10.31)$$

where we have used Equation 10.8b and the integral definition of the Dirac delta function to arrive at the second equality. Inserting from Equation 10.21, we have

$$J^{(0)}(\boldsymbol{\rho}_1, \boldsymbol{\rho}_2) = 2\pi \sum_{m=0}^{\infty} \alpha_m \int_{-\infty}^{\infty} e_m^{(0)*}(\boldsymbol{\rho}_1; \omega) e_m^{(0)}(\boldsymbol{\rho}_2; \omega) \mathrm{d}\omega. \qquad (10.32)$$

Especially, if we are not considering few-cycle pulses or wideband SC light, the laser pulses can be assumed quasimonochromatic with center frequency ω_0 and we may write, to a good approximation, the modal fields in space–frequency-separable form

$$e_m^{(0)}(\boldsymbol{\rho}; \omega) = g_m(\omega)\, e_m^{(0)}(\boldsymbol{\rho}; \omega_0), \qquad (10.33)$$

where $g_m(\omega)$ represent the spectral (amplitude) line shapes of the modes. In this case, Equation 10.32 reduces to

$$J^{(0)}(\boldsymbol{\rho}_1, \boldsymbol{\rho}_2) = \sum_{m=0}^{\infty} \alpha'_m e_m^{(0)*}(\boldsymbol{\rho}_1; \omega_0) e_m^{(0)}(\boldsymbol{\rho}_2; \omega_0), \qquad (10.34)$$

where we have defined

$$\alpha'_m = 2\pi \alpha_m \int_{0}^{\infty} |g_m(\omega)|^2 \, \mathrm{d}\omega. \qquad (10.35)$$

The result is thus formally analogous to Equation 10.30.

The spatial Mercer-type coherent-mode representations introduced above for the CSD are completely general and are often natural representations of fields radiated by lasers, especially if the modes can be determined by knowledge of the cavity structure. Then the only question that remains is the number and the weight distribution of the modes present in the beam. This problem may be resolved by performing explicit spatial coherence measurements [14,15] or, at least if the modes are of the Hermite–Gaussian form, simply by measuring the intensity distribution across the beam and applying the algorithm described in Refs. [16–18].

10.4.2 Spectral and Temporal Mercer-Type Coherent-Mode Representations

Let us consider the coherence properties of nonstationary fields at a single spatial point $\mathbf{r}_1 = \mathbf{r}_2 = \mathbf{r}$ and neglect the explicit dependence on $\mathbf{r}$ for brevity of notation. Now, the CSD and MCF may be written in the forms $W(\omega_1, \omega_2)$ and $\Gamma(t_1, t_2)$. In view of Equations 10.8a and 10.8b, these functions are related by

$$\Gamma(t_1, t_2) = \iint_0^\infty W(\omega_1, \omega_2) \exp\left[i\left(\omega_1 t_1 - \omega_2 t_2\right)\right] d\omega_1 d\omega_2, \qquad (10.36)$$

and its inverse. The correlation functions are globally of this form if we consider plane-wave pulses with no spatial dependence. Moreover, often the CSD at the source plane $z = 0$ is, to a good approximation, separable in space and frequency coordinates and the MCF is, therefore, separable in space and time coordinates [19,20]. In this case, the correlation functions can again be expressed like indicated above. Generally, even if these assumptions are true at $z = 0$, complex space–time coupling will be introduced by propagation. However, for quasimonochromatic pulsed beams, this coupling is typically negligible.

The spectral density and the instantaneous intensity now take the forms $S(\omega) = W(\omega, \omega)$ and $I(t) = \Gamma(t, t)$, and one can define the normalized forms of the CSD and MCF in analogy with Equations 10.7 and 10.3, respectively. It is also useful to define the so-called effective or integrated degree of coherence $\bar{\gamma}$ of nonstationary fields as [10]

$$\bar{\gamma}^2 = \frac{\displaystyle\iint_{-\infty}^{\infty} |\Gamma(t_1, t_2)|^2 \, dt_1 \, dt_2}{\left[\displaystyle\int_{-\infty}^{\infty} I(t) dt\right]^2} \qquad (10.37)$$

in analogy with the corresponding definition in the spatial domain [21]. Using the generalized Wiener–Khintchine theorem, one can readily establish that

$$\bar{\mu}^2 = \frac{\displaystyle\iint_0^\infty |W(\omega_1, \omega_2)|^2 \, d\omega_1 \, d\omega_2}{\left[\displaystyle\int_0^\infty S(\omega) d\omega\right]^2} = \bar{\gamma}^2, \qquad (10.38)$$

that is, we have the same value of the effective degree of coherence in both space–time and space–frequency domains.

If the spatial dependence is neglected, the spectral coherent-mode representation in Equation 10.21 takes the form [10]

$$W(\omega_1, \omega_2) = \sum_{m=0}^{\infty} \alpha_m e_m^*(\omega_1) e_m(\omega_2) \qquad (10.39)$$

and the integral equation 10.18 simplifies into

$$\int_0^\infty W(\omega_1, \omega_2) e_m(\omega_1) d\omega_1 = \alpha_m e_m(\omega_2). \qquad (10.40)$$

Furthermore, the temporal coherent-mode representation in Equation 10.24 assumes the form

$$\Gamma(t_1, t_2) = \sum_{m=0}^{\infty} \alpha_m e_m^*(t_1) e_m(t_2), \tag{10.41}$$

where again the spectral and temporal coherent modes $e_m(\omega)$ and $e_m(t)$ form a Fourier-transform pair. In general, the effective degree of coherence defined by Equation 10.37 or by Equation 10.38 has the expression [10]

$$\bar{\gamma} = \bar{\mu} = \frac{\left[\sum_{m=0}^{\infty} \alpha_m^2 \right]^{1/2}}{\sum_{m=0}^{\infty} \alpha_m}. \tag{10.42}$$

If only one mode has a nonzero eigenvalue α_m, we evidently have $\bar{\gamma} = \bar{\mu} = 1$. However, if many modes have significant eigenvalues, the effective degree of coherence is reduced and the field eventually becomes *quasistationary*. This means that its effective coherence time is significantly shorter than the pulse duration.

10.4.3 Some Examples of Coherent-Mode Expansions

Let us study in some detail the effects of superposition of several transverse modes in the spatial coherence properties of a continuous-wave (stationary) laser beam, assuming that the modes are of the usual Hermite–Gaussian form (we consider, for simplicity, the fields in the x-direction only but note that this is actually not a restriction since the two-dimensional [2D] modes are separable in x and y). Then the orthogonal modes in the coherent-mode expansion are (we assume that they all oscillate at $\omega_m \approx \omega_0$)

$$e_m^{(0)}(x; \omega_0) = (2/\pi)^{1/4} \left(2^m m! \, w_c \right)^{-1/2} H_m \left(\frac{\sqrt{2}x}{w_c} \right) \exp \left(-\frac{x^2}{w_c^2} \right), \tag{10.43}$$

where $H_m(x)$ is a Hermite polynomial of order m and w_c is the transverse scale factor of the coherent modes.

First, assume that $a_m(\omega_0) = 1$ for $m = 0, \ldots, M$ and $a_m(\omega_0) = 0$ for $m > M$. We can now readily calculate the transverse intensity profile $S^{(0)}(x)$ at $\omega = \omega_0$ and the normalized distribution of the complex degree of spatial coherence between the axial point $x_1 = 0$ and an off-axis point $x_2 = x$, that is, the function $\mu^{(0)}(0, x)$. Some results for different values of M are illustrated in Figure 10.1, where the curves are shown only for positive values of x since they are all symmetric with respect to the origin. Clearly, since the higher-order modes extend further away from the axis, the intensity distribution of the partially coherent field expands when more modes are added. At the same time, the central peak of the spatial coherence profile gets narrower, which indicates a decreased

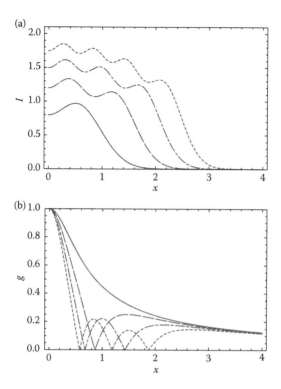

FIGURE 10.1 Plots of (a) the spatial intensity profile and (b) the absolute value of the complex degree of spatial coherence between an axial and an off-axis point for incoherent superpositions of $M + 1$ lowest-order Hermite–Gaussian modes. Solid lines: $M = 1$. Dashed lines: $M = 3$. Dash-dotted lines: $M = 5$. Dashed lines: $M = 7$.

spatial coherence width especially if one compares it with the width of the intensity profile. When the ratio of the effective widths of the coherence function and the intensity profile is much smaller than unity, one speaks of a *quasihomogeneous* field. This is already the case with the largest values of M shown in Figure 10.1.

It is worth noting that the appropriate incoherent superpositions of Hermite–Gaussian modes can lead to flattop intensity distributions that remain as such while the beam propagates in free space since all the Hermite–Gaussian modes expand at the same rate. An algorithm for determining the required modal weights is given in Ref. [22].

A particularly important example of spatially partially coherent fields obtained by superpositions of Hermite–Gaussian modes is the class of (stationary) Gaussian–Schell-model fields described by Gaussian distributions of the spatial intensity profile and the complex degree of spatial coherence [23–26]. Such fields have the functional form (here, we again look only at the x-dependent part for brevity of notation)

$$W^{(0)}(x_1, x_2) = \left[S^{(0)}(x_1) S^{(0)}(x_2) \right]^{1/2} \mu^{(0)}(x_1, x_2) \qquad (10.44)$$

with

$$S^{(0)}(x) = S_0 \exp\left(-\frac{2x^2}{w^2}\right)$$
(10.45)

and

$$\mu^{(0)}(x_1, x_2) = \exp\left[-\frac{(x_1 - x_2)^2}{2\sigma^2}\right].$$
(10.46)

Hence, these sources are characterized by a Gaussian spatial intensity profile with scale factor w and a Gaussian distribution of the complex degree of spatial coherence with scale factor σ. We stress that, strictly speaking (e.g., in spherical-mirror resonators), both w and σ generally depend on ω, but we have omitted this because we assume a quasimonochromatic field. A fully coherent Gaussian beam is obtained in the limit $\sigma \to \infty$.

It has been shown [25,27] that a Gaussian–Schell-model field characterized by the parameters w and σ is obtained as a superposition of Hermite–Gaussian modes if their modal scale factor is chosen as

$$w_c = w\sqrt{\beta},$$
(10.47)

where

$$\beta = \left[1 + (w/\sigma)^2\right]^{-1/2},$$
(10.48)

and if the modal weights assume the form

$$\alpha_m = S_0 \frac{\sqrt{2\pi}\, w}{1 + 1/\beta} \left(\frac{1 - \beta}{1 + \beta}\right)^m.$$
(10.49)

To represent a Gaussian–Schell-model field properly, all modes with significant values of α_m need to be retained. In view of Equation 10.49, the required number of modes increases as the effective degree of spatial coherence of the field is reduced, that is, as the ratio σ/w (or the value of β) decreases.

Since all Hermite–Gaussian modes expand at the same rate when the field propagates in free space, the beams radiated by Gaussian–Schell-model sources are of Gaussian–Schell-model type at any plane $z =$ constant. The coherent modes and the total beam width thus obey the laws

$$w_c(z) = w_c\sqrt{1 + \left(\frac{z}{z_R}\right)^2} \to \frac{w_c z}{z_R},$$
(10.50)

$$w(z) = w\sqrt{1 + \left(\frac{z}{z_R}\right)^2} \to \frac{wz}{z_R},$$
(10.51)

where the last forms apply in the far-zone limit $z \gg z_R$. In these expressions

$$z_R = \frac{\omega_0 w_c^2}{2c} = \frac{\omega_0 w^2 \beta}{2c}$$
(10.52)

represents the (common) Rayleigh range of the modes and the total field. The characteristic (paraxial) far-field divergence angles of the coherent modes and the entire Gaussian–Schell-model beam are

$$\theta_c = \lim_{z \to \infty} \frac{w_c(z)}{z} = \frac{2c}{\omega_0 w_c} \tag{10.53a}$$

and

$$\theta = \lim_{z \to \infty} \frac{w(z)}{z} = \frac{2c}{\omega_0 w \beta}, \tag{10.53b}$$

respectively. The coherence width of the total beam satisfies the propagation law

$$\sigma(z) = \frac{\sigma}{w} w(z). \tag{10.54}$$

Both the coherent modes and the total field are characterized by the same propagation-dependent parameter

$$R(z) = z + \frac{z_R}{z} \to z, \tag{10.55}$$

where the last form again applies to large values of z. This parameter represents the common radius of curvature of all the coherent modes and is associated with the phase of the CSD of the total field [25,26]. The parameters introduced above are illustrated in Figure 10.2.

The results presented above show that any two Gaussian–Schell-model beams with the same value of the product $w\beta$ have the same far-zone diffraction angle θ. In particular, we may assume that one of the fields is fully coherent, that is, that it only contains the lowest-order coherent mode in the Mercer expansion. Then one can readily see that a coherent Gaussian beam characterized by w_G and a partially coherent beam characterized by w and σ produce equally directional beams provided that the equivalence relation

$$\frac{1}{w_G^2} = \frac{1}{w^2} + \frac{1}{\sigma^2} \tag{10.56}$$

holds [23].

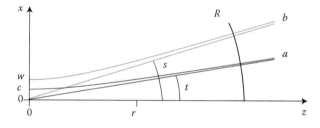

FIGURE 10.2 Propagation parameters and divergence characteristics of Gaussian–Schell-model beams, and of the coherent modes associated with them.

Anisotropic Gaussian–Schell-model fields [28–30] can be constructed if we let the parameters w and σ assume different values in the x- and y-directions. Then, in general, the far-field divergence angles are also different in the x- and y-directions. Using the subscripts x and y, we see that if the condition $w_x \beta_x = w_y \beta_y$ holds, $\theta_x = \theta_y$, that is, an anisotropic Gaussian–Schell-model source produces a rotationally symmetric far-field intensity pattern. If, on the other hand, the condition $w_x^2 \beta_x = w_y^2 \beta_y$ holds, we have $w_x(z)/w_y(z) = w_x/w_y$ for all z and also $\theta_x/\theta_y = w_x/w_y$. Thus, the beam retains the same ellipticity at all propagation distances including the far field. By choosing the parameters w_x, σ_x, w_y, and σ_y properly, one can readily construct fields with large values of w_x/w_y as noted by Gori and Guattari [31], who called this type of shape-invariant fields "blade-like."

The spatial M^2 factor is defined as the ratio of the effective beam divergence θ of an arbitrary field and the divergence θ_G of a fully coherent Gaussian beam with a matching scale factor w. Hence, a beam with a given value of the M^2 factor may be called M^2 times diffraction limited. For Gaussian–Schell-model beams, the M^2 factor is uniquely defined and (in the isotropic case), it depends on the spatial coherence of the source according to

$$M^2 = \frac{\theta}{\theta_G} = \frac{1}{\beta} = \left[1 + (w/\sigma)^2\right]^{1/2}. \tag{10.57}$$

In the quasihomogeneous limit $\sigma \ll w$, we thus have $M^2 \approx w/\sigma \gg 1$.

The Gaussian–Schell model is convenient and also useful in the characterization of pulsed fields. By neglecting the spatial dependence as in Section 10.4.2, we may write the CSD in the form $W(\omega_1, \omega_2) = [S(\omega_1)S(\omega_2)]^{1/2} \mu(\omega_1 - \omega_2)$ with

$$S(\omega) = S_0 \exp\left[-\frac{2}{\Omega^2}(\omega - \omega_0)^2\right] \tag{10.58}$$

and

$$\mu(\omega_1, \omega_2) = \exp\left[-\frac{(\omega_1 - \omega_2)^2}{2\Omega_\mu^2}\right], \tag{10.59}$$

where Ω and Ω_μ are the characteristic widths of the spectrum and spectral coherence, respectively, and ω_0 is the center frequency of the spectrum. We can express the CSD in the form of Equation 10.39 by employing spectral Hermite–Gaussian coherent modes [10]

$$e_m(\omega) = (2/\pi)^{1/4} \left(2^m m! \, \Omega_c\right)^{-1/2} H_m\left[\frac{\sqrt{2}(\omega - \omega_0)}{\Omega_c}\right] \exp\left[-\frac{(\omega - \omega_0)^2}{\Omega_c^2}\right] \tag{10.60}$$

with

$$\Omega_c = \Omega\sqrt{\beta} \tag{10.61}$$

if we define the spectral β parameter as

$$\beta = \left[1 + \left(\Omega/\Omega_\mu\right)^2\right]^{-1/2} \tag{10.62}$$

and choose the expansion coefficients as

$$\alpha_m = S_0 \frac{\sqrt{2\pi}\,\Omega}{1 + 1/\beta} \left(\frac{1-\beta}{1+\beta}\right)^m \tag{10.63}$$

in analogy with Equation 10.49 in the spatial domain.

By applying the generalized Wiener–Khintchine theorem, we obtain the MCF of the total field in the form $\Gamma(t_1, t_2) = [I(t_1)I(t_2)]^{1/2}\,\gamma(t_1 - t_2)$ with [32]

$$I(t) = I_0 \exp\left(-\frac{2t^2}{T^2}\right) \tag{10.64}$$

and

$$\gamma(t_1, t_2) = \exp\left[-\frac{(t_1 - t_2)^2}{2T_\gamma^2}\right] \exp\left[i\omega_0(t_2 - t_1)\right], \tag{10.65}$$

provided that expressions

$$\frac{1}{\Omega_\mu^2} + \frac{1}{\Omega^2} = \frac{T^2}{4} \tag{10.66}$$

and

$$\frac{1}{T_\gamma^2} + \frac{1}{T^2} = \frac{\Omega^2}{4} \tag{10.67}$$

connect the spectral and temporal characteristics of the pulses.

The MCF can of course be expressed as a superposition of fully coherent temporal modes in the form of Equation 10.41. The time-domain coherent modes are also of the Hermite–Gaussian type

$$e_m(t) = (-i)^m (2/\pi)^{1/4} \left(2^m m!\, T_c\right)^{-1/2} H_m\left(\frac{\sqrt{2}t}{T_c}\right) \exp\left(-\frac{t^2}{T_c^2}\right) \exp\left(-i\omega_0 t\right), \tag{10.68}$$

where

$$T_c = 2/\Omega_c. \tag{10.69}$$

It then follows, for example, that $T_c = T\sqrt{\beta}$ in analogy with Equation 10.61, and that we may define the temporal β parameter also as

$$\beta = \left[1 + \left(T/T_\gamma\right)^2\right]^{-1/2} \tag{10.70}$$

since $\Omega_c/\Omega = T_\gamma/T$.

The effective degree of coherence for Gaussian–Schell-model pulses, defined by Equation 10.37, can be straightforwardly cast in the forms

$$\bar{\gamma} = \bar{\mu} = \sqrt{\frac{2}{\Omega T}} = \sqrt{\beta}. \tag{10.71}$$

In analogy with the spatial case, we may also define the temporal M^2 factor as the ratio of the pulse duration T of an arbitrary nonstationary field and the pulse length T_G of a fully spectrally coherent Gaussian pulse with a matching spectral width Ω. For Gaussian–Schell-model pulses also, this quantity has a uniquely defined value

$$M^2 = \frac{T}{T_G} = \frac{1}{\beta} = \left[1 + (\Omega/\Omega_\mu)^2\right]^{1/2} = \left[1 + (T/T_\gamma)^2\right]^{1/2}. \tag{10.72}$$

In the quasistationary limit $\Omega \gg \Omega_\mu$, we now have $M^2 \approx \Omega/\Omega_\mu \gg 1$. Finally, an equivalence theorem concerning fully and partially coherent Gaussian pulses can be easily obtained. The two pulses have identical spectra (of width Ω) provided that the relation

$$\frac{1}{T_G^2} = \frac{1}{T^2} + \frac{1}{T_\gamma^2} \tag{10.73}$$

holds for their temporal properties. Thus, a temporally partially coherent Gaussian pulse can have the same spectrum as a coherent laser pulse if its duration is extended. In the case of a fully stationary field with $T \to \infty$, we have simply $T = T_\gamma$, that is, the spectra are the same if the duration of the coherent pulse is equal to the effective coherence time of the stationary field.

10.4.4 Elementary-Field Representations of Partially Coherent Fields

The coherent-mode representation introduced above is completely general, but it is not the only possible expansion of the correlation functions MCF and CSD in terms of fully coherent fields. Let us next assume that the MCF of a nonstationary field at the plane $z = 0$ can be written as

$$\Gamma^{(0)}(\boldsymbol{\rho}_1, \boldsymbol{\rho}_2; t_1, t_2) = \iint_{-\infty}^{\infty} p(\boldsymbol{\rho}_e; t_e) f^{(0)*}(\boldsymbol{\rho}_1 - \boldsymbol{\rho}_e; t_1 - t_e) f^{(0)}$$
$$\times (\boldsymbol{\rho}_2 - \boldsymbol{\rho}_e; t_2 - t_e)\, d^2\rho_e dt_e \tag{10.74}$$

so that the spatiotemporal intensity distribution takes the form

$$I^{(0)}(\boldsymbol{\rho}; t) = \iint_{-\infty}^{\infty} p(\boldsymbol{\rho}_e; t_e) \left|f^{(0)}(\boldsymbol{\rho} - \boldsymbol{\rho}_e; t - t_e)\right|^2 d^2\rho_e dt_e. \tag{10.75}$$

In these expressions, $f^{(0)}(\rho;t)$ is a fully spatially and temporally coherent (pulsed) field. The integrals contain replicas of this same field, which are shifted spatially (in the transverse direction) into new center positions ρ_e and their temporal origins are shifted into time instants t_e. The real and positive function $p(\rho_e;t_e)$ represents the spatiotemporal weights of these shifted replicas in the incoherent summation.

The type of field representations described by Equation 10.74 (and other closely related representations), known as elementary-field representations, are investigated thoroughly in Ref. [33]. They appear to be first introduced into the theory of spatially partially coherent light some 30 years ago to represent Gaussian–Schell-model fields as superpositions of coherent Gaussian beams [34,35] and later extended to more general spatially [36,37] and temporally [38,39] partially coherent fields. However, in contrast to the Mercer-type coherent-mode representation, the elementary-field representation is only applicable to a limited class of partially coherent fields [33].

Instead of considering the general implications of Equation 10.74, we assume for simplicity that both the weight function and the elementary field can be expressed in space–time-separable forms $p(\rho_e;t_e) = p(\rho_e)p(t_e)$ and $f^{(0)}(\rho;t) = f^{(0)}(\rho)f^{(0)}(t)$. These are in general only approximations, but they are in fact good approximations if we consider reasonably directional and quasimonochromatic laser beams and pulses. Clearly, the MCF then also assumes a space–time-separable form

$$\Gamma^{(0)}(\rho_1,\rho_2;t_1,t_2) = \Gamma^{(0)}(\rho_1,\rho_2)\Gamma^{(0)}(t_1,t_2), \tag{10.76}$$

where

$$\Gamma^{(0)}(\rho_1,\rho_2) = \int_{-\infty}^{\infty} p(\rho_e)f^{(0)*}(\rho_1 - \rho_e)f^{(0)}(\rho_2 - \rho_e)\, d^2\rho_e \tag{10.77}$$

and

$$\Gamma^{(0)}(t_1,t_2) = \int_{-\infty}^{\infty} p(t_e)f^{(0)*}(t_1 - t_e)f^{(0)}(t_2 - t_e)\, dt_e \tag{10.78}$$

represent the spatial and temporal contributions, respectively.

By applying the generalized Wiener–Khintchine theorem and Equation 10.8a to Equations 10.76 through 10.78, we see that

$$W^{(0)}(\rho_1,\rho_2;\omega_1,\omega_2) = W^{(0)}(\rho_1,\rho_2)W^{(0)}(\omega_1,\omega_2), \tag{10.79}$$

where $W^{(0)}(\rho_1,\rho_2) = \Gamma^{(0)}(\rho_1,\rho_2)$,

$$W^{(0)}(\omega_1,\omega_2) = 2\pi F^{(0)*}(\omega_1)F^{(0)}(\omega_2)P(\omega_2 - \omega_1), \tag{10.80}$$

and we have defined the Fourier transforms of the weight function and the temporal elementary field as

$$P(\omega) = \frac{1}{2\pi} \int_{-\infty}^{\infty} p(t)\exp(i\omega t)\, dt \tag{10.81}$$

and

$$F^{(0)}(\omega) = \frac{1}{2\pi} \int_{-\infty}^{\infty} f^{(0)}(t) \exp{(\mathrm{i}\omega t)}\, \mathrm{d}t, \tag{10.82}$$

respectively. In view of these results, the spectral density across the plane $z = 0$ is of the form $S^{(0)}(\boldsymbol{\rho}, \omega) = S^{(0)}(\boldsymbol{\rho})S^{(0)}(\omega)$, where $S^{(0)}(\boldsymbol{\rho}) = W^{(0)}(\boldsymbol{\rho}, \boldsymbol{\rho}) = \Gamma^{(0)}(\boldsymbol{\rho}, \boldsymbol{\rho})$ and $S^{(0)}(\omega) = W^{(0)}(\omega, \omega) = 2\pi P(0)\left|F^{(0)}(\omega)\right|^2$. The complex degree of spectral coherence takes the form $\mu^{(0)}(\boldsymbol{\rho}_1, \boldsymbol{\rho}_2; \omega_1, \omega_2) = \mu^{(0)}(\boldsymbol{\rho}_1, \boldsymbol{\rho}_2)\mu^{(0)}(\omega_1, \omega_2)$ with

$$\mu^{(0)}(\boldsymbol{\rho}_1, \boldsymbol{\rho}_2) = \frac{W^{(0)}(\boldsymbol{\rho}_1, \boldsymbol{\rho}_2)}{\sqrt{S^{(0)}(\boldsymbol{\rho}_1)S^{(0)}(\boldsymbol{\rho}_2)}} = \frac{\Gamma^{(0)}(\boldsymbol{\rho}_1, \boldsymbol{\rho}_2)}{\sqrt{S^{(0)}(\boldsymbol{\rho}_1)S^{(0)}(\boldsymbol{\rho}_2)}} \tag{10.83}$$

and

$$\begin{aligned} \mu^{(0)}(\omega_1, \omega_2) &= \frac{W^{(0)}(\omega_1, \omega_2)}{\sqrt{S^{(0)}(\omega_1)S^{(0)}(\omega_2)}} \\ &= \frac{P(\omega_2 - \omega_1)}{P(0)} \exp\left\{ \mathrm{i}\arg\left[F^{(0)}(\omega_2)\right] - \mathrm{i}\arg\left[F^{(0)}(\omega_1)\right] \right\}. \end{aligned} \tag{10.84}$$

Thus, we can conclude that the spectrum $S^{(0)}(\omega)$ of the entire partially coherent field is directly proportional to the spectrum of the elementary field and the absolute value of the complex degree of spectral coherence is fully determined by the weight function of the temporally shifted elementary fields. Finally, we may write the mutual intensity of the nonstationary field, defined by Equation 10.31, in the form

$$J^{(0)}(\boldsymbol{\rho}_1, \boldsymbol{\rho}_2) = J^{(0)}\Gamma^{(0)}(\boldsymbol{\rho}_1, \boldsymbol{\rho}_2) \tag{10.85}$$

with $J^{(0)} = (2\pi)^2 P(0) \int_0^{\infty} \left|F^{(0)}(\omega)\right|^2 \mathrm{d}\omega$.

In the special case of stationary fields, we express the noncorrelation of different frequency components by writing the spectral weight function in the form $P(\omega_2 - \omega_1) = \delta(\omega_2 - \omega_1)$. This leads to $W^{(0)}(\boldsymbol{\rho}_1, \boldsymbol{\rho}_2; \omega) = 2\pi \left|F^{(0)}(\omega)\right|^2 \Gamma^{(0)}(\boldsymbol{\rho}_1, \boldsymbol{\rho}_2)$ and the mutual intensity defined by Equation 10.27 is given by Equation 10.85 with $J^{(0)} = 2\pi \int_0^{\infty} \left|F^{(0)}(\omega)\right|^2 \mathrm{d}\omega$.

The propagated version of Equation 10.79 is obtained as described in Section 10.3. The spatially shifted elementary fields may be propagated independently and superposed incoherently to get the propagated version of the CSD. Let us consider, as an example, the free-space propagation of a narrowband field into an arbitrary plane $z_1 = z_2 = z$ in the far zone. Using Equation 10.15 we then get

$$\begin{aligned} W^{(\infty)}(\boldsymbol{\rho}_1, z, \boldsymbol{\rho}_2, z; \omega_1, \omega_2) &= \left(\frac{\omega_0}{2\pi c z}\right)^2 W^{(0)}(\omega_1, \omega_2)\exp\left[-\frac{\mathrm{i}\omega_0}{2cz}\left(|\boldsymbol{\rho}_1|^2 - |\boldsymbol{\rho}_2|^2\right)\right] \\ &\times \iint_{-\infty}^{\infty} W^{(0)}(\boldsymbol{\rho}_1', \boldsymbol{\rho}_2')\exp\left[\frac{\mathrm{i}\omega_0}{cz}\left(\boldsymbol{\rho}_1 \cdot \boldsymbol{\rho}_1' - \boldsymbol{\rho}_2 \cdot \boldsymbol{\rho}_2'\right)\right]\mathrm{d}^2\boldsymbol{\rho}_1'\mathrm{d}^2\boldsymbol{\rho}_2'. \end{aligned} \tag{10.86}$$

Defining the 2D spatial Fourier transforms of the spatial elementary-field function and the weight function as

$$F^{(0)}(\boldsymbol{\kappa}) = \frac{1}{(2\pi)^2} \int_{-\infty}^{\infty} f^{(0)}(\boldsymbol{\rho}') \exp\left(-i\boldsymbol{\kappa} \cdot \boldsymbol{\rho}'\right) d^2\rho' \tag{10.87}$$

and

$$P(\boldsymbol{\kappa}) = \frac{1}{(2\pi)^2} \int_{-\infty}^{\infty} p(\boldsymbol{\rho}_e) \exp\left(-i\boldsymbol{\kappa} \cdot \boldsymbol{\rho}_e\right) d^2\rho_e, \tag{10.88}$$

respectively, we obtain (after some straightforward calculation) the result

$$W^{(\infty)}(\boldsymbol{\rho}_1, z, \boldsymbol{\rho}_2, z; \omega_1, \omega_2) = \left(\frac{4\pi^2 \omega_0}{cz}\right)^2 W^{(0)}(\omega_1, \omega_2) \exp\left[-\frac{i\omega_0}{2cz}\left(|\boldsymbol{\rho}_1|^2 - |\boldsymbol{\rho}_2|^2\right)\right]$$
$$\times P\left[\frac{\omega_0}{cz}\left(\boldsymbol{\rho}_2 - \boldsymbol{\rho}_1\right)\right] F^{(0)*}\left(\frac{\omega_0}{cz}\boldsymbol{\rho}_1\right) F^{(0)}\left(\frac{\omega_0}{cz}\boldsymbol{\rho}_2\right). \tag{10.89}$$

Therefore, the spectral density and the complex degree of spectral coherence assume the forms

$$S^{(\infty)}(\boldsymbol{\rho}, z; \omega) = \left(\frac{4\pi^2 \omega_0}{cz}\right)^2 S^{(0)}(\omega) P(0) \left|F^{(0)}\left(\frac{\omega_0}{cz}\boldsymbol{\rho}\right)\right|^2 \tag{10.90}$$

with $S^{(0)}(\omega) = W^{(0)}(\omega, \omega)$ and

$$\mu^{(\infty)}(\boldsymbol{\rho}_1, z, \boldsymbol{\rho}_2, z; \omega_1, \omega_2) = \mu^{(0)}(\omega_1, \omega_2) \exp\left[-\frac{i\omega_0}{2cz}\left(|\boldsymbol{\rho}_1|^2 - |\boldsymbol{\rho}_2|^2\right)\right]$$
$$\times \frac{P\left[(\omega_0/cz)\left(\boldsymbol{\rho}_2 - \boldsymbol{\rho}_1\right)\right]}{P(0)} \exp\left\{i \arg\left[F^{(0)}\left(\frac{\omega_0}{cz}\boldsymbol{\rho}_2\right)\right]\right.$$
$$\left. -i \arg\left[F^{(0)}\left(\frac{\omega_0}{cz}\boldsymbol{\rho}_1\right)\right]\right\}, \tag{10.91}$$

respectively. Clearly, the total spatial distribution of the spectral density in the far zone is proportional to that of an individual elementary field, while the spatial form of the absolute value of the complex degree of spectral coherence at $\omega_1 = \omega_2 = \omega_0$ is determined by the weight function of the elementary fields.

10.4.5 Some Examples of Elementary-Field Representations

Let us consider again only the x-dependence of the field for brevity and choose the spatial weight function and the elementary-field mode to assume the Gaussian forms

$$p(x_e) = p_0 \exp\left(-\frac{2x_e^2}{w_P^2}\right) \tag{10.92}$$

and

$$f^{(0)}(x) = f_0 \exp\left(-\frac{x^2}{w_e^2}\right). \tag{10.93}$$

Then, by inserting these expressions in the one-dimensional (1D) form of Equation 10.77 and carrying out the integration, we again obtain a Gaussian–Schell-model field with parameters w and σ, provided that the transverse scales of the weight function and the elementary-field mode are given by

$$w_p = w\sqrt{1 - \beta^2} \tag{10.94}$$

and

$$w_e = w\beta, \tag{10.95}$$

respectively. In view of Equations 10.47 and 10.94, $w_e/w_c = \sqrt{\beta}$. Hence, $w_e < w_c$ for all partially spatially coherent fields, that is, if w and σ are fixed, the elementary-field mode is always narrower than the lowest-order coherent mode. However, it is evident by considering Equations 10.53a and 10.53b that the far-zone divergence angles of the elementary field and the total Gaussian–Schell-model field are the same.

Figure 10.3 illustrates the elementary-field representation of Gaussian–Schell-model sources. The curves $g(x) = S^{(0)}(x)$, shown by thick solid lines, represent the intensity profile of the entire partially coherent field. The thin solid curves illustrate the function $g(x) = \mu^{(0)}(0, x)$, that is, the degree of spatial coherence between an axial and an off-axis point. In Figure 10.3a, we consider a nearly coherent case with $\sigma = 2w$ and in Figure 10.3b, we consider a less-coherent case with $\sigma = w/3$. The intensity profile $g(x) = |f^{(0)}(x)|^2$ of the elementary field is in each case shown by a dash-dotted line, and the corresponding weight function $g(x) = p(x)$ is illustrated by a dotted line. For comparison, we also show the intensity profile $g(x) = |e_0^{(0)}(x)|^2$ of the lowest-order coherent mode in the Mercer expansion (dashed line). In the nearly coherent case, the intensity profiles of both the elementary field and the lowest-order coherent mode are almost as wide as the intensity profile of the entire partially coherent field, and the weight function in the elementary-field representation is relatively narrow as one would expect. In the less coherent case, the elementary field as well as the lowest-order coherent mode become substantially narrower while the spatial width of the weight function increases. In the quasihomogeneous limit $\sigma \ll w$, the weight function in fact becomes proportional to the intensity profile of the partially coherent field. This is generally true; hence, in the quasihomogeneous case, we may cast the space-dependent part of Equation 10.79 in the form [37]

$$W^{(0)}(\boldsymbol{\rho}_1, \boldsymbol{\rho}_2) = S^{(0)}(\bar{\boldsymbol{\rho}})\mu^{(0)}(\Delta\boldsymbol{\rho}), \tag{10.96}$$

where

$$\mu^{(0)}(\Delta\boldsymbol{\rho}) = \frac{\displaystyle\int_{-\infty}^{\infty} f^{(0)*}(\boldsymbol{\rho}' - \Delta\boldsymbol{\rho})f^{(0)}(\boldsymbol{\rho}')\mathrm{d}^2\rho'}{\displaystyle\int_{-\infty}^{\infty} |f^{(0)}(\boldsymbol{\rho}')|^2 \mathrm{d}^2\rho'}, \tag{10.97}$$

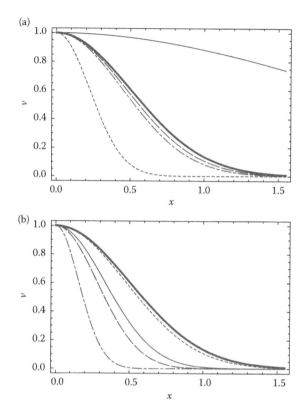

FIGURE 10.3 Representation of Gaussian–Schell-model fields by coherent modes and elementary fields. Thick solid lines: intensity profile of the partially coherent field. Thin solid lines: degree of spatial coherence of the partially coherent field. Dashed lines: intensity of the lowest-order coherent mode. Dash-dotted lines: intensity of the elementary field. Dashed lines: weight function of the elementary fields. (a) Nearly coherent case $\sigma = 2w$. (b) Intermediate coherence case $\sigma = w/3$.

and we have introduced the average and difference spatial coordinates $\bar{\rho} = \frac{1}{2}\left(\rho_1 + \rho_2\right)$ and $\Delta\rho = \rho_2 - \rho_1$.

Let us conclude this section by considering the temporal elementary-field representation of Gaussian–Schell-model pulses [38]. To this end, we write

$$p(t_e) = p_0 \exp\left(-\frac{2t_e^2}{T_p^2}\right) \tag{10.98}$$

and

$$f^{(0)}(t) = f_0 \exp\left(-\frac{t^2}{T_e^2}\right) \exp\left(-i\omega_0 t\right). \tag{10.99}$$

An MCF characterized by Equations 10.58 and 10.59 follows if we choose $T_p = T\sqrt{1-\beta^2}$ and $T_e = T\beta$, where β is given by Equation 10.70. By evaluating Equations 10.81 and 10.82, we obtain a CSD characterized by Equations 10.58, with $\Omega = 2/T_e$ and $\Omega_\mu = 2/T_p$.

10.5 Focusing, Imaging, and Shaping Partially Coherent Fields

We now turn our attention to beam propagation in paraxial optical systems that may contain classical lenses and also more complex optical elements. It will be seen that the spatial coherence properties of light generally have a profound influence an beam propagation in such systems.

10.5.1 Partially Coherent Beams in Paraxial Optical Systems

In paraxial ray optics, rotationally symmetric systems are often characterized by matrices

$$\mathbf{M}(\omega) = \begin{bmatrix} A(\omega) & B(\omega) \\ C(\omega) & D(\omega) \end{bmatrix}, \tag{10.100}$$

which connect the input-plane and output-plane values of a vector $[y(\omega), \theta(\omega)]^{\mathsf{T}}$. Here, $y(\omega)$ is the local ray height and $\theta(\omega)$ is the ray angle with respect to the optical axis. The elements of this so-called ABCD matrix generally depend on ω because of dispersion. The Fresnel formula 10.14 can be extended to cover wave propagation in paraxial systems characterized by the matrix $\mathbf{M}(\omega)$ [40]. If we denote by $E^{(0)}(\rho'; \omega)$ and $E(\rho; \omega)$ the fields in the input and output planes, respectively, this lens-system diffraction formula reads as

$$E(\rho; \omega) = \frac{\omega}{i2\pi cB(\omega)} \exp\left[i\omega L(\omega)/c\right] \exp\left[\frac{i\omega D(\omega)}{2cB(\omega)} |\rho|^2\right]$$
$$\times \int_{-\infty}^{\infty} E^{(0)}(\rho'; \omega) \exp\left[\frac{i\omega A(\omega)}{2cB(\omega)} |\rho'|^2\right] \exp\left[-\frac{i\omega}{cB(\omega)} \rho \cdot \rho'\right] d^2\rho'. \tag{10.101}$$

Here, $L(\omega)$ is the axial optical path length through the system and we have assumed that the refractive indices in the input and output media are equal to unity, in which case, $\det \mathbf{M}(\omega) = A(\omega)D(\omega) - B(\omega)C(\omega) = 1$.

A particularly interesting class of an optical system is that described by ABCD matrices with $A(\omega) = 0$. Clearly, in view or ray theory, systems with such a property either focus a parallel incident bundle of rays into the output plane or collimate light from a point source in the input plane into a parallel bundle of rays in the output plane. If we denote $B(\omega) = F(\omega)$, the condition $\det \mathbf{M}(\omega) = 1$ gives $C(\omega) = -1/F(\omega)$ and we may interpret

$F(\omega)$ as the focal length of the system. It now follows from Equation 10.101 that

$$
E(\boldsymbol{\rho};\omega) = \frac{\omega}{i2\pi\, cF(\omega)} \exp\left[i\omega L(\omega)/c\right] \exp\left[\frac{i\omega D(\omega)}{2cF(\omega)}|\boldsymbol{\rho}|^2\right]
$$
$$
\times \int_{-\infty}^{\infty} E^{(0)}(\boldsymbol{\rho}';\omega) \exp\left[-\frac{i\omega}{cF(\omega)}\boldsymbol{\rho}\cdot\boldsymbol{\rho}'\right] d^2\rho'.
\tag{10.102}
$$

Thus, apart from a multiplicative factor that includes a parabolic phase term, the system performs a scaled Fourier transformation of the input field (for a given ω). Another system of special interest is an imaging system with focal length $F(\omega)$ and linear magnification $m(\omega)$. Its matrix elements are $A(\omega) = m(\omega)$, $B(\omega) = 0$, $C(\omega) = -1/F(\omega)$, and $D(\omega) = 1/m(\omega)$. In this case, Equation 10.101 becomes formally indefinite but one can show (e.g., by applying the method of stationary phase) that

$$
E(\boldsymbol{\rho};\omega) = \frac{\operatorname{sign}\left[m(\omega)\right]}{m(\omega)} E^{(0)}\left[\frac{\boldsymbol{\rho}}{m(\omega)};\omega\right] \exp\left[i\omega L(\omega)/c\right] \exp\left[-\frac{i\omega}{cm(\omega)F(\omega)}|\boldsymbol{\rho}|^2\right].
\tag{10.103}
$$

Hence, apart from phase scaling factors, we indeed obtain an image of the input field distribution.

Lens-system diffraction integrals for stationary and nonstationary spatially partially coherent fields may again be obtained simply by inserting the results for coherent fields into the appropriate definitions of the CSD. As with propagation in uniform media, this generally leads to 4D integrals. For example, if we consider an ideal Fourier-transforming system with $D(\omega) = 0$ and $F(\omega) = F$, and assume a quasimonochromatic beam, the space-dependent part of the CSD (or the mutual intensity) at the output plane is related to that in the input plane by the expression

$$
W(\boldsymbol{\rho}_1,\boldsymbol{\rho}_2) = \left(\frac{\omega_0}{2\pi\, cF}\right)^2 \iint_{-\infty}^{\infty} W^{(0)}(\boldsymbol{\rho}_1',\boldsymbol{\rho}_2') \exp\left[\frac{i\omega_0}{cF}\left(\boldsymbol{\rho}_1\cdot\boldsymbol{\rho}_1' - \boldsymbol{\rho}_2\cdot\boldsymbol{\rho}_2'\right)\right] d^2\rho_1' d^2\rho_2'.
\tag{10.104}
$$

However, we can again make use of either the Mercer-type coherent-mode representations or the elementary-field representations to reduce the dimensionality of the integrals. Considering, for example, the elementary-field expansion, Equation 10.104 reduces to

$$
W(\boldsymbol{\rho}_1,\boldsymbol{\rho}_2) = \left(\frac{4\pi^2\omega_0}{cF}\right)^2 P\left[\frac{\omega_0}{cF}\left(\boldsymbol{\rho}_2 - \boldsymbol{\rho}_1\right)\right] F^{(0)*}\left(\frac{\omega_0}{cF}\boldsymbol{\rho}_1\right) F^{(0)}\left(\frac{\omega_0}{cF}\boldsymbol{\rho}_2\right),
\tag{10.105}
$$

where $F^{(0)}(\kappa)$ and $P(\kappa)$ are defined by Equations 10.87 and 10.88, respectively.

10.5.2 Some Examples of Partially Coherent Beam Propagation in Lens Systems

In addition to reducing the dimensionality of the propagation integrals, both the coherent-mode and elementary-field approaches can provide considerable physical insight into the coherence dependence of phenomena such as focal shifts of beam-like fields in paraxial systems. The general laws for propagation of the CSD have been applied to study such phenomena for Gaussian–Schell-model beams theoretically [41–43]. In this case, the integrations can be performed analytically and the results appear in the form that is a direct generalization of the results for spatially coherent Gaussian beams. The theoretical predictions have also been confirmed by experiments [44,45]. Let us now illustrate, by some simple examples, the usefulness of the coherent-mode representation in the interpretation of such results.

In the theory of fully coherent Hermmite–Gaussian laser beams, one often introduces a so-called complex q-parameter as a compact way of describing the beam width and radius of curvature simultaneously. For the coherent modes of a Gaussian–Schell-model beam, the definition of the q-parameter reads as

$$\frac{1}{q(z)} = \frac{1}{R(z)} - i\frac{2c}{\omega_0 w_c^2(z)}. \tag{10.106}$$

It is well known that the values of the q-parameter at the input plane $z = 0$ and the output plane $z = $ constant of an arbitrary $ABCD$ system are connected by

$$q(z) = \frac{Aq(0) + B}{Cq(0) + D}. \tag{10.107}$$

In view of Equation 10.47 and the definition 10.106, the q-parameter of an entire Gaussian–Schell-model beam with any value of β is the same as that of its lowest-order coherent mode. Hence, the transformation of the (generalized) q-parameter in optical systems also obeys Equation 10.107 for these partially coherent beams.

It is of particular interest to consider systems that transform a beam with its waist at the input plane into a beam with a waist at the output plane, that is, $R(0) = R(z) = \infty$. Let us consider for simplicity a system that consists of a single thin lens of focal length F, with the input and output planes at distances z_1 and z_2 from the lens. Considering a fully coherent Gaussian beam with Rayleigh range z_R, such as the fundamental coherent mode of a Gaussian–Schell-model beam, we again have well-known expressions for the distance z_2 and the magnification of the waist $m = w_c(z)/w_c$:

$$\frac{1}{z_2} = \frac{1}{z_1 + z_R^2/(z_1 - F)} - \frac{1}{F}, \tag{10.108}$$

$$m = \left[(1 - z_1/F)^2 + (z_R/F)^2\right]^{-1/2}. \tag{10.109}$$

In view of Equation 10.52, the Rayleigh range of the entire Gaussian–Schell-model beam is the same as that of its coherent modes, and hence, we may conclude that Equations 10.108 and 10.109 also apply to the entire beam if the magnification of the waist is redefined as $m = w(z)/w$.

Clearly, Equations 10.108 and 10.109 reduce to the familiar geometrical imaging laws of paraxial optics if we let $z_R \to 0$. Obviously, for a fully coherent beam, this means that the input waist size is reduced toward zero (point source in the input plane). In the case of partially coherent beams with an arbitrary value of w, geometrical imaging laws are also obtained by letting $\beta \to 0$, which represents the incoherent limit. Thus, we obtain the expected result that, in the spatially incoherent limit, the output waist is located at the geometrical image plane of the lens and its size is determined by the geometrical magnification of the system.

10.5.3 Spatial Shaping of Partially Coherent Fields

Let us proceed to consider problems related to shaping the transverse intensity distributions of spatially partially coherent beams using aspheric optical elements, which can be either refractive or diffractive. In both cases, we assume that the element is placed immediately behind the plane $z = 0$ and that its optical action can be described by a deterministic complex-amplitude transmission function

$$t(\boldsymbol{\rho}'; \omega) = |t(\boldsymbol{\rho}'; \omega)| \exp\left[i\phi(\boldsymbol{\rho}'; \omega)\right], \tag{10.110}$$

which generally depends on the optical frequency ω. When using the formulas for fully coherent light, such as Equations 10.14 and 10.15 for propagation in homogeneous media and Equations 10.101 and 10.102 for transmission through lens systems, the field $E^{(0)}(\boldsymbol{\rho}'; \omega)$ inside the integral in question should then be replaced with the field

$$E(\boldsymbol{\rho}'; \omega) = t(\boldsymbol{\rho}'; \omega)E^{(0)}(\boldsymbol{\rho}'; \omega) \tag{10.111}$$

transmitted by the element. When applying the propagation laws for partially coherent light such as Equation 10.16a, 10.16b, or 10.104, one should replace the original CSD by a modified CSD, which is of the form

$$W(\boldsymbol{\rho}_1', \boldsymbol{\rho}_2'; \omega_1, \omega_2) = t^*(\boldsymbol{\rho}_1'; \omega_1)t(\boldsymbol{\rho}_2'; \omega_2) W^{(0)}(\boldsymbol{\rho}_1', \boldsymbol{\rho}_2'; \omega_1, \omega_2) \tag{10.112}$$

in the case of nonstationary fields and of the form

$$W(\boldsymbol{\rho}_1', \boldsymbol{\rho}_2'; \omega) = t^*(\boldsymbol{\rho}_1'; \omega)t(\boldsymbol{\rho}_2'; \omega) W^{(0)}(\boldsymbol{\rho}_1', \boldsymbol{\rho}_2'; \omega) \tag{10.113}$$

in the case of stationary fields. If narrowband fields are of concern, we may again neglect the frequency dependence of the transmission function especially if the element is refractive.

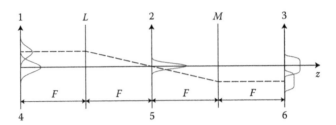

FIGURE 10.4 Schematic illustration of a 4F imaging system, including a pupil with a complex-amplitude transmittance $t(\rho'')$ in the Fourier plane. The propagation of two elementary field modes through the system is shown: one is the the axial elementary field and the other is its laterally shifted replica, which travels around a principal ray illustrated with the dashed line.

The transmission function $t(\rho')$ may describe, for example, an object with specified amplitude transmittance $|t(\rho')|$, or a refractive or diffractive element characterized by a phase function $\phi(\rho')$. On the other hand, we may consider the field at the exit plane of an arbitrary paraxial system as a (new) secondary source, modulate this field by a transmittance function, and apply the lens-system diffraction formula again to evaluate the new output field. This approach allows us to analyze, for example, imaging systems with hard or apodized apertures and aberrations, as well as the effects of spatial filtering in partially coherent optics. Let us specifically assume that the system generating the secondary source is an ideal Fourier-transforming system with focal length F as illustrated in Figure 10.4 and that the CSD in the input plane of this first system may be represented as a superposition of elementary fields in the form

$$W^{(0)}(\rho_1', \rho_2') = \int_{-\infty}^{\infty} p(\rho_e) f^{(0)*}(\rho_1' - \rho_e) f^{(0)}(\rho_2' - \rho_e) \, \mathrm{d}^2\rho_e. \qquad (10.114)$$

By introducing an arbitrary transmission function $t(\rho'')$ in the Fourier plane, we may write the CSD of the transmitted field at the Fourier plane in the form

$$W(\rho_1'', \rho_2'') = \left(\frac{4\pi^2\omega_0}{cF}\right)^2 P\left[\frac{\omega_0}{cF}(\rho_2'' - \rho_1'')\right] F^*\left(\frac{\omega_0}{cF}\rho_1''\right) F\left(\frac{\omega_0}{cF}\rho_2''\right), \qquad (10.115)$$

where

$$F\left(\frac{\omega_0}{cF}\rho''\right) = t\left(\frac{\omega_0}{cF}\rho''\right) F^{(0)}\left(\frac{\omega_0}{cF}\rho''\right) \qquad (10.116)$$

in analogy with Equation 10.111.

We further assume that the field described by Equation 10.115 enters a second ideal Fourier-transforming system of focal length F, so that in the absence of modulation $t(\rho'')$, we have an imaging system with magnification -1 (the discussion could be straightforwardly extended to systems with arbitrary linear magnification m). Replacing $W^{(0)}(\rho_1', \rho_2')$ by $W(\rho_1'', \rho_2'')$ in Equation 10.102 and carrying out the integrations leads to the result

$$W(\rho_1, \rho_2) = \int_{-\infty}^{\infty} p(\rho_e) f^* \left(-\rho_1 - \rho_e\right) f\left(-\rho_2 - \rho_e\right) d^2\rho_e, \tag{10.117}$$

which implies that the spatial intensity profile at the output plane takes the form

$$S(\rho) = \int_{-\infty}^{\infty} p(\rho_e) \left| f\left(-\rho - \rho_e\right) \right|^2 d^2\rho_e. \tag{10.118}$$

Here, the function $f(-\rho)$ can be interpreted as the complex-amplitude response of the system to illumination by the (axial) elementary field $f^{(0)}\left(\rho'\right)$, that is, as a generalization of the conventional impulse response of the system. Furthermore, $\left| f(-\rho) \right|^2$ provides the intensity response of the system illuminated by an axial elementary field. On the other hand, the weight function $p(\rho_e)$ takes the role of the source-plane spatial intensity distribution if we compare Equation 10.118 with the conventional expression for the image of an incoherent source.

Let us next consider the problem of converting a Gaussian–Schell-model beam in to an approximately flat-top beam in a system of Figure 10.4 using an aspheric refractive or diffractive element with the transmission function designed, for example, by the map-transform method (see, e.g., Ref. [46] and the references cited therein). We assume that the element is designed for the elementary field defined in Equation 10.93 (with $f_0 = 1$) and that the resulting intensity response is of the super-Gaussian form

$$|f(x)|^2 = \exp\left(-\frac{2x^Q}{w_f^Q}\right). \tag{10.119}$$

The element can be designed to approximate this distribution very well if w_f is several times larger than w_e and the value of Q is not too large; the practical criterion is that the edge of the super-Gaussian profile should not be steeper than the diffraction limit provided by the image of the elementary field, $\left| f^{(0)}(x) \right|^2$. Figure 10.5 illustrates this image and one suitable choice of the super-Gaussian function with $w_f = 5w_e$ and $Q = 12$ (see the curve marked 1.0, which gives the coherent response of the beam shaping system). Figure 10.5 also illustrates the system response to partially coherent illumination with Gaussian–Schell-model beams having different values of w, thus leading to weight functions of the form of Equation 10.92 with w_p determined from Equations 10.94 and 10.95. We see that decreasing the degree of spatial coherence smooths out the edges of the (approximately) flattop profile. This can be understood qualitatively from the convolution-type result expressed in Equation 10.118, and also from the geometrical construction of Figure 10.4. In the partially coherent case, the relative steepness of the edges compared to the width of the flattop region can, however, be improved by increasing the ratio w_f/w_e.

Apart from elements designed using the map-transform method, various types of diffusers are often used to shape laser beams (see, e.g., Refs. [47,48]). With coherent laser beams, such diffusers easily lead to the formation of speckle patterns in the shaped beam.

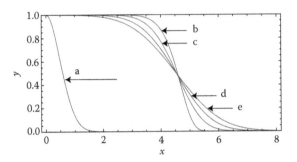

FIGURE 10.5 Effect of decreasing spatial coherence in shaping Gaussian–Schell-model beams: the numbers associated with the arrows indicate the values $w/w_e = 1/\beta$.

One of the general advantages of using partially coherent light is automatic speckle reduction. This is also the case in beam shaping [49–53], which can be easily appreciated by reconsideration of Figure 10.4. First, assume that the element in the Fourier plane produces a coherent intensity response with a flattop envelop but with sharply varying speckle substructure. Then, note that the speckle patterns produced by the off-axis elementary fields are shifted spatially in the image plane and when a sufficient number of such shifted copies is added incoherently, the speckle is effectively washed out at the expense of a slight broadening of the target profile.

We illustrate the speckle problem and the effect of partial coherence in remedying it with the aid of a simple example, in which we assume that a periodic object (grating) with a transmittance function

$$t(x'') = \exp\left[i\Delta\phi \sin\left(2\pi x''/d\right)\right] = \sum_{m=-\infty}^{\infty} J_m(\Delta\phi) \exp\left(i2\pi m x''/d\right) \qquad (10.120)$$

is inserted in the Fourier plane. Here, J_m is a Bessel function of the first kind and order m and d is the grating period. The coherent response to illumination with the fundamental elementary mode now reads as

$$f(x) = \sum_{m=-\infty}^{\infty} J_m(\Delta\phi) \exp\left[-\frac{(x - m\Delta x)^2}{w_e^2}\right], \qquad (10.121)$$

where Δx is the lateral separation of the harmonic images of the elementary field, centered at the center positions given by the paraxial grating equation. Hence, $\Delta x = F\lambda_0/d$ with $\lambda_0 = 2\pi c/\omega_0$.

If we choose $\Delta\phi \approx 1.435$ and if $\Delta x \gg w_e$, the grating acts as a "triplicator" with ~90% of light appearing in the three central peaks as shown by the solid line in Figure 10.6a, where we have chosen $\Delta x = 4w_e$. However, if we attempt to reduce Δx to make the harmonic images overlap and thus form a flattop beam, interference takes place as shown by

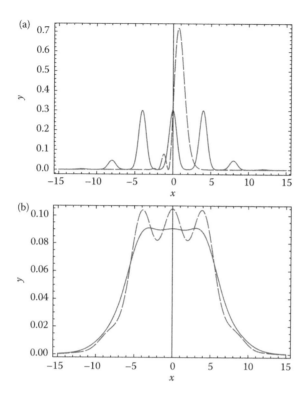

FIGURE 10.6 Beam shaping with periodic elements. The effect of speckle (a) and a decrease of spatial coherence (b) in shaping coherent and partially coherent Gaussian beams.

the dashed line in Figure 10.6a, where $\Delta x = w_e$. This is a simple effect of speckle. Considering Gaussian–Schell-model beams in the input plane of the system and increasing the waist size w is an alternative way to make the harmonic images overlap. This is illustrated in Figure 10.6b. The dashed and solid lines show the results for $w = 3w_e$ and $w = 4w_e$, respectively, with no significant interference effect (we have slightly adjusted the values of $\Delta\phi$ to 1.44 and 1.47, respectively, to optimize the uniformity of the residual peaks). Finally, it should be noted that triplicators for partially coherent beams can also be realized using gratings if one just increases Δx sufficiently to resolve the partially coherent harmonic images. On the other hand, gratings that produce a larger number of harmonic images with equal efficiency can be used to widen the flattop region further [51,52].

10.6 Modeling Some Practical Laser Sources

In the discussion presented above, we have considered light fields with varying degrees of spatial, spectral, or temporal coherence from the general viewpoint of second-order

optical coherence theory, with just a few remarks on the applicability of this theory to modeling the actual low-coherence lasers. In this section, an attempt is made
to rectify matters: the theoretical models will be applied to a number of realistic
lasers.

10.6.1 Excimer Lasers

Excimer lasers are high-power pulsed lasers with applications in, for example, materials
processing and medical sciences. They operate in the ultraviolet region; spectral center
wavelengths of 193 nm (ArF) and 248 nm (KrF) are the most common. The spectral line
widths are in the picometer wavelength range and, therefore, we can model excimer lasers
as quasimonochromatic sources. Excimer lasers typically generate nanosecond pulses,
which means that the temporal M^2 factor is large and, therefore, we can consider excimer
lasers as temporally quasistationary sources. The beam emitted from the cavity is usually elliptical and so is the far-field radiation pattern. Moreover, often, the eccentricities
of the output beam and the far-field pattern are nearly equal and the spatial M^2 factor
is large.

Let us substantiate the general arguments presented above by considering an excimer
with some representative spatial and angular beam characteristics, as well as spectral and
temporal pulse properties. Similar techniques can of course be applied to any excimer
laser with different specifications. We assume that the beam at $z = 0$ is Gaussian with
an anisotropic cross section defined by $w_x = 10$ mm and $w_y = 25$ mm. Furthermore,
the field in the far zone is also assumed to be Gaussian with far-field divergence angles
$\theta_x = 1$ mrad and $\theta_y = 3$ mrad. Thus, we model the spatial field to be of the anisotropic
Gaussian–Schell-model form. Assuming that the center wavelength of the spectrum is
$\lambda_0 = 193$ nm, the elementary field at $z = 0$ has an anisotropic Gaussian distribution with
scale factors $w_{ex} = \lambda_0/(\pi\theta_x) \approx 61$ μm and $w_{ey} = \lambda_0/(\pi\theta_y) \approx 20.5$ μm. Hence, the field
is definitely quasihomogeneous in both directions and we may approximate $w_{px} \approx w_x$,
$w_{py} \approx w_y$, $\sigma_x \approx w_{ex}$, and $\sigma_y \approx w_{ey}$. Thus, from Equation 10.48, $\beta_x \approx w_{ex}/w_{px} \approx 0.0061$
and $\beta_y \approx w_{ey}/w_{py} \approx 0.00082$. In view of Equation 10.57, the spatial M^2 factors in the
x- and y-directions are $M_x^2 \approx 160$ and $M_y^2 \approx 1200$. Finally, the widths of the lowest-
order coherent mode are $w_{cx} \approx 0.78$ mm and $w_{cy} \approx 0.72$ mm. Although the lowest-order
coherent mode happens to be almost circularly symmetric with these parameters, the
number of modes required to properly represent the excimer beam is much greater in
the y-direction than in the x-direction.

Turning our attention to the temporal and spectral properties of the excimer field,
we consider Gaussian pulses with duration $T = 20$ ps and assume that the $1/e^2$ width of
a Gaussian spectrum in wavelength scale is 2 pm (representative for an excimer with
an intracavity etalon to reduce the line width). This corresponds to $\Omega \approx 10$ THz at
$\lambda_0 = 193$ nm. The temporal width of the elementary pulse is then $T_e = 2/\Omega \approx 0.2$ ps,
which means that the pulses are highly quasistationary. Hence, $T_p \approx T$, the tempo-
ral $\beta \approx T_e/T_p \approx 0.01$, the temporal M^2 factor is $M^2 \approx 100$, and the pulse width of the
lowest-order temporal coherent mode is $T_c \approx 2$ ps.

10.6.2 Free-Electron Lasers

FELs can produce light from the x-ray region (wavelengths of the order of 10 nm or even less) to the terahertz region. Particularly in the interesting extreme ultraviolet (EUV) and x-ray regimes [54], the generation of radiation is based on self-amplified spontaneous emission (since it is not possible to construct an optical cavity) and, therefore, the light is expected to be partially spatially and temporally coherent (see Ref. [55] and the references cited therein).

Recent direct measurements on the spatial coherence at the free-electron laser in Hamburg (FLASH) facility [56–58] using Young's interferometer show conclusively that the FEL radiation can be very well described by the Gaussian–Schell model. In fact, it is sufficient to include only a few transverse modes in the Mercer expansion to obtain a highly accurate model. Hence, the elementary mode description is also possible and techniques based on the same principles as above for excimers (near-field and far-field intensity measurements) can in principle be applied to quickly estimate the spatial coherence of FEL radiation without direct measurements.

The characterization of the temporal coherence properties of FEL radiation is a somewhat more difficult task. Autocorrelation measurements have been performed recently [57,59] but they do not alone provide sufficient information for the determination of the two-time MCF or the complex degree of temporal coherence $\gamma(t_1, t_2)$. Nevertheless, because of the fact that the single-shot spectra of FEL pulses differ strongly from each other [59], the radiation must be spectrally (and, therefore, temporally) partially coherent. Moreover, the FEL theory suggests that the radiation within any single pulse can be described as a superposition of "temporally coherent spikes" that may be considered mutually uncorrelated [54]. Such a description has a striking similarity with the temporal elementary-field model introduced above and indicates that the temporal M^2 factor could be on the order of $M^2 = 100$, making even femtosecond FEL quasistationary sources in the sense defined earlier in this chapter.

10.6.3 Multimode Edge-Emitting Semiconductor Lasers

The number of spatial modes radiated by edge-emitting semiconductor lasers depends on the transverse characteristics of the cavity; if more than one mode is generated, the radiation is spatially partially coherent [60] and may be described by the techniques introduced in Section 10.4.1. Most often, only a single mode is supported by the cavity in the direction perpendicular to the junction (fast axis), while several modes may oscillate in the parallel direction (slow axis). This is particularly true in high-power broad-area laser diodes (BALDs) with cavity dimensions in the orthogonal directions differing by orders of magnitude.

The longitudinal and transverse mode structure of BALDs depends on the cavity geometry and the functional forms of the transverse coherent modes can, therefore, be calculated. The modal weights can then be determined directly by high-resolution spectral imaging [61,62]. Such an analysis reveals that a large number of transverse modes are present in BALDs, and hence, they can be well described as quasihomogeneous sources.

As a result, the spatial elementary-field description is readily available [63]: the spatial weight function is obtained from near-field intensity measurements while the functional form of the elementary field is obtained from the far-zone radiation pattern as indicated in Equations 10.96 and 10.97. This approach also works if the cavity has imperfections such as aberrations due to a nonuniform gain profile, which may yield nonsymmetric far-zone intensity profiles that cannot be predicted; be considering superpositions of ideal coherent modes (which are real at the exit plane $z = 0$ of the cavity); then the elementary field at the exit plane simply acquires a spatial phase.

The elementary-field description is particularly natural if we consider diode bars that consist of an array of identical, uncoupled, and single-mode cavities. In this case, each cavity generates its own elementary field, which all overlap in the far zone.

10.6.4 VCSELs and Arrays

The multi-quantum-well structures in VCSELs, surrounded by Bragg reflectors on both sides, allow one to easily achieve single-longitudinal-mode operation. The number of transverse modes and, therefore, the spatial coherence of the emitted light depends on the emitting area and the actual laser structure (see, e.g., Refs. [64,65] and the references cited therein) and sometimes also on the operating conditions of the device [66].

High-power operation can also be achieved by using 2D arrays of single-mode-emitting VCSEL units, in analogy of using edge-emitting lasers in an arrayed (bar) format. Many array configurations are possible since the individual emitters are defined by lithographic patterning of the multilayer structure. If the neighboring emitters are sufficiently closely spaced, evanescent coupling between them leads to array modes of wide spatial extent [67]. Then the spatial coherence properties of the radiation depend on the number of excited modes as in any multimode laser.

If the individual VCSEL emitters are far enough from each other, there is no evanescent coupling and each (small) emitter acts as a spatially coherent source with its characteristic far-zone radiation pattern. Now, if the independent emitters are identical, each emitter can again be considered as a source of an elementary-field mode, and the far-field pattern of the entire array is identical to that of a single emitter. In the image plane of the array (provided that there is no aperture to truncate the beam), we of course obtain an array of discrete image spots. However, beam-shaping elements could be inserted in the Fourier plane as explained in Section 10.5.3 to homogenize the output pattern. At least two options are available. First, one could use an element with an elementary-field response that just fills the space between neighboring VCSEL images. Second, since the individual emitters are uncorrelated, one could use a beam-shaping element with a wider elementary-field response, extending over several array positions, to make the system less sensitive to positioning and other errors.

10.6.5 SC Light Generated in Microstructured Fibers

Light fields with extremely wide spectral content, known as SC fields, can be generated by illuminating many materials with intense laser pulses; the broadening of the spectrum is due to various nonlinear optical processes that take place in the medium. In particular, light fields with octave-wide spectra can be generated in a controlled way by feeding the pump pulse in a relatively short section of a microstructured nonlinear optical fiber [68,69]. Typically, fibers with a single spatial mode are used, which results in nearly spatially coherent SC output. However, since the pump pulses are never completely identical, different SC pulse realizations also differ from each other both spectrally and temporally. Hence, if we consider ensemble averages, the SC light is in general both spectrally and temporally partially coherent. Understanding the coherence properties of SC light is of crucial importance: in applications such as frequency metrology [70–72], a high degree of spectral coherence is required, but in other applications, including white-light interferometry or optical coherence tomography, the spectral shape of the field rather than the coherence is of primary interest.

The realizations of SC light generated in nonlinear microstructured (or tapered) optical fibers can be constructed numerically by solving a nonlinear wave equation that contains appropriate contributions from several nonlinear processes that occur in the fiber; this enables one to compute ensemble averages and thereby construct correlation functions. Dudley and Coen [73,74] applied this approach to characterize pulse-to-pulse correlations using a measure for spectral coherence that directly corresponds to a practical experimental approach for measuring these correlations with a delayed-time interferometer. It was established that, with appropriate random quantum noise associated with the individual input pulses, numerical simulations correspond to the experimental results at a remarkable accuracy. It was also found that the spectral coherence properties of SC light strongly depend on factors such as input pulse power and fiber length. Remarkably, frequency-resolved optical gating allows the measurement of single realizations [75,76] and thereby, in principle, the experimental construction of second-order correlation functions.

Figure 10.7 illustrates a typical infrared SC spectrum obtained by simulations as outlined above and the corresponding ensemble-averaged picosecond pulse. The remainder of the pump pulse at $\lambda = 1060$ nm is surrounded by a relatively smooth spectrum extending over a wavelength range from ~900 to ~1250 nm. The simulations were made assuming the following parameters: pump peak power 100 W, pump pulse duration 5 ps, and fiber length 20 m. In the intensity curve, the arrival time of the pump pulse at the fiber end is the zero temporal position; the SC pulse has spread temporally on both sides and has a double-peaked form.

The ability to numerically simulate realizations of SC pulses in both spectral and temporal domains also allows one to construct the CSD and MCF directly by ensemble averaging [77,78]. This approach reveals that SC light fields can be approximately resolved into two distinct parts: one is a nearly coherent field in both a spectral and temporal sense, while the other is effectively a quasistationary contribution. Thus, we may write, to a good approximation

$$\Gamma(t_1, t_2) = \langle E^*(t_1)E(t_2)\rangle = E_c^*(t_1)E_c(t_2) + I_s(\bar{t})\gamma_s(\Delta t) \qquad (10.122)$$

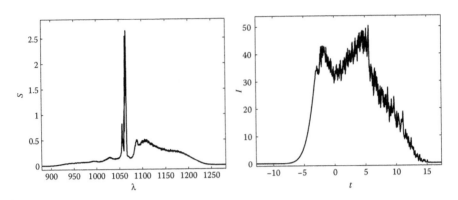

FIGURE 10.7 Wavelength dependence of the spectral density (left) and the temporal intensity distribution (right) for a simulated ensemble of SC pulses. (Courtesy of G. Genty and M. Korhonen.)

and

$$W(\omega_1, \omega_2) = \langle E^*(\omega_1)E(\omega_2)\rangle = E_c^*(\omega_1)E_c(\omega_2) + S_s(\bar{\omega})\mu_s(\Delta\omega), \qquad (10.123)$$

where $E(t)$ and $E(\omega)$ are the time- and frequency-domain field realizations, the subscripts c and s refer to the quasicoherent and quasistationary contributions, respectively, and we have neglected the spatial dependence of the SC field. The relative strengths of the two contributions depend crucially on the experimental parameters.

Figure 10.8 illustrates graphically the partition of the MCF into the two contributions. Here, the square represents the quasicoherent part of the pulse train, which has a high-coherence temporal extent of approximately 7 ps, that is, of the same order of magnitude as the temporal duration of the pump pulse. This part is mainly responsible for the first peak in the intensity profile in Figure 10.7. However, it is not temporally fully coherent because of its "wings" that are nevertheless approximately separable in t_1 and t_2 and, therefore, belong to the quasicoherent contribution in Equation 10.122. The line around $\Delta t = 0$ represents the quasistationary part with low temporal coherence (since the line is narrow in the Δt direction). This part extends temporally over the entire ensemble-averaged intensity distribution. In this example, the integrated degree of coherence $\bar{\gamma}$, defined in Equation 10.37, has the value $\bar{\gamma} = 0.1420$; hence the quasistationary part is already dominant. Graphs of similar nature are also obtained for the CSD [77,78].

The temporal and spectral coherent-mode expansions of SC light, as defined in Section 10.4, can be computed directly from the MCF and CSD and the number of coherent modes required to represent the field properly depends critically on the degree of spectral/temporal coherence [79]. Finally, if one uses the lowest-order coherent mode to represent the quasicoherent part of the SC field in Equations 10.122 and 10.123, the temporal and spectral elementary-field representations (Section 10.4.4) provide good approximations for the quasistationary contribution [80].

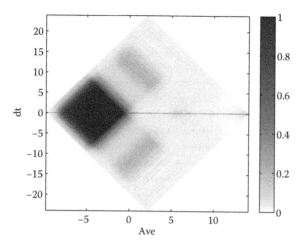

FIGURE 10.8 Graph of the absolute value of the normalized MCF, $|\gamma(\bar{t}, \Delta t)|$, illustrating the partition of the temporal coherence function into distinct quasicoherent and quasi-stationary contributions. (Courtesy of G. Genty and M. Korhonen.)

10.6.6 Random Lasers

Random lasers are based on a mixture of strongly scattering media and materials with gain [81], such as powders made of active media, suspensions of scattering particles in dye solutions, and porous solids with the pores filled with gain media. Early studies of such disordered gain media, illuminated by short laser pulses, revealed simultaneous narrowing of the emission spectrum and shortening of output pulses when a threshold was exceeded [82]. These features suggest laser-like output, which, however, was found to have many characteristics of spectrally and spatially low-coherence light. Later, discrete "lasing spikes" with narrow spectral width were observed, indicating coherent feedback in wavelength-scale regions of disordered amplifying media [83]. Several different (and competing) theoretical approaches have been developed to model the physical mechanisms that occur in random lasers (see Ref. [81] and the work cited therein for details), but to date, the coherence properties of random-laser emission seem to be understood only on a qualitative level.

Typically, the spectral positions of the laser spikes and their spectral strengths change from one excitation pulse to another, which is a clear indication of low overall spectral and temporal coherence in the sense discussed above (the emitted light is quasi-monochromatic and quasistationary). If only a small area (a few square micrometers) of the disordered medium is illuminated at a time and the disordered medium is rigid, the spectral positions have been observed to stay the same while the spectral strengths still change from pulse to pulse [84]. If the emission from a random laser were to originate from modes localized in spatially separate areas of the disordered medium, the illumination of just one of them at a time would result in a spatially nearly coherent speckle pattern with a large angular divergence (because of the small size and randomness of the "effective resonator" and the quasimonochromatic nature of the emission). The illumination of several effective resonators (with separate resonance frequencies and different speckle patterns) would then lead to a smooth far-field distribution of mutual intensity. The results reported in Ref. [84], however, also show evidence of mode competition; this indicates that the effective resonators overlap partially, leading to a multimode operation at each spatial point in the exit face of a random-laser source.

With large-area illumination of the disordered medium, evidence has been obtained of the coexistence of strongly localized and more extended modes [85]. With many such modes oscillating independently and overlapping partially, the complete field would be spatially quasihomogeneous, with the coherence area roughly given by the average spatial extent of the modes. Despite the interesting observations made in Refs. [84,85] and elsewhere, the subject of spatial, spectral, and temporal coherence of random-laser radiation is still well worth more detailed studies.

References

1. Mandel, L. and Wolf, E. 1995. *Optical Coherence and Quantum Optics*. Cambridge: Cambridge University Press.
2. Goodman, J.W. 2000. *Statistical Optics*. New York: Wiley.
3. Bertolotti, M., Ferrari, A., and Sereda, L. 1995. Coherence properties of nonstationary polychromatic light sources, *J. Opt. Soc. Am. B* 12: 341–47.
4. Bertolotti, M., Sereda, L, and Ferrari, A. 1997. Application of the spectral representation of stochastic processes to the study of nonstationary light radiation: A tutorial, *Pure Appl. Opt.* 6: 153–71.
5. Sereda, L., Bertolotti, M., and Ferrari, A. 1998. Coherence properties of nonstationary light wave fields, *J. Opt. Soc. Am. A* 15: 695–705.
6. Wolf, E. 1981. New spectral representation of random sources and the partially coherent fields they generate, *Opt. Commun.* 38: 3–6.
7. Wolf, E. 1982. New theory of partial coherence in the space-frequency domain. Part I: Spectra and cross-spectra of steady-state sources, *J. Opt. Soc. Am.* 72: 343–51.
8. Wolf, E. 1986. New theory of partial coherence in the space-frequency domain. Part II: Steady-state field and higher-order correlations, *J. Opt. Soc. Am. A* 3: 76–85.
9. Gori, F. 1983. Mode propagation of the fields generated by Collett–Wolf–Schell-model sources, *Opt. Commun.* 46: 149–54.
10. Lajunen, H., Tervo, J., and Vahimaa, P. 2004. Overall coherence and coherent-mode expansion of spectrally partially coherent plane-wave pulses, *J. Opt. Soc. Am. A* 21: 2117–23.
11. Lajunen, H., Tervo, J., and Vahimaa, P. 2005. Theory of spatially and spectrally partially coherent pulses, *J. Opt. Soc. Am. A* 22:1536–45.
12. Lajunen, H. and Friberg, A.T. 2006. Quasi-stationary plane-wave optical pulses and the van Cittert–Zernike theorem in time, *J. Opt. Soc. Am. A* 23: 2530–37.

13. Wolf, E. and Agarwal, G.S. 1984. Coherence theory of laser resonator modes, *J. Opt. Soc. Am. A* 1: 541–46.

14. Turunen, J., Tervonen, E., and Friberg, A.T. 1989. Coherence-theoretic algorithm to determine the transverse mode structure of lasers, *Opt. Lett.* 14: 627–29.

15. Tervonen, E., Turunen, J., and Friberg, A.T. 1989. Transverse laser mode structure determination from spatial coherence measurements: Experimental results, *Appl. Phys. B* 49: 409–14.

16. Gori, F., Santarsiero, M., Borghi, R., and Guattari, G. 1998. Intensity-based modal analysis for partially coherent beams with Hermite–Gaussian modes, *Opt. Lett.* 23: 989–91.

17. Borghi, R., Guattari, G., de la Torre, L., Gori, F., and Santarsiero, M. 2003. Evaluation of the spatial coherence of a light beam through transverse intensity measurements, *J. Opt. Soc. Am. A* 20: 1763–70.

18. Rydberg, C. and Bengtsson, J. 2007. Numerical algorithm for the retrieval of spatial coherence properties of partially coherent beams from transverse intensity measurements, *Opt. Express* 15: 13613–23.

19. Christov, I.P. 1986. Propagation of partially coherent pulses, *Opt. Acta* 33:64–77.

20. Wang, L., Lin, Q., Chen, H., and Zhu, S. 2003. Propagation of partially coherent optical pulsed beams in the spatiotemporal domain, *Phys. Rev. E* 67: 056612.

21. Vahimaa, P. and Tervo, J. 2006. Unified measures for optical fields: Degree of polarization and effective degree of coherence, *J. Opt. A: Pure Appl. Opt.* 6: S41–S44.

22. Borghi, R. and Santarsiero, M. 1998. Modal decomposition of partially coherent flat-topped beams produced by multimode lasers, *Opt. Lett.* 23: 313–15.

23. Wolf, E. and Collett, E. 1978. Partially coherent sources which produce the same far-field intensity distribution as a laser, *Opt. Commun.* 25: 293–96.

24. Collett, E. and Wolf, E. 1979. New equivalence theorems for planar sources that generate the same distributions of radiant intensity, *J. Opt. Soc. Am.* 61: 942–50.

25. Gori, F. 1980. Collett–Wolf sources and multimode lasers, *Opt. Commun.* 34: 301–5.

26. Friberg, A.T. and Sudol, R.J. 1982. Propagation parameters of Gaussian–Schell-model beams, *Opt. Commun.* 41: 383–87.

27. Starikov, A. and Wolf, E. 1979. Coherent-mode representation of Gaussian–Schell-model sources and their radiation fields, *J. Opt. Soc. Am.* 72: 923–27.

28. Li, Y. and Wolf, E. 1982. Radiation from anisotropic Gaussian–Schell-model sources, *Opt. Lett.* 7: 256–58.

29. Simon, R. 1984. A new class of anisotropic Gaussian beams, *Opt. Commun.* 55: 381–85.

30. DeSantis, F., Gori, F., Guattari, G., and Palma, C. 1986. Anisotropic Gaussian–Schell-model sources, *Opt. Acta* 33: 315–26.

31. Gori, F. and Guattari, G. 1983. A new type optical fields, *Opt. Commun.* 48: 7–12.

32. Pääkkönen, P., Turunen, J., Vahimaa, P., Friberg, A.T., and Wyrowski, F. 2002. Partially coherent Gaussian pulses, *Opt. Commun.* 204: 53–58.

33. Turunen, J. 2011. Elementary-field representations in partially coherent optics, *J. Mod. Opt.* 58: 509–27.

34. Gori, F. and Palma, C. 1978. Partially coherent sources which give rise to highly directional laser beams, *Opt. Commun.* 27: 185–87.

35. Gori, F. 1980. Directionality and partial coherence, *Opt. Acta* 27: 1025–34.

36. Vahimaa, P. and Turunen, J. 2006. Finite-elementary-source model for partially coherent radiation, *Opt. Express* 14: 1376–82.

37. Tervo, J., Turunen, J., Vahimaa, P., and Wyrowski, F. 2010. Shifted-elementary-mode representation of partially coherent vectorial fields, *J. Opt. Soc. Am. A* 24: 2014–47.

38. Vahimaa, P. and Turunen, J. 2006. Independent-elementary-pulse representation for nonstationary fields, *Opt. Express* 14: 5007–12.

39. Saastamoinen, K., Turunen, J., Vahimaa, P., and Friberg, A.T. 2009. Spectrally partially coherent propagation-invariant fields, *Phys. Rev. A* 80: 053804.

40. Collins, S. A. 1969. Lens-system diffraction integral written in terms of matrix optics, *J. Opt. Soc. Am.* 69: 1168–77.

41. Simon, R., Sudarshan, E.C.G., and Mukunda, N. 1984. Generalized rays in first-order optics: Transformation properties of Gaussian–Schell-model fields, *Phys. Rev. A* 29: 3273–79.

42. Turunen, J. and Friberg, A.T. 1986. Matrix representation of Gaussian–Schell-model beams in optical systems, *Opt. Laser Technol.* 18: 259–67.

43. Friberg, A.T. and Turunen, J. 1988. Imaging of Gaussian–Schell-model sources, *J. Opt. Soc. Am. A* 5: 713–20.

44. He, Q., Turunen, J., and Friberg, A.T. 1988. Propagation and imaging experiments with Gaussian–Schell-model sources, *Opt. Commun.* 67: 245–50.

45. Lavi, S., Prochaska, R., and Keren, E. 1988. Generalized beam parameters and transformation laws for partially coherent light, *Appl. Opt.* 27: 3696–703.

46. Romero, L.A. and Dickey, F.M. 1996. Lossless laser beam shaping, *J. Opt. Soc. Am. A* 13: 751–60.

47. Deng, X., Liang, X., Chen, Z., Yu, W., and Ma, R. 1986. Uniform illumination of large targets using a lens array, *Appl. Opt.* 25: 377–81.

48. Dixit, S.N., Thomas, I.M., Woods, B.W., Morgan, A.J., Henesian, M.A., Wegner, P.J., and Powell, H.T. 1993. Random phase plates for beam smoothing on the Nova laser, *Appl. Opt.* 32: 2543–54.

49. Ozaki, Y. and Takamoto, K. 1989. Cylindrical fly's eye lens for intensity redistribution of an excimer laser beam, *Appl. Opt.* 28: 106–10.

50. Holmér, A.-K. and Hård, S. 1995. Laser-machining experiment with an excimer laser and a kinoform, *Appl. Opt.* 34: 7718–23.

51. Turunen, J., Pääkkönen, P., Kuittinen, M., Laakkonen, P., Simonen, J., Kajava, T., and Kaivola, M. 2000. Diffractive shaping of excimer laser beams, *J. Mod. Opt.* 47: 2467–75.

52. Kajava, T., Hakola, A., Elfström, H., Simonen, J., Pääkkönen, P., and Turunen, J. 2006. Flat-top profile of an excimer-laser beam generated using diffractive gratings, *Opt. Commun.* 268: 289–93.

53. Zhou, A.F. 2011. UV excimer laser beam homogenization for micromachining applications, *Opt. Photonics Lett.* 4: 75–81.

54. McNeil, B.W.J. and Thompson, N.R. 2010. X-ray free-electron lasers, *Nat. Photonics* 4: 814–21.

55. Vartanyats, I.A. and Singer, A. 2011. Temporal and spatial coherence properties of free-electron laser pulses in the extreme ultraviolet region, *Phys Rev. Spec. Tops. — Accel. Beams* 14: 080701.

56. Singer, A., Vartanyats, I.A., Kuhlmann, M., D'lusterer, S., Treusch, R., and Feldhaus, J. 2008. Transverse-coherence properties of the free-electron-laser FLASH at DESY, *Phys. Rev. Lett.* 101: 254801.

57. Vartanyats, I.A., Mancuso, A.P. , Singer, A., Yefanov, O.M., and Gulden, J. 2010. Coherence measurements and coherent diffractive imaging at FLASH, *J. Phys. B: At. Mol. Opt. Phys.* 43: 194016.

58. Vartanyats, I.A. et al. 2011. Coherence properties of individual femtosecond pulses of an x-ray free-electron laser, *Phys. Rev. Lett.* 107: 144801.

59. Mitzner, R. et al. 2008. Spatio-temporal coherence of free electron laser pulses in the soft x-ray regime, *Opt. Express* 16: 19909–19.

60. Spano, P. 1980. Connection between spatial coherence and mode structure in optical fibers and semiconductor lasers, *Opt. Commun.* 33: 265–70.

61. Stelmakh, N. and Flowers, M. 2006. Measurement of spatial modes of broad-area diode lasers with 1-GHz resolution grating spectrometer, *IEEE Photon. Technol. Lett.* 18: 1618–20.

62. Stelmakh, N. 2007. Harnessing multimode broad-area laser-diode emission into a single-lobe diffraction-limited spot, *IEEE Photon. Technol. Lett.* 19: 1392–94.

63. Partanen, H., Tervo, J., and Turunen, J. 2013. Spatial coherence of broad area laser diodes, *Appl. Opt.* 52: 3221–28.

64. Valle, A. and Pesquera, L. 2002. Analytical calculation of transverse mode characteristics in vertical-cavity surface emitting lasers, *J. Opt. Soc. Am. B* 19: 1549–57.

65. Gensty, T., Becker, K., Fischer, I., ElslaSSer, W., Degen, C., Debernardi, P., and Bava, G.P. 2005. Wave chaos in real-world vertical-cavity surface-emitting laser, *Phys. Rev. Lett.* 94: 233901.

66. Peeters, M., Verschaffelt, G., Thienpont, H., Mandre, S.K., and Fischer, I. 2005. Spatial decoherence of pulsed broad-area vertical-cavity surface-emitting lasers. *Opt. Express* 13: 9337–45.

67. Yoo, H.-J., Hayes, J.R., Paek, E.G., Scherer, A., and Kwon, Y.-S. 1990. Array mode analysis of two-dimensional phased arrays of vertical-cavity surface-emitting lasers, *IEEE J. Quantum Electron.* 26: 1039–51.

68. Dudley, J.M., Genty, G., and Coen, S. 2006. Supercontinuum generation in photonic crystal fiber, *Rev. Mod. Phys.* 78: 1135–84.

69. Genty, G., Coen, S., and Dudley, J.M. 2007. Fiber supercontinuum sources, *J. Opt. Soc. Am. B* 24: 1771–85.

70. Bellini, M. and Hänsch, T.W. 2000. Phase-locked white-light continuum pulses: Toward a universal optical frequency-comb synthesizer, *Opt. Lett.* 25: 1049–51.

71. Holzwarth, R., Zimmermann, M., Udem, Th., Hänsch, T.W. et al. 2001. White-light frequency comb generation with a diode-pumped Cr:LiSAF laser. *Opt. Lett.* 1376–78.

72. Stenger, J., Schnatz, H., Tamm, C., and Telle, H.R. 2002. Ultraprecise measurement of optical frequency ratios, *Phys. Rev. Lett.* 88: 073601.

73. Dudley, J.M. and Coen, S. 2002. Coherence properties of supercontinuum spectra generated in photonic crystal and tapered optical fibers, *Opt. Lett.* 27: 1180–82.

74. Dudley, J.M. and Coen, S. 2002. Numerical simulations and coherence properties of supercontinuum generation in photonic crystal and tapered optical fibers, *IEEE J. Sel. Top. Quantum Electron.* 8: 651–59.

75. Gu, X., Kimmel, M., Zeek, E., OŠShea, P., Shreenath, A.P., Trebino, R., and Windeler, R.S. 2002. Frequency-resolved optical gating and single-shot spectral measurements reveal fine structure in microstructure-fiber supercontinuum, *Opt. Lett.* 27: 1174–76.

76. Gu, X., Kimmel, M., Shreenath, A.P., Trebino, R., Dudley, J.M., Coen, S., and Windeler, R.S. 2003. Experimental studies of the coherence of microstructure-fiber supercontinuum, *Opt. Express* 11: 2697–703.

77. Genty, G., Surakka, M., Turunen, J., and Friberg, A.T. 2010. Second-order coherence of supercontinuum light, *Opt. Lett.* 35: 3057–59.

78. Genty, G., Surakka, M., Turunen, J., and Friberg, A.T. 2011. Complete characterization of supercontinuum coherence, *J. Opt. Soc. Am. B* 28: 2301–9.

79. Erkintalo, M., Genty, G., Surakka, M., Turunen, J., and Friberg, A.T. 2012. Coherent-mode representation of supercontinuum light, *Opt. Lett.* 37: 169–71.

80. Korhonen, M., Friberg, A.T., Turunen, J., and Genty, G. 2013. Elementary field representation of supercontinuum, *J. Opt. Soc. Am. B* 30: 21–26.

81. Wiersma, D.S. 2008. The physics and applications of random lasers. *Nat. Phys.* 4: 359–67.

82. Gouedard, C., Husson, D., Sauteret, C., Auzel, F., and Migus, A. 1993. Generation of spatially incoherent short pulses in laser-pumped neodymium stoichiometric crystals and powders, *J. Opt. Soc. Am. B* 12: 2358–63.

83. Cao, H., Zhao, Y.G., Ho, S.T., Seelig, E.W., Wang, Q.H., and Chang, R.P.H. 1999. Random laser action in semiconductor powder, *Phys. Rev. Lett.* 82: 2278–81.

84. van der Molen, K.L., Tjerkstra, R.W., Mosk, A.P., and Lagendijk, A. 2007. Spatial extent of random laser modes, *Phys. Rev. Lett.* 98: 143901.

85. Fallert, J., Dietz, R.J.B., Sartor, J., Schneider, D., Klingshirn, C., and Kalt, H.. 2009. Co-existence of strongly and weakly localized random laser modes. *Nat. Photonics* 3: 279–82.

11

Orbital Angular Momentum of Light

Miles Padgett
University of Glasgow

Reuben Aspden
University of Glasgow

11.1 Introduction

That light beams carry a momentum in addition to their energy is firmly described by Maxwell's equations. Whereas linear momentum has been understood since the original formulation of Maxwell's equations, what was only much later definitively ascertained is that Maxwell's formulation also describes an angular momentum associated with light. This angular momentum was first mathematically described by Allen et al. in 1992 and has since found applications not just using light but with all wave phenomena.

315

11.2 Historical Introduction

From Maxwell's equations, it is easy to show that the linear-momentum-to-energy ratio of a light beam is $1/c$, and it follows that, for light in a vacuum, the radiation pressure exerted by a beam of intensity I is given as $P = I/c$. Interestingly, the precise form of this linear momentum within a dielectric was and to some extent still is debated [1–3]. Also contained in Maxwell's formulation is that light can carry an angular momentum. The first clear expression of light's angular momentum was by Poynting, who used a mechanical analogy to a revolving shaft to reason that a circular polarized light beam carried an angular momentum. He showed that the angular-momentum-to-energy ratio was $1/\omega$, where ω is the angular frequency of light [4].

We note that all of this early work pre-dates the concept of the photon and is not dependent upon quantization. When considering optical momentum today, it is natural to express both linear and angular momentum in terms of h/λ and $\hbar$ per photon, respectively, but, recognizing that the energy of the photon is $\hbar/\omega$, we see that these per photon values are entirely consistent with the momentum/energy ratios expressed above. For the remainder of this chapter, we will adopt the photon concept while stressing that, for the majority of the discussion, the physics we invoke will be classical.

Although recognizing the presence of angular momentum within light beams, Poynting also felt that, due to its small size, a mechanical observation of this angular momentum was extremely unlikely. However, in the 1930s, an experiment by Beth showed that optical angular momentum could rotate a physical object [5]. Upon transmission through a half-wave plate, the handedness of circularly polarized light is reversed, reversing also its angular momentum. This change in the angular momentum of light requires an equal but opposite reaction torque on the waveplate, potentially causing it to rotate. By suspending the waveplate from a quartz fiber, Beth was able to reduce the frictional forces to a point where the mechanical action of light could be observed.

The angular momentum that arises from the circular polarization of light is equivalent to the spin of individual photons and is referred to as the spin angular momentum. However, from the original considerations of optical momentum and certainly from the 1930s, this spin component of light's angular momentum was recognized as not being the whole story. Whereas the mechanical action of light's momentum is difficult to observe, its interaction with atoms is self-evident. A dipole transition has associated with it a change in the angular momentum of the electronic state of $\hbar$. If the emitted photon is circularly polarized, then angular momentum is conserved. But what happens for higher-order transitions? For example, a quadrupole transition requires an angular momentum exchange of $2\hbar$. This question was considered by Darwin [6] (grandson of the pioneer of evolutionary science!). He reasoned that the additional angular momentum could be transferred if the light was emitted a short radius away from the center of mass of the atom. This short radius, combined with the linear momentum of light, gives a torque acting on the center of mass. This additional contribution is what we would now call orbital angular momentum (OAM).

Beyond an association with high-order transitions, and hence rare events, it was not until 1992 when the work of Allen et al. realized that whole light beams could be generated

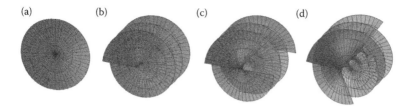

(a)   (b)   (c)   (d)

FIGURE 11.1 Phase profile of light beams with phase cross section of $\exp(i\ell\theta)$ and angular-momentum-to-energy ratio of $\ell\hbar$, where (a) $\ell = 0$, (b) $\ell = 1$, (c) $\ell = 2$, and (d) $\ell = 3$.

in specific OAM states [7]. They reasoned that any light beam with helical phase fronts described by a phase cross section of $\exp(i\ell\theta)$ has an OAM-to-energy ratio of ℓ/ω, that is, $\ell\hbar$ per photon (Figure 11.1). A consequence of the helical phase fronts is that the beam axis is a phase singularity, necessarily of zero intensity.

Although not considered in terms of their angular momentum, fields containing phase singularities had been recognized for some time. Immediately prior to the 1992 work, Soskin and coworkers had generated helically phased laser beams using diffraction gratings containing a forked dislocation [8]. Earlier work still had identified helical phase fronts as being a property of certain classes of laser modes [9,10]. The presence of phase singularities within more complicated field distributions had been the subject of much research in the 1970s for both acoustic [11] and optical [12] fields. However, none of this earlier work had recognized the implications such phase features had on transverse momentum and angular momentum in particular. The modern interest in the OAM of light stems from the studies in Leiden by Allen, Woerdman, and coworkers in the early 1990s.

11.3 OAM in Paraxial Light Beams

Allen, Woerdman, and coworkers first investigated OAM within the context of the paraxial approximation and a sketch of this derivation will be presented here [7]. Using the Lorentz gauge, a linearly polarized laser mode can be represented by

$$\mathbf{A} = \mathbf{x}u(x, y, z)e^{-ikz}, \tag{11.1}$$

where $\mathbf{x}$ is the unit vector in the x-direction, and $u(x, y, z)$ is the complex scalar function describing the distribution of the field amplitude, which satisfies the wave equation. Using this vector potential gives the following expressions for the electric and magnetic fields:

$$\mathbf{B} = \mu_0 \mathbf{H} = ik\left[u\mathbf{y} + \frac{i}{k}\frac{\partial u}{\partial y}\mathbf{z}\right]e^{ikz} \tag{11.2a}$$

and

$$\mathbf{E} = ik \left[u\mathbf{x} + \frac{i}{k} \frac{\partial u}{\partial x} \mathbf{z} \right] e^{ikz}. \tag{11.2b}$$

From these fields, the real part of the time-averaged Poynting vector, $\varepsilon_0 \mathbf{E} \times \mathbf{B}$, can be calculated as

$$\varepsilon_0 \mathbf{E} \times \mathbf{B} = \frac{\varepsilon_0}{2} \left[(\mathbf{E}^* \times \mathbf{B}) + (\mathbf{E} \times \mathbf{B}^*) \right]$$

$$= i\omega \frac{\varepsilon_0}{2} (u\nabla u^* - u^*\nabla u) + \omega k \varepsilon_0 |u|^2 \mathbf{z}. \tag{11.3}$$

It should be noted that the term of order zero, $\partial u / \partial z$, in the expression for the magnetic field $\mathbf{B}$ (Equation 11.2a) has been retained, allowing the grad terms to be written concisely. It should be clear that this result also applies when expressing the scalar function in cylindrical coordinates, $u(r, \phi, z)$. The cycle-averaged momentum density is the linear momentum density in the beam. The angular momentum density along the beam axis, j_z, is found by taking the cross product between the ϕ component in Equation 11.3 and the radius vector, r. If we consider any field of the form

$$u(r, \phi, z) = u_0(r, z) \exp(il\phi), \tag{11.4}$$

which obeys the conditions of the paraxial approximation, we see that the ϕ component of the linear momentum density simply becomes $\varepsilon_0 \langle \mathbf{E} \times \mathbf{B} \rangle_\phi = \varepsilon_0 \omega l |u|^2 / \mathbf{r}$ and so taking the cross product with r gives an angular momentum density of magnitude

$$j_z = \varepsilon_0 \omega l |u|^2. \tag{11.5}$$

Multiplying the linear momentum density by the speed of light gives the energy density of the beam, $w = c\varepsilon_0 \omega k |u|^2 = \varepsilon_0 \omega^2 |u|^2$ and thus the ratio between angular momentum density and energy density in a beam is shown to be

$$\frac{j_z}{w} = \frac{l}{\omega}. \tag{11.6}$$

Integrating the angular momentum and energy densities over the x–y plane gives a ratio of angular momentum to energy per unit length in the beam of

$$\frac{J_z}{W} = \frac{\iint r \, dr \, d\phi \, (\mathbf{r} \times \langle \mathbf{E} \times \mathbf{B} \rangle)_z}{c \iint r \, dr \, d\phi \, \langle \mathbf{E} \times \mathbf{B} \rangle_z} = \frac{l}{\omega}. \tag{11.7}$$

It is trivial to see that the ratio of angular to linear momentum densities is

$$\frac{\omega l}{\omega k} = \frac{l\lambda}{2\pi}, \tag{11.8}$$

showing that the beam possesses an OAM of l per photon. As the beam is not polarized, this angular momentum cannot be due to spin and so is evidence of the OAM of light.

This derivation can be simply extended to include the case of nonparaxial beams and polarized light beams.

11.4 Generation of Light Beams Carrying OAM

The original generation of OAM-carrying beams in 1992 was proposed and demonstrated using a subtle combination of cylindrical lenses to efficiently convert the high-order Hermite–Gaussian modes emitted by a laser into Laguerre–Gaussian modes. Similar lens-induced mode transformations of this kind had previously been studied [13], but again without any reference to their angular momentum. The complex amplitudes of the Hermite–Gaussian and Laguerre–Gaussian modes are respectively described by (see Chapter 1, Section 1.7)

$$
u_{mn}^{HG}(x, y, z) = C_{nm}^{HG}\left(\frac{1}{w}\right)\exp\left(-ik\frac{x^2+y^2}{2R}\right)\exp\left(-\frac{x^2+y^2}{w^2}\right)
$$

$$
\times \exp(-i(n+m+1)\psi)H_n\left(\frac{x\sqrt{2}}{w}\right)H_m\left(\frac{y\sqrt{2}}{w}\right), \tag{11.9}
$$

$$
u_{mn}^{LG}(r, \phi, z) = C_{nm}^{LG}\left(\frac{1}{w}\right)\exp\left(-ik\frac{r^2}{2R}\right)\exp\left(\frac{r^2}{w^2}\right)\exp(-i(n+m+1)\psi)
$$

$$
\times \exp(-i(n-m)\phi)L_{\min(n,m)}^{|n-m|}\left(\frac{2r^2}{w^2}\right), \tag{11.10}
$$

where k is the wavenumber, R is the radius of curvature of the phase fronts, w is the beam waste of the Gaussian term, n and m are the HG mode indices, $\min(n, m) = p$, $n - m = \ell$, with ℓ and p as the usual LG indices (note that $n + m = 2p + \ell = N$, the mode order), $\psi = \arctan z/z_r$ gives the Gouy phase for a beam with Rayleigh range z_r, $H_n(x)$ is the Hermite polynomial of order n, and $L_p^\ell(r)$ is the generalized Laguerre polynomial. The key term in these expressions is the $\exp(-i(n - m)\phi)$ azimuthal phase that gives rise directly to the OAM. From these equations, one can derive the expressions for the energy and momentum flow within the beam shown above. Unsurprisingly, for the helically phased beams, the Poynting vector has an azimuthal component equal to $\ell/k_0 r$ times the axial component. This gives a local ray direction with respect to the optical axis of $\beta = \ell/k_0 r$.

The transformation between Hermite–Gaussian and Laguerre–Gaussian modes is possible because both are complete basis sets, where any mode of order N can be synthesized by the summation of N modes of the other set. The ability to transform also depends upon the surprising fact that the rotation by 45° of an HG mode around the beam axis creates exactly the modal magnitude weightings required to create the corresponding LG mode. Furthermore, a specific combination of cylindrical lenses introduces a mode-dependent

Gouy phase to rephase the constituent HG modes to form the LG beam. Most usefully, the same pair of cylindrical lenses will convert any HG mode, oriented at 45°, to the corresponding LG mode, where $\ell = n - m$ and $p = \min(n, m)$ [14].

Although it is one of the few methods capable of producing pure LG modes with near 100% conversion efficiency, the cylindrical lenses approach is practically limited by lens tolerances, meaning that great care is required if residual aberrations are to be avoided [15]. Upon propagation, astigmatism causes any $\ell > 1$ singularity to break up into ℓ unity singularities.

Another method for producing helically phased beams pioneered by the Leiden group was the use of spiral phaseplates [16] (Figure 11.2). Conceptually, these are simple transmissive elements with an optical thickness that increases with azimuthal angle, such that a plane wave input beam acquires a helical phase upon transmission. The complication is that the optical thickness needs to be controlled to sub-wavelength precision. In the original work, this was accomplished by immersing the spiral phaseplate in a near index match fluid bath, the temperature of which could then be adjusted to give the desired phase step. Subsequently, with microfabrication techniques, it has been possible to make plates of the correct thickness directly [17]. One big advantage of this approach is that it transforms the output of a conventional laser into a helically phased beam, albeit one that comprises a superposition of multiple LG modes of the same ℓ indices but differing p. The superposition is weighted in favor of the $p=0$ mode, which for $\ell=1$ can reach

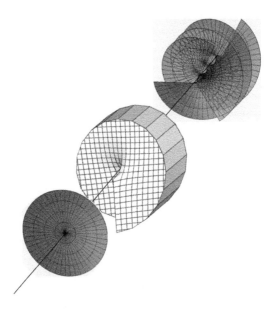

FIGURE 11.2   Spiral phase plates have an increasing optical thickness with azimuthal angle, transforming the beam, in this instance, from one with an $\ell = 0$ phase profile to a beam with a phase profile of $\ell = 3$.

80%. For higher values of ℓ, the mode weighting between different p values becomes more uniform [18].

An interesting feature of the spiral phaseplate is that it makes clear the link between the linear and angular momentum of light. When illuminated by a plane wave, the inclined surface of the phaseplate causes refraction of light in an azimuthal direction, making an angle, β, with respect to the optical axis. The resulting azimuthal component of the light's linear momentum, $\sin \beta \times \hbar k_0 = \ell \hbar / r$, when multiplied by the radius vector gives exactly the OAM of light [19].

As discussed, in parallel with the early studies of OAM, helically phased beams were already under study using diffraction gratings with a fork dislocation. These diffractive optical components are in essence simple holograms, where the first-order diffracted beam has a desired intensity and phase structure. As a general recipe, the desired phase profile is added modulo 2π to a linear phase ramp. For a helically phased beam, the result is a diffraction grating with an ℓ-pronged fork dislocation centered on the beam axis [2,8] (Figure 11.3). When illuminated with a Gaussian beam from a conventional laser, the resulting diffraction pattern creates a helically phased diffracted first order. For a perfect imaging system, the corresponding intensity distribution matches that of the illuminating beam. However, the finite numerical aperture of the intervening lenses means that the very center of the beam (where the skew angle of the rays is greatest) is attenuated, transforming the Gaussian illuminating intensity into an annular distribution (Figure 11.4). These forked diffraction gratings are the holographic equivalent of the spiral phase plates discussed above, but in most cases they are easier to fabricate. Overall, the diffractive approach has proved to be an extremely popular way of producing helically phased beams, not least because of the emergence of computer-controlled phase-only spatial light modulators, which allow easy implementation and real-time update of diffractive optical components [20]. The use of spatial light modulators is, of course, not just restricted to the generation of Laguerre–Gaussian beams; any complex beam can be generated, including superpositions of Laguerre–Gaussian [21] and Bessel beams [22] (see Chapter 6 for further details).

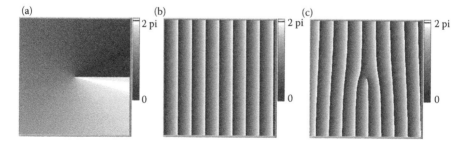

FIGURE 11.3 Phase profiles of holograms used to generate helically phased beams. (a) Spiral phase plate phase profile for $\ell = 1$, (b) linear phase ramp, and (c) the combined phase plate and phase ramp hologram.

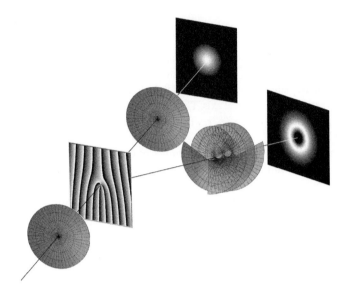

FIGURE 11.4   A forked hologram producing a beam with $\ell = 3$ and annular intensity profile in the first diffracted order.

Another approach to the generation of helically phased beams are the so-called Q-plates [23], liquid-crystal-based devices that perform a transformation between spin and orbital angular momentum. These components are half-wave plates but with a direction to the optic axis, which changes with spatial position. An incident circularly polarized plane wave is converted to the opposite handedness but acquires a geometric phase that depends upon the orientation of the optic axis. By an appropriate spatial dependence, the transmitted beam has a helical phase. Such devices are useful not only for beam generation but also as devices in quantum optics, allowing entanglement exchange between spin and orbital angular momentum [24].

Although the early work on OAM was in relation to light beams that were both spatially and temporally coherent, neither is a fundamental requirement. For polarization and spin angular momentum, achromatic waveplates can produce a broadband polarization state. Similarly, a forked diffraction grating produces the same ℓ index for all wavelengths, albeit with an angular dispersion. However, by reimaging the grating onto a second dispersive element, this angular dispersion can be compensated to create a white light optical vortex [25,26], which, as with its monochromatic equivalent, can set microscopic particles into rotation [27]. Spatial coherence is certainly not a requirement for a pure polarization state, but since it depends upon spatial phase structure, OAM is more complicated. If a forked grating or spiral phase plate is illuminated by a beam with only partial spatial coherence, the result is no longer a pure OAM state but rather an incoherent superposition. These Rankin vortices have a finite on axis intensity but still carry a nonzero OAM [28].

11.5 Optical Manipulation and Optical Spanners

Following the recognition of OAM as a phenomenon within laser beams created in the laboratory, an obvious question was "could this orbital angular momentum be transferred to objects in a similar fashion to that shown for spin angular momentum by Beth [5]?" Rather than following the approach of Beth, the first successful demonstration of this transfer was performed using optical tweezers by Rubinstein-Dunlop and coworkers [29]. Optical tweezers use tightly focused beams of light to trap microscopic particles, attracted to the beam focus by the gradient force [30]. When the tweezing beam is helically phased, an object is then both trapped on the beam axis and subject to a torque that results from a slight absorption of the light's angular momentum (Figure 11.5). Absorption of light works equally for the transfer of both spin and orbital angular momentum, hence a circularly polarized helically phased beam with $\ell = 1$ has a total angular momentum of $2\hbar$ per photon or, if combined with opposite handedness, no angular momentum. By flipping the polarization state, this equivalence allows the rotation of the trapped particle to be stopped and started again without losing hold of the particle; an optical spanner [31].

When the particle is held on the beam axis, the equivalence of the spin and orbital angular momentum seems a natural property, but this is not generally so. For particles held away from the beam axis, the polarization state and phase structure of the beam have different symmetries. Polarization is a local property, and a circularly polarized beam

FIGURE 11.5 A particle trapped in the beam focus of the optical spanner by the gradient force can be rotated using the angular momentum of the beam.

has a spin angular momentum that is the same at all positions in its cross section. By contrast, helical phase fronts have an associated ray direction that changes with transverse position. When considered locally, the skew rays give a transverse momentum that varies with both radius and azimuthal angle. A small particle confined in the annular intensity of a circularly polarized, helically phased beam will experience both a torque, inducing a spin about its own axis, and an azimuthal scattering force, causing an orbiting around the beam axis [32,33].

Beyond fundamental studies into the character of light's angular momentum, the use of helically phased beams in optical tweezers and the resulting rotation of the trapped particles suggests various applications [34]. These applications range from the use of rotating particles to create microscopic pumps [35,36] and their associated drag to measure viscosity [37,38] through to more generic, optically driven machines [39].

11.6 Beam Transformations

Beam transformations are fundamental to the operation of the cylindrical lens mode converters used to transform between Hermite–Gaussian and Laguerre–Gaussian modes [14]. The analogy with polarization is both useful and accurate. For transformations in the polarization state of a beam, a useful construct is the Poincaré sphere [40], which is a specific example of a Bloch sphere for the representation of a two-state system. For polarization, the north and south poles of the sphere represent right- and left-handed circularly polarized light. An equally weighted superposition of the states gives linearly polarized states, with an orientation depending upon their relative phase (Figure 11.6a). In general, the longitude on the sphere represents the relative phase of the superposition and the latitude the relative weighting. All possible polarization states can be represented as a point on the sphere or a vector direction. The operation of any waveplate corresponds to a rotation of this vector about a direction depending upon the orientation of the plate. For example, a halfwave plate rotates the state to invert the sense of circular polarization, and a quarter-wave plate can transform a linear polarization state to a circular one.

But the use of a Bloch sphere is not restricted to polarization. Similar spheres can be created for the addition of two orthogonal modes. The simplest example is for the Laguerre–Gaussian modes $\ell = 1$, and $\ell = 1$, $p = 0$ [41]. As with the polarization case, this gives polar states with angular momentum of $\pm\hbar$. Adding these two Laguerre–Gaussian modes together gives an Hermite–Gaussian $m = 1$, $n = 0$ mode with an orientation that depends upon their relative phase (Figure 11.6b). The equivalent transformations produced by waveplates are possible for these Laguerre–Gaussian modes using the cylindrical lens mode converters developed by the Leiden group [14]. However, unlike polarization, OAM state space is not limited to two orthogonal states. For Laguerre–Gaussian modes of order $N = \ell + 2p$, there are N modes, and rather than a sphere it is necessary to represent them as an N-element vector. The associated transformations produced by the cylindrical lenses are similarly represented by $N \times N$ matrices [42]. This equivalence between spin and orbital angular momentum transformations

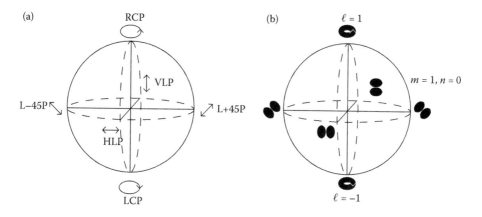

FIGURE 11.6 (a) Poincare sphere showing positions of common polarization states, (b) Bloch sphere for LG modes with $\ell = \pm 1$ and associated HG modes with $m = 1$ and $n = 0$.

is particular relevant to the understanding of the role of OAM in tests of quantum optics [43].

Leaving aside the precise mathematical description, one can recognize two primary types of transformation. First, there is a coordinate transform corresponding to a rotation of the state around the optical axis and no change in angular momentum, which for polarization might arise from optical activity or Faraday type effects (this can also be considered to arise from a phase difference between the circular states). Secondly, there is a rephasing of the linear states that, in general, does change the angular momentum and for polarization might arise from birefringence. For OAM, the rotation of the state corresponds to a rotation of the resulting mode or image (formed by the superposition of modes) around the optical axis, corresponding to a change in phase between positive and negative OAM states. The rephasing of the Hermite–Gaussian (cf. linear polarization) is the transformation made possible with the cylindrical lenses.

The equivalence between the rotation of the polarization state and the rotation of an image [44] is one that is particularly relevant to work in photon drag. In the 1970s, Jones reported several experimental realizations of Fresnel, or aether, drag. That the speed of light is independent of the frame is only true in vacuum. Upon transmission through a moving medium, the light can be dragged in either the transverse or longitudinal direction [45]. Jones showed that not only could a light beam be displaced by transverse motion of the medium [46], but that a spinning medium slightly rotates the plane of transmitted polarization [47], a phenomena later called the mechanical Faraday effect [48]. However, while it was recognized that transverse drag in a rotating medium would lead to the rotation of a transmitted mode or image, it has now been shown to move through the same angle as the polarization [49]. The equality of these two angles can be interpreted as an equivalence of aether drag acting on the spin and orbital angular momentum of light. Observing this image rotation is hard but has been accomplished using a slow light medium to increase the transmit time to give rotations of a few degrees [50].

11.7 Phenomena Associated with Orbital Angular Momentum

Underpinning any modal description of the light field are Maxwell's equations, and consequently all modal sets are equally valid for analyzing spatial effect. However, although OAM modes are, therefore, not a unique description of the phenomena, they often lead to a simple understanding, especially if the system is rotationally symmetric and/or contains optical vortices.

The continual rotation, at a frequency Ω, of a circularly polarized beam with angular momentum $\sigma\hbar$ per photon is equivalent to the frequency shift of the beam by $\Delta\omega = \sigma\Omega$ [51], where $\sigma = \pm 1$ for right- and left-handed circularly polarized light, respectively. One method of rotating the polarization state of the beam is by transmission through a rotating half-wave plate [52]. The rotation of a helically phased beam possessing a phase cross section $\phi(r,\theta) = \exp(i\ell\theta)$ also brings about a similar frequency shift, where the ℓ-fold rotational symmetry gives a frequency shifted by $\Delta\omega = \ell\Omega$ [53,54]. Such frequency shifts can be introduced by passing the beam through an image-inverting (e.g., Dove) prism that, when rotated, rotates the transmitted beam by twice the angular rotation rate of the prism. Similar frequency shifts also occur in atomic interactions where the atom has an azimuthal component to its velocity [55].

An interesting question arises when the beam carries both spin and orbital angular momentum, that is, when the helically phased beam is also circularly polarized. In this case, a careful examination of the vectorial nature of the beam cross section reveals that the beam has a $J = \ell + \sigma$-fold rotational symmetry, giving a frequency shift of $\Delta\omega = J\Omega$, which is proportional to the total angular momentum of light.

None of these frequency shifts fundamentally require OAM as an explanation, but treatment in the OAM basis set leads to an intuitive understanding much simpler than possible in a Cartesian reference frame.

Another area where OAM leads to an intuitive understanding of a generic concept is the angular uncertainty relationship. For linear variables, one can consider the finite transverse aperture of the beam as resulting in an uncertainty in the transverse momentum, leading to the finite size of the focused far-field spot [56,57]. For a beam with no angular restriction, the cyclic nature of the azimuthal phase leads to a pure angular momentum state with, in principle, no uncertainty in OAM. Restricting the aperture, blocking the light over an azimuthal range, means the transmitted light field is inherently described by a spread of OAM states (Figure 11.7). The relationship between the open angle of the aperture and the spread of OAM values is a form of uncertainty relationship and has been the subject of both experimental and theoretical study [58–60]. What complicates the formulation of this form of the uncertainty relationship is the periodicity of the angular aperture. Considering the case where the angular position is in no

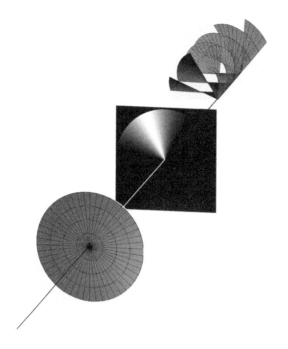

FIGURE 11.7 Restricting the azimuthal range of the incoming beam results in a beam described by a spread of OAM states. The relationship between the aperture angle and resultant OAM states follows a form of uncertainty relationship.

way restricted and so $\Delta L = 0$, the periodicity of the angular position means it still has a finite distribution and so has a readily calculated uncertainty of $\Delta\phi = \pi/\sqrt{3}$. To take account of this periodicity in angular position, the uncertainty principle for states with finite uncertainty in angular momentum, can be stated as

$$\Delta L_z \Delta\phi_\theta \geq \frac{1}{2}\hbar|1 - 2\pi P(\theta)|, \tag{11.11}$$

where $\Delta L_z = \Delta l\hbar$ and $P(\theta)$ is the angular probability density at the boundary of the chosen angular range. For narrow Gaussian profiled apertures of width $\Delta\phi$, the corresponding spread in OAM states, ΔL, is given by $\Delta L \Delta\phi = \hbar/2$ [59].

The Goos–Hänchen shift is normally associated with the apparent displacement of a beam upon reflection from a dielectric surface. The helical nature of an OAM-carrying beam and the inherent variation in the local ray direction and their associated Fresnel reflection coefficients also results in beam shifts both in displacement and angle [61–64]. These shifts can scale with the OAM and hence be much larger than normally encountered, although they often result in beam distortion and hence modification of the OAM content.

11.8 Measuring Beams with OAM

OAM is associated with helical phase fronts. It follows that any interferometric measurement of a beam that reveals the phase structure can be used to deduce the OAM spectrum of the beam. Perhaps most obvious is the interference of an unknown helically phased beam with a colinear plane wave to give the characteristic spiral interference fringes, where the number of fringes gives the ℓ of the beam [65–67]. Similarly, the diffraction pattern from any aperture gives a unique signature of the illuminating beam and can be used to deduce the OAM spectrum [68]. Specific examples of diffraction studies have included single [69] and double [70] slits but also more complicated arrays [71]. Perhaps the most unexpected of these diffraction type experiments is the diffraction of a helical beam by a triangular aperture that yields a triangular array of $\ell(\ell+1)/2$ spots, revealing both the magnitude and, from their orientation, the sign of the OAM [72] (Figure 11.8). However, all of these approaches require the interpretation of an extended optical pattern that requires an imaging detector and many photons in the same state.

A completely different approach to the measurement of OAM is based on the forked diffraction gratings that we have previous considered for generating helically phased beams [8]. For generation, one might consider the collimated output from a single-mode fiber illuminating a forked hologram to create a helically phased vortex beam in the far-field. Optical systems are reciprocal, such that the same vortex beam in the far-field will be converted by the forked grating into a fundamental Gaussian mode and coupled into the fiber. However, the coupling is only efficient if the converted mode matches the mode of the fiber, which requires the OAM of the incoming beam to match that of the diffraction grating [73]. By implementing the diffraction grating on an SLM, it is possible to sequentially test the incident light for different values of ℓ [74]. However, although this does allow the measurement of low light levels, and even single photons, each individual measurement has only a binary outcome and cannot be used to realize the higher information content of the photon encoded using OAM. More complicated designs of diffraction grating are possible in which the same grating can couple to different detectors [75,76], but the probabilistic nature of the photon distribution does not overcome this restriction.

The rotational frequency shift discussed previously might also offer a way of measuring the OAM of the beam, where every value of ℓ gives a distinct frequency sideband [77].

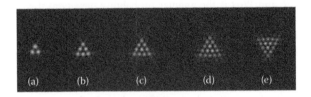

FIGURE 11.8 Triangular array interference patterns obtained when a helical beam is diffracted through a triangular aperture where the incident beam has (a) $\ell = 1$, (b) $\ell = 2$, (c) $\ell = 3$, (d) $\ell = 4$, and (e) $\ell = -4$.

However, both the precise beam rotation and the frequency measurement of low light levels are difficult. As an alternative, rather than continually rotating the beam to give a frequency shift, a fixed rotation gives a static, ℓ-dependent phase shift. Using image-rotating Dove prisms, this type of phase shift had been built into Mach Zehnder [78,79] and Sagnac [80] interferometers. The approach can also be adapted to measure the total (spin + orbital) AM of the beam rather than just the OAM [81]. Unfortunately, to measure N states requires $N-1$ interferometers, thus limiting the technique to a small number of states.

The general OAM measurement problem has remained a challenge since the 1990s. Ideally, just as with polarization, one would like an OAM beam splitter, but unlike polarization, the OAM equivalent needs many ports. The problem of mode separation is not insurmountable; for example, a lens can separate plane wave states dependent upon their direction. This function of a lens suggests an approach to OAM sorting. By mapping the azimuthal coordinate of the input helical beam onto the transverse coordinate, every helically phased input mode is transformed to a tilted plane wave, which can be separated with a lens. The required transformation needs two bespoke optical elements in conjugate planes [82]. Initially, these were implemented using spatial light modulators but now the two components have been specifically manufactured [83]. The overall system resembles a beam telescope, but one for which each helically phase input beam is focused to a spot with a transverse position proportional to ℓ.

11.9 Quantum Properties of OAM

The precursor to much of the work on the use of OAM in quantum optics was the observation of second-harmonic generation of beams containing OAM [84]. It was observed that the OAM was conserved such that the second-harmonic output had twice the OAM of the input beam. These second-order processes work in both directions, so the natural question was how does the OAM of the pump beam divide into the signal and idler photons? We start with the state generated by spontaneous parametric down-conversion as described in Ref. [85] and follow the presentation in Ref. [86]:

$$|\psi\rangle = |vac\rangle + const \times \int d\mathbf{k}_s \int d\mathbf{k}_i \sin c \frac{1}{3} (\omega_s + \omega_i - \omega_p) t$$
$$\times F(\mathbf{k}_s, \mathbf{k}_i) |1, \mathbf{k}_s\rangle |1, \mathbf{k}_i\rangle, \tag{11.12}$$

where $|1, \mathbf{k}_s\rangle |1, \mathbf{k}_i\rangle$ denotes the creation of one signal and one idler photon with wavevector $\mathbf{k}_s$ and $\mathbf{k}_i$, respectively, and $F(\mathbf{k}_s, \mathbf{k}_i)$ is as follows:

$$F(\mathbf{k}_s, \mathbf{k}_i) = \int d\mathbf{q}_p E(\mathbf{q}_p) \left[\frac{\omega_s \omega_i \omega_p}{n^2(\mathbf{k}_s) n^2(\mathbf{k}_i) n^2(\mathbf{k}_p)} \right]^{\frac{1}{2}} \times \prod_{j=1}^{3} \sin c \frac{1}{2} (\mathbf{k}_s + \mathbf{k}_i - \mathbf{k}_p)_j L_j \tag{11.13}$$

$\mathbf{q}_p$ is the transverse wave vector of the pump, $E(\mathbf{q}_p)$ is the angular spectrum of the pump, $n(\mathbf{k})$ are the respective refractive indices, and L_j denotes the length of the crystal in $\{x, y, z\}$. We consider only the monochromatic case where $\omega_s + \omega_i = \omega_p$ such that the time dependence in Equation 11.12 disappears. We also assume that the pump beam has a narrow angular spectrum and the photon pairs are observed close to the z-axis, a collinear phase matching regime, so that $|\mathbf{q}| \ll |\mathbf{k}|$. By ensuring that the transverse size of the crystal is large, the transverse momentum of the pump beam is conserved, $\mathbf{q}_p = \mathbf{q}_s + \mathbf{q}_i$. Using these approximations, the refractive indexes can be considered as constants and so Equation 11.13 becomes

$$F(\mathbf{k}_s, \mathbf{k}_i) = E(\mathbf{q}_s + \mathbf{q}_i) \sin c \left| \frac{1}{2}(k_{p,z} - k_{s,z} - k_{i,z})L_z \right|. \tag{11.14}$$

When considering the thin crystal approximation, the sinc term above tends toward unity and so it can be seen that the angular spectrum of the pump beam is transferred to the two-photon state. Assuming the phase matching condition embodied in Equation 11.14 and the fact that the angular spectrum of the pump beam is transferred to the down-converted photons, for collinear to near collinear SPDC OAM is conserved, that is

$$m = l_s + l_i, \tag{11.15}$$

where $m\hbar$ is the OAM per photon in the pump beam and $l\hbar$ is the OAM in the signal and idler photons, respectively.

Zeilinger and coworkers used forked holograms to measure the OAM of single photons, showing not only that the OAM was still conserved, but that although the OAM of one of the down-converted beams could take on a wide range of different values [87], the sum of the two beams was conserved as per Equation 11.15, indicative of quantum entanglement [73] (Figure 11.9). The extent of the entanglement was clarified with experiments by themselves [88] and others [89].

In this early work, the measurements of each state were made by fixed holograms or precision-machined spiral phase plates [90], but the approach becomes more flexible when programmable spatial light modulators are used instead [74,91]. The ease of switching the measurement state makes possible many more and arbitrary measurements. By direct analogy through the Poincaré sphere, with experiments on the nonlocal correlation of two polarization states, tests of Bell [43] and Leggett [92] inequalities have been made. The generality with which the spatial light modulators can be programmed extends beyond purely helically phased beam to include full control of the complex amplitude [93]. This control allows for more general studies of angular momentum and angle states enabling a test of the angular form of the EPR paradox [94], the spatial entanglement of complex modal superpositions [95], and radial states [96].

Other quantum-type properties of OAM and vortex states include light–matter interactions. The interaction of vortex beams with cold atoms via four-wave mixing is constrained by the conservation of OAM [97,98]. Vortex beams have also been used to store information in Ref. [99] or stir [100] Bose–Einstein condensates.

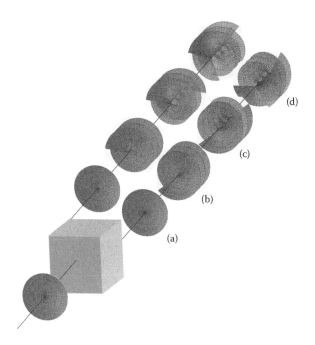

FIGURE 11.9 Measuring the OAM of the photon pairs generated by down-conversion shows that while each individual photon can take any value of OAM, the sum of the OAM of the pair always equals that of the pump beam, possible pair values shown here are (a) $\ell = 0$ pair, (b) $\ell = 1$ and $\ell = -1$, (c) $\ell = 2$ and $\ell = -2$, and (d) $\ell = 3$ and $\ell = -3$.

11.10 Other Applications with OAM

One of the key differences between spin and orbital AM is that, whereas the spin (polarization) has only two orthogonal states, OAM is unbounded. This larger state space immediately suggests the use of OAM as a variable for encoding either classical or quantum information [77]. Using spatial light modulators to both encode and decode the beam, it was demonstrated that superpositions of OAM states could provide multiplexing within a free-space communication system [76]. However, unlike polarization, the decomposition of a beam into its constituent OAM eigenstates depends upon both the transverse and angular alignment of the beam [101]. Furthermore, like any other extended beam, the fidelity of phase fronts and hence the OAM spectrum is degraded by atmospheric turbulence [102–104]. Despite these problems, recent work has shown OAM multiplexing over short transmission distances yielding extremely high data rates [105].

As an alternative to shaping an outgoing laser beam, the same optical elements can be used for detection (as there were for the single photon experiments discussed in the previous section). However, when used at a classical level, this gives rise to new forms of imaging. Placing a spiral phase plate within a telescope gives a phase filter blocking any on-axis light [106], applied with a degree of added sophistication means that the light from one astronomical object, such as a star, can be attenuated to reveal the light from a neighboring yet fainter object [107], potentially an extrasolar planet. Similar techniques can be applied to microscopy, with the inclusion of a forked hologram using unidirectional edge enhancement of the image for both classical [108] and quantum [109] regimes. Slight variation of the technique gives spiral interferometry, where a height feature gives spiral rather than circular fringes in the interferogram, the sense of which breaks the up/down degeneracy of traditional interferometric images [110].

11.11 Conclusions

It did not take long for OAM to become firmly rooted within the mainstream of optics research after Allen et al. wrote their seminal paper. However, OAM is not a solely optical property. Being associated with the phase cross section of a beam, it applies to all wave phenomenon, and OAM has been discussed for mm- [19], radio wave [111], and x-ray [112] frequencies. Beyond electromagnetism, OAM is finding application within electron beams for imaging [113,114], acoustics [115,116], and ultrasonics [117]. However, besides the insights that OAM provides and its potential applications, its most significant contribution to optics is perhaps to reinforce the importance of phase in determining properties beyond those apparent in the intensity distribution alone. OAM has given us optical spanners, angular entanglement and new forms of imaging. No doubt it still has a few more twists to perform.

References

1. Abraham, M., 1914. On the issue of the symmetry of electromagnetic tension tensors, *Annalen der Physik* 44(12): 537–544.
2. Bazhenov, V.Y., Soskin, M.S., and Vasnetsov, M.V. 1992. Screw dislocations in light wavefronts, *J. Mod. Phys.* 39(5): 985–990.
3. Barnett, S.M. and Loudon, R., 2010. The enigma of optical momentum in a medium. *Philos.Trans. R. Soc. A Math. Phys. Eng. Sci*, 368(1914), 927–39.
4. Poynting, J.H., 1909. The wave motion of a revolving shaft, and a suggestion as to the angular momentum in a beam of circularly polarized light. *Proc. R. Soc. Lond. A* 82: 560–67.
5. Beth, R.A., 1936. Mechanical detection and measurement of the angular momentum of light, *Phys Rev* 50(2): 115–125.
6. Darwin, C.G. 1932. Notes on the theory of radiation. *Proc. R. Soc. Lond. A* 136(829): 36–52.
7. Allen, L. et al. 1992. Orbital angular-momentum of light and the transformation of Laguerre-Gaussian laser modes. *Phys Rev A* 45(11): 8185–89.

8. Bazhenov, V.Y., Vasnetsov, M.V., and Soskin, M.S. 1990. Laser-beams with screw dislocations in their wave-fronts *Jetp Lett.*, 52(8): 429–431.

9. Coullet, P., Gil, G., and Rocca, F. 1989. Optical vortices. *Opt. Commun.* 73(5): 403–8.

10. Tamm, C., 1988. Frequency locking of 2 transverse optical modes of a laser. *Phys Rev A* 38(11): 5960–63.

11. Nye, J.F. and Berry, M.V. 1974. Dislocations in wave trains, *Proc. R. Soc. Lond. A*, 336(1605): 165–190.

12. Berry, M.V., Nye, J.F., and Wright, F. 1979. The elliptic umbilic diffraction catastrophe. *Philos. Trans. R. Soc. Lond.* 291: 453–84.

13. Abramochkin, E. and Volostnikov, V. 1991. Beam transformations and non-transformed beams, *Opt. Commun.* 83(1–2): 123–35.

14. Beijersbergen, M.W. et al. 1993. Astigmatic laser mode converters and transfer of orbital angular momentum, *Opt. Commun.* 96(1–3): 123–32.

15. Courtial, J. and Padgett, M.J. 1999. Performance of a cylindrical lens mode converter for producing Laguerre-Gaussian laser modes, *Opt. Commun.* 159(1–3): 13–18.

16. Beijersbergen, M.W. et al. 1994. Helical-wavefront laser beams produced with a spiral phaseplate, *Opt. Commun.* 112(5–6): 321–27.

17. Oemrawsingh, S.S.R. et al. 2005. Experimental demonstration of fractional orbital angular momentum entanglement of two photons, *Phys Rev Lett.*, 95(24): 240501.

18. Dennis, M.R., O'Holleran, K., and Padgett, M.J. 2009. Singular optics: Optical vortices and polarization singularities, *Prog. Optics* 53: 293–363.

19. Turnbull, G.A. et al., 1996. Generation of free-space Laguerre-Gaussian modes at millimetre-wave frequencies by use of a spiral phaseplate, *Opt. Commun.* 127(4–6): 183–88.

20. Curtis, J.E., Koss, B.A., and Grier, D.G. 2002. Dynamic holographic optical tweezers, *Opt. Commun.* 207(1–6), 169–75.

21. Leach, J. et al. 2005. Vortex knots in light, *New J. Phys.* 7: 55–55.

22. Vasilyeu, R. et al., 2009. Generating superpositions of higher–order Bessel beams, *Opt. Express* 17: 23389–3395.

23. Marrucci, L., Manzo, C., and Paparo, D. 2006. Optical spin-to-orbital angular momentum conversion in inhomogeneous anisotropic media, *Phys. Rev. Lett.*, 96(16): 163905.

24. Nagali, E. et al., 2009. Optimal quantum cloning of orbital angular momentum photon qubits through Hong-Ou-Mandel coalescence, *Nat. Photonics* 3(12): 720–23.

25. Leach, J. and Padgett, M.J., 2003. Observation of chromatic effects near a white-light vortex, *New J. Phys.* 5: 154.

26. Mariyenko, I.G., Strohaber, J., and Uiterwaal, C., 2005. Creation of optical vortices in femtosecond pulses, *Opt. Express* 13: 7599–608.

27. Wright, A. et al. 2008. Transfer of orbital angular momentum from a super-continuum, white-light beam, *Opt. Express*, 16(13): 9495–500.

28. Swartzlander, G. and Hernandez-Aranda, R., 2007. Optical rankine vortex and anomalous circulation of light, *Phys. Rev. Lett.* 99(16): 163901.

29. He, H. et al. 1995. Direct observation of transfer of angular momentum to absorptive particles from a laser beam with a phase singularity, *Phys. Rev. Lett.*, 75(5): 826–29.

30. Ashkin, A. et al. 1986. Observation of a single-beam gradient force optical trap for dielectric particles, *Opt. Lett.* , 11(5): 288–90.

31. Simpson, N.B. et al. 1997. Mechanical equivalence of spin and orbital angular momentum of light: An optical spanner, *Opt. Lett.* 22(1): 52–54.

32. O'Neil, A.T. et al. 2002. Intrinsic and extrinsic nature of the orbital angular momentum of a light beam, *Phys. Rev. Lett.* 88(5): 053601.

33. Garces-Chavez, V. et al., 2003. Observation of the transfer of the local angular momentum density of a multiringed light beam to an optically trapped particle, *Phys. Rev. Lett.* 91(9): 093602.

34. Padgett, M. and Bowman, R. 2011. Tweezers with a twist, *Nat. Photonics* 5: 343–48.

35. Ladavac, K. and Grier, D. 2004. Microoptomechanical pumps assembled and driven by holographic optical vortex arrays, *Opt. Express* 12(6): 1144–49.

36. Leach, J. et al. 2006. An optically driven pump for microfluidics, *Lab Chip* 6(6): 735–39.

37. Bishop, A.I. et al. 2004. Optical microrheology using rotating laser-trapped particles, *Phys. Rev. Lett.* 92(19): 198104.

38. Parkin, S. et al. 2007. Picoliter viscometry using optically rotated particles, *Phys. Rev. E*, 76(4), 041507.

39. Knoner, G. et al. 2007. Integrated optomechanical microelements, *Opt. Express*, 15(9): 5521–30.

40. Born, M. and Wolf, E., 1999. *Principles of Optics* 7(null) ed, Cambridge University Press, Cambridge, UK.

41. Padgett, M.J. and Courtial, J. 1999. Poincare-sphere equivalent for light beams containing orbital angular momentum, *Opt. Lett.* 24(7): 430–32.

42. Allen, L., Courtial, J., and Padgett, M.J. 1999. Matrix formulation for the propagation of light beams with orbital and spin angular momenta, *Phys. Rev. E*, 60(6): 7497–503.

43. Leach, J. et al. 2009. Violation of a Bell inequality in two-dimensional orbital angular momentum state-spaces, *Opt. Express* 17(10): 8287–93.

44. Allen, L. and Padgett, M. 2007. Equivalent geometric transformations for spin and orbital angular momentum of light, *J. Mod. Opt.* 54(4), 487–91.

45. Leonhardt, U. and Piwnicki, P. 2000. Light in moving media, *Contemp. Phys.* 41(5): 301–308.

46. Jones, R.V., 1975. Ether-drag in a transversely moving medium, *Proc. R. Soc. Lond. A.* 345(1642): 351–64.

47. Jones, R.V. 1976. Rotary aether drag, *Proc. R. Soc. Lond. A.* 349(1659): 423–39.

48. Nienhuis, G., Woerdman, J.P., and Kuscer, I., 1992. Magnetic and mechanical Faraday effects, *Phys. Rev. A* 46(11): 7079–92.

49. Padgett, M. et al., 2006. Polarization and image rotation induced by a rotating dielectric rod: an optical angular momentum interpretation, *Opt. Lett.* 31(14): 2205–07.

50. Franke-Arnold, S. et al., 2011. Rotary photon drag enhanced by a slow-light medium, *Science* 333(6038): 65.

51. Garetz, B.A. 1981. Angular dopper effect, *J. Opt. Soc. Am.* 71(5): 609–11.

52. Garetz B.A. and Arnold, S. 1979. Variable frequency-shifting of circularly polarized laser-radiation via a rotating half-wave retardation plate, *Opt. Com mun.* 31(1): 1–3.

53. Courtial, J. et al. 1998. Measurement of the rotational frequency shift imparted to a rotating light beam possessing orbital angular momentum, *Phys. Rev. Lett.* 80(15): 3217–19.

54. Basistiy, I.V. et al. 2003. Manifestation of the rotational Doppler effect by use of an off-axis optical vortex beam, *Opt. Lett.* 28(14): 1185–87.

55. Allen, L., Babiker, M., and Power, W.L. 1994. Azimuthal Dopper-shift in light-beams with orbital angular momentum, *Opt. Commun.* 112(3–4): 141–44.

56. Stelzer, E.H.K. and Grill, S., 2000. The uncertainty principle applied to estimate focal spot dimensions, *Opt. Commun.* 173(1–6): 51–56.

57. Padgett, M., 2008. On the focussing of light, as limited by the uncertainty principle, *J. Mod. Opt.* 55(18): 3083–89.

58. Forbes, G.W. and Alonso, M.A. 2001. Consistent analogs of the fourier uncertainty relation. *Am. J. Phys.* 69: 1091.

59. Franke-Arnold, S. et al. 2004. Uncertainty principle for angular position and angular momentum, *New. J. Phys.* 6: 103.

60. Øeháèek, J. et al. 2008. Experimental test of uncertainty relations for quantum mechanics on a circle, *Phys. Rev. A* 77(3): 13.

61. Fedoseyev, V.G. 2001. Spin-independent transverse shift of the centre of gravity of a reflected and of a refracted light beam, *Opt. Commun.* 193(1–6): 9–18.

62. Bliokh, K.Y., Shadrivov, I.V., and Kivshar, Y.S. 2009. Goos–Hänchen and Imbert–Fedorov shifts of polarized vortex beams, *Opt. Lett.*, 34(3): 389–391.

63. Aiello, A. and Woerdman, J.P. 2011. Goos–Hänchen and Imbert-Fedorov shifts of a nondiffracting Bessel beam, *Opt. Lett.* 36(4): 543–45.

64. Dennis, M.R. and Götte, J.B. 2012. The analogy between optical beam shifts and quantum weak measurements, *New. J. Phys.*, 14(7): 073013.

65. Harris, M. et al. 1994. Laser modes with helical wave-fronts, *Phys. Rev. A* 49(4): 3119–22.

66. Padgett, M. et al. 1996. An experiment to observe the intensity and phase structure of Laguerre-Gaussian laser modes. *Am. J. Phys.* 64(1): 77–82.

67. Soskin, M.S. et al. 1997. Topological charge and angular momentum of light beams carrying optical vortices, *Phys. Rev. A*, 56(5): 4064–75.

68. Mazilu, M. et al. 2012. Simultaneous determination of the constituent azimuthal and radial mode indices for light fields possessing orbital angular momentum, *Appl. Phys. Lett.* 100(23): 231115.

69. Ghai, D., Senthilkumaran, P., and Sirohi, R.S. 2009. Single-slit diffraction of an optical beam with phase singularity, *Opt. Laser Eng.* 47(1), 123–26.
70. Sztul, H.I. and Alfano, R.R. 2006. Double-slit interference with Laguerre-Gaussian beams, *Opt. Lett.* 31(7), 999–1001.
71. Berkhout, G.C.G. and Beijersbergen, M.W. 2008. Method for probing the orbital angular momentum of optical vortices in electromagnetic waves from astronomical objects, *Phys. Rev. Lett.* 101(10): 100801.
72. Hickmann, J.M. et al. 2010. Unveiling a truncated optical lattice associated with a triangular aperture using light's orbital angular momentum, *Phys. Rev. Lett.* 105: 053904.
73. Mair, A. et al. 2001. Entanglement of the orbital angular momentum states of photons, *Nature* 412(6844): 313–316.
74. Yao, E. et al. 2006. Observation of quantum entanglement using spatial light modulators, *Opt. Express* 14(26), 13089–94.
75. Khonina, S.N. et al., 2000. Gauss-Laguerre modes with different indices in prescribed diffraction orders of a diffractive phase element, *Opt. Commun.* 175(4–6): 301–8.
76. Gibson, G. et al., 2004. Free-space information transfer using light beams carrying orbital angular momentum, *Opt. Express* 12(22): 5448–56.
77. Molina-Terriza, G., Torres, J.P., and Torner, L., 2001. Management of the angular momentum of light: preparation of photons in multidimensionals vector states of angular momentum, *Phys. Rev. Lett.* 88(1): 013601.
78. Leach, J. et al. 2002. Measuring the orbital angular momentum of a single photon, *Phys. Rev. Lett.* 88(25): 257901.
79. Lavery, M.P.J. et al. 2011. Robust interferometer for the routing of light beams carrying orbital angular momentum, *New J. Phys.* 13(9): 093014.
80. Slussarenko, S. et al. 2010. The polarizing Sagnac interferometer: A tool for light orbital angular momentum sorting and spin-orbit photon processing, *Opt. Express*, 18(26): 27205–16.
81. Leach, J. et al., 2004. Interferometric methods to measure orbital and spin, or the total angular momentum of a single photon, *Phys. Rev. Lett.* 92(1): 013601.
82. Berkhout, G.C.G. et al. 2010. Efficient sorting of orbital angular momentum states of light, *Phys. Rev. Lett.* 105(15): 153601.
83. Lavery, M.P.J. et al. 2012. Refractive elements for the measurement of the orbital angular momentum of a single photon, *Opt. Express*, 20(3): 2110–15.
84. Dholakia, K. et al. 1996. Second-harmonic generation and the orbital angular momentum of light, *Phys. Rev. A*, 54(5): R3742–R3745.
85. Hong, C.K. and Mandel, L. 1985. Theory of parametric frequency down conversion of light, *Phys. Rev. A*, 31(4): 2409.
86. Monken, C.H., Ribeiro, P.H.S., and Pádua, S. 1998. Transfer of angular spectrum and image formation in spontaneous parametric down-conversion, *Phys. Rev. A*, 57(4): 3123.
87. Torres, J.P., Alexandrescu, A., and Torner, L. 2003. Quantum spiral bandwidth of entangled two-photon states, *Phys. Rev. A*, 68(5): 050301.

88. Vaziri, A. et al. 2003. Concentration of higher dimensional entanglement: Qutrits of photon orbital angular momentum, *Phys. Rev. Lett.* 91(22): 227902.

89. Barreiro, J.T. et al. 2005. Generation of hyperentangled photon pairs. *Phys. Rev. Lett.* 95(26): 260501.

90. Oemrawsingh, S.S.R. et al. 2005. Experimental demonstration of fractional orbital angular momentum entanglement of two photons, *Phys. Rev. Lett.* 95(24): 240501.

91. Stuetz, M. et al. 2007. How to create and detect N-dimensional entangled photons with an active phase hologram, *Appl. Phys. Lett.* 90(26): 261114.

92. Romero, J. et al. 2010. Violation of Leggett inequalities in orbital angular momentum subspaces, *New J. Phys.* 12: 123007.

93. Kirk, J.P. and Jones, A.L. 1971. Phase-only complex-valued spatial filter, *JOSA* 61(8): 1023–28.

94. Leach, J. et al. 2010. Quantum correlations in optical angle-orbital angular momentum variables, *Science*, 329(5992): 662–65.

95. Romero, J. et al. 2011. Entangled optical vortex links, *Phys. Rev. Lett.* 106(10): 100407.

96. Salakhutdinov, V.D., Eliel, E.R., and Löffler, W., 2012. Full-field quantum correlations of spatially entangled photons, *Phys. Rev. Lett.*, 108: 173604.

97. Barreiro, S. and Tabosa, J. 2003. Generation of light carrying orbital angular momentum via induced coherence grating in cold atoms, *Phys. Rev. Lett.* 90(13): 133001.

98. Walker, G., Arnold, A.S., and Franke-Arnold, S. 2012. Trans-spectral orbital angular momentum transfer via four-wave mixing in Rb vapor, *Phys. Rev. Lett.*, 108: 243601.

99. Dutton, Z. and Ruostekoski, J. 2004. Transfer and storage of vortex states in light and matter waves, *Phys. Rev. Lett.* 93(19): 193602.

100. Andersen, M.F. et al. 2006. Quantized rotation of atoms from photons with orbital angular momentum, *Phys. Rev. Lett.* 97(17): 170406.

101. Vasnetsov, M.V., Pas'ko, V.A., and Soskin, M.S. 2005. Analysis of orbital angular momentum of a misaligned optical beam, *New J. Phys.* 7, 46–46.

102. Paterson, C., 2005. Atmospheric turbulence and orbital angular momentum of single photons for optical communication, *Phys. Rev. Lett.* 94(15): 153901.

103. Pors, B.J. et al. 2011. Transport of orbital-angular-momentum entanglement through a turbulent atmosphere, *Opt. Express*, 19(7): 6671–83.

104. Malik, M. et al. 2012. Influence of atmospheric turbulence on optical communications using orbital angular momentum for encoding, *Opt. Express*, 20(12): 13195–200.

105. Wang, J. et al. 2012. Terabit free-space data transmission employing orbital angular momentum multiplexing, *Nat. Photoics*, 6(7): 488–96.

106. Swartzlander, G.A. 2001. Peering into darkness with a vortex spatial filter, *Opt. Lett.* 26(8): 497–99.

107. Swartzlander, G. et al. 2008. Astronomical demonstration of an optical vortex coronagraph, *Opt. Express*, 16(14): 10200–207.

108. Fürhapter, S., Jesacher, A., Bernet, S., and Ritsch-Marte, M., 2005b. Spiral phase contrast imaging in microscopy, *Opt. Express* 13(3): 689–94.

109. Jack, B. et al. 2009. Holographic ghost imaging and the violation of a bell inequality, *Phys. Rev. Lett.* 103(8): 083602.

110. Fürhapter, S., Jesacher, A., Bernet, S., and Ritsch-Marte, M., 2005a. Spiral interferometry, Opt. Lett. 30(15): 1953–55.

111. Thidé, B. et al. 2007. Utilization of photon orbital angular momentum in the low-frequency radio domain. *Phys. Rev. Lett.* 99(8): 087701.

112. Sasaki, S. and Mcnulty, I. 2008. Proposal for generating brilliant x-ray beams carrying orbital angular momentum, *Phys. Rev. Lett.* 100(12): 124801.

113. Verbeeck, J., Tian, H., and Schattschneider, P. 2010. Production and application of electron vortex beams, *Nature* 467(7313): 301–4.

114. McMorran, B.J. et al. 2011. Electron vortex beams with high quanta of orbital angular momentum, *Science* 331(6014): 192–95.

115. Volke-Sepúlveda, K., Santillán, A.O., and Boullosa, R.R. 2008. Transfer of angular momentum to matter from acoustical vortices in free space, *Phys. Rev. Lett.* 100(2): 024302.

116. Skeldon, K.D. et al. 2008. An acoustic spanner and its associated rotational Doppler shift, *New J. Phys.* 10: 013018.

117. Demore, C. et al. 2012. Mechanical evidence of the orbital angular momentum to energy ratio of vortex beams, *Phys. Rev. Lett.* 108(19): 194301.

Index

Printed and bound by CPI Group (UK) Ltd, Croydon, CR0 4YY

24/10/2024

01778302-0010